# RETIRING WILD

## BOOK EIGHT OF THE ALASKA OFF GRID SURVIVAL SERIES

### MILES MARTIN

ALASKA DREAMS PUBLISHING

**Retiring Wild**
*By Miles Martin*

Book Eight of The Alaska Off Grid Survival Series
©2021 Miles Martin
Artwork, Photos, Original Poetry ©2021 by Miles Martin - All rights reserved

This Book may not be re-sold or given away to others. All rights reserved, including the rights to reproduce this Book.

No part of this text may be reproduced, transmitted, downloaded, scanned, or copied in any form or means, whether electronic or mechanical, without the express written permission of the publisher.

Any distribution of this Book without the permission of the publisher is illegal and punishable by law. Please purchase only authorized editions of this book and do not take part in or encourage piracy of copyrighted materials.

Published by:
Alaska Dreams Publishing
www.alaskadp.com
2nd ADP Edition December 2021
PRINT PAPERBACK ISBN: 978-1-956303-12-4
PRINT HARDCOVER ISBN: 978-1-956303-13-1
*This book was previously published by Miles of Alaska*

Visit www.milesofalaska.com to find a bio of Miles, additional photos, stories, how-to videos, handmade artwork, and raw materials for sale.

*Camped in a field of mint.*

*Nenana cabin in winter.*

# CONTENTS

*Introduction*   7

Chapter 1   11
*Illegal ivory laws, The truth will not set us free*

Chapter 2   41
*Homestead problems, Land dispute, Gift shop*

Chapter 3   64
*Stuff not taught in school, Mayor ideas shared, I am the source, The unprotected*

Chapter 4   89
*Tucson cop, ANILCA agreement, Subsistence laws*

Chapter 5   109
*Heart attack*

Chapter 6   134
*Mammoth tusk hunting, Meet Papa and Flower in a time travel.*

Photos   161

Chapter 7   184
*Mammoth tusks aftermath*

Chapter 8   208
*A big art deal, Visit friends, Try to collect money owed*

Chapter 9   222
*Patriots, My rabbit cycle theory of civilization, My son and the Ex*

Chapter 10   247
*Wolves, Bears, Bees*

Chapter 11   266
*Lawyer in land dispute*

Chapter 12   288
*Banned from shop, Computer compromised*

Chapter 13   304
*Lose web site, Lost emails and order shipments*

Chapter 14   313
*Federal government takes over state, Reality TV comes to Nenana*

Chapter 15   334
*Snow machine trip, Get a moose*

Chapter 16   345
*Ancient life hunting mammoths*

| | |
|---|---|
| *About The Alaska Off Grid Survival Series* | 353 |
| *Magazine and News Stories* | 356 |
| *Other Titles Available From Alaska Dreams Publishing* | 358 |
| *Notes* | 359 |

# INTRODUCTION

Alaskan wilderness survival is defined by your ability to stay warm at fifty below, work with sled dogs, handle the twenty-four hours of dark in winter, and the same hours of light in summer. When older I realize survival in the long term is about understanding your surroundings and being prepared. This includes prospering and being happy, maybe even the ability to get along with others, for man is a social being. Even the most primitive man cannot live in the Alaskan wilderness without the northern lights.

Early years have houseboat adventures, dog trips, and meeting mail order women. A whole lifestyle is described in detail through daily events. I help raise two children with the mother on a remote homestead for a few years. I end up in the village of Nenana, Alaska. My wilderness lifestyle is in conflict with civilized ways. Some misunderstandings are funny! I thought a 'tanning salon' was a place to sell furs. I pull out my library card for ID, trying to cash my first check. *No one mentioned photo ID!* Other misunderstandings are not so funny. I'm the only suspect in a murder in a community that has vigilante justice. I begin getting high profile. I'm smart. I know how to make money and run a business. I'm number five in the world on the internet selling wilderness garbage, animal by-products of living off the land, bones, skulls, teeth, claws.

I begin to feel I live a catch 22. Society offers up catchy phrases, 'Off the Grid', 'Lone Wolf', and, 'Think outside the box'. The reality of these terms is not supported by society as expressed in its laws. Doing well and prospering is the same as bragging about being on welfare as a dumpster diver. Few fans who admire a survival life understand what it means. Few who can read and write volunteer to step down

# INTRODUCTION

in class, and actually do it. Many who try have a one-time adventurous event that puts them in the news, never to return. Those born and raised in the life are not able or willing to communicate this lifestyle to the civilized world.

'Off grid' is usually a highly successful civilized person who reads Waldon Pond, and wishes to avoid Civilizations issues, yet still have many of its benefits. You're most likely in a borough (county in many states,) getting mail; meaning jury duty, taxes, permits, needing to know what day it is, paying attention to burn bans and such. Not really a wilderness survival life. Keep in mind, as Waldon Pond was being written, Thoreau's mother was darning his socks and bringing him cookies, paying the taxes on the cabin and property. Off-grid should be about independence; subsistence off the land, not society. The lifestyle may not be easier (or harder), just different.

If you truly understand, and are good at wilderness survival, the biggest danger you face is civilization. The same danger your peers, the wild animals and entire environment, faces. You can shoot bears, dress for the cold, but you cannot escape pollution, taxes, fees, permits, worse, not qualifying for these permits. Civilization does not want you to be gone, not plugged in and paying. You will pay, or be a criminal. Few civilized believe this. If you ever arrive, you will understand.

INTRODUCTION

*112 ft. mammoth tusk, 40,000 years old. Where did I find it?*

9

# CHAPTER ONE

## ILLEGAL IVORY LAWS, THE TRUTH WILL NOT SET US FREE

"Miles, is your power back on?" My log cabin is dark, no electric sounds. This is an isolated community of 300 people, Nenana, Alaska. Supplies for us are fifty miles away in Fairbanks, serving the entire interior of Alaska, 50,000 people, covering an area bigger than most states. We are supposed to be prepared for 'off the grid' existence. Nenana is defined as a subsistence community. Meaning we live off the land, not the government.

I'm a ham radio operator.

"KL1VC, Miles here, hey, Bean! No power yet. It's been a couple of days now, are you ok?" I'm supposed to give his call sign, but I am not good with names and numbers, and out remote like this the rules are not held to civilized standards. Ham radio can be run on low voltage and power, so many ham operators around the world including Bean and I, run our system with twelve volt power. I have a deep cycle golf cart battery with a solar power charger. Bean and I can communicate all over the world in an emergency. Such as find out how large this power outage is, get an idea what happened and when it might get fixed. There is a system in place, discussed ahead of time, for a communication line world-wide. In a worse case, we may find out where it is safe to head for in the world.

Bean lives only a few blocks from me, and has been a friend for over twenty years. He's an accountant, knows money, numbers, laws, civilized things that elude me. He's tall, always happy smile on a gnome face. I think he likes me because he loves the outdoors, and I am his go-to wilderness expert. We are both seniors now. I'm not much over five feet tall, bald, with Einstein white tuffs of hair on the side. I have an over-all elf look, living in my low ceiling greenery and flower covered hobbit house, that most people have to bend over to get into, and cannot stand up in.

Bean and I have been in communication since the outage, and it looks like even Fairbanks is out of power. In isolated Nenana, how would we know? The whole world could be dark. A nuclear bomb took out most of civilization, an asteroid, a blast of solar flare got the world grid? Certainly some local residents are in a panic thinking the sky is falling. Some residents, like new urban school teachers, do not comprehend subsistence, and what it means.

"No electric, no life." There is no going back to 'before there were cell phones'. Bean and I would chuckle, except this is serious. Panicked people running down the street can be dangerous; more so than the power outage. I know from local ham two meter communication that the outage will get fixed. Something about a power intertie issue between Anchorage and Fairbanks involving one of the military bases.

"WL7BDO, KL1VC, Yes, Miles the power is already coming back in Fairbanks so no big deal, assume we will be back on line in a day or two."

I decide to use my four wheeler to go visit in person to discuss how this is going for our neighbors. We are part of WIN, meaning 'Wellness in Nenana'. Whatever is good for our community we discuss at weekly meetings. We address community drug issues, local economy, food distribution, how to run a community garden-farmers market. We have discussed what to do in the event of various disasters like forest fires, floods, power outages, anything that might affect our well-being. We have the equivalent of fire drills. The city office has a list of who owns rescue boats, and who has various skills that may be needed. Such as, who has communication with the outside world, like Bean and I.

Tosjacket, the name of the hill across the river means 'A good place to meet between two rivers.' In Athabascan Indian, 'na' means river. The Nenana and Tanana rivers meet here. We are all about water—one of the necessities of life. I'm a 'river rat', as the New York Times called me. After a lifetime of living in the Alaska wilderness, I know every river in interior Alaska, and am known in every village.

The Tanana River runs downstream 300 miles to the Yukon. The Yukon runs another 800 miles to the ocean. The last road and bridge on the entire route is right here in Nenana. This is the end of civilization, all the way to the ocean. We are the hub for the barge line taking supplies 1,000 miles to those with no roads. Nenana remaining operative is crucial for supply distribution for all downriver villages.

The unpaved dirt road to Bean's house is familiar to me. I pass our two room post-office. It is odd to see no lights on. Not open in the middle of the week; middle of a work-day. There will be no mail going in or out till power comes back. Mail is not one of the critical necessities of life. Unless you are a senior waiting for meds to arrive in the mail that you will die without. Our Senior Center has a list of those who need medicines and has a van that can go to Fairbanks for pick up - all part of our community preparedness plan.

My own area of expertise and interest is in personal survival. I try to be a people person, but cannot quite pull it off. I spent too many years alone in the wilderness living on my houseboat as a mountain man- trapper-artist. However, many skills required for an individual to survive also applies to a group. Food, shelter, water, fire, self-defense. For the group to survive, the individual has to survive. Right now the community does not need to be in emergency mode. Three days without power is getting to be almost common.

I see Al way up ahead of me staggering up a grassy footpath. He is headed for the liquor store which he will discover is closed. Alcohol can be a necessity if you are addicted. He can die without his drink. He looks and walks like a zombie. He is a zombie. He could not tell me his name right now. Part of my lack of social skills is not being very interested in what these zombies will do to stay alive. In fact, I take a side street to avoid going by Al. He will for sure try to stop me and ask for money for a drink. If he could get away with it, try to rob me.

There is talk of gangs going round taking advantage of the power outage knowing security cameras and lights are not working. Thieves take advantage of life being out of routine. In the chaos, they will try to step in and make the goods of the community disappear. The good news is, this has not happened yet, and may not. Nenana is not like the big city where there will be guaranteed looting and burning. The streets are, for the most part, quiet and devoid of people. Here and there I see candles or kerosene lights in the windows as dawn arrives and these lights get blown out.

Bean greets me at the door with his usual round happy face. This power outage is not bothering him a bit. He is enjoying a change of routine, not having to go to work, sitting by the warmth of the wood stove and by kerosene lantern light reading a good book for a few days.

I smile taking all this in; I nod toward the wood stove. "Where would our community be now, if the Fairbanks North Star Borough got its way and absorbed us into its folds?"

He knows what I am talking about. There is a ban on wood stoves in order to comply with Federal air quality requirements. We'd all be 100% dependent on oil and electricity. In fact, even now it is hard to find wood stoves for sale. To buy one, you have to prove you live remote and are exempt from the rules. Many remote people cannot prove such; no address, no ID. Bean has ten cords of split birch wood in the yard. There is a 100 year supply of wood for the entire community in the wilderness behind us.

"Yes, freezing… and broken, frozen water pipes!"

"Maybe not, Bean, there might be no water to freeze." I thought he'd get my point, but Bean looks puzzled so I explain, "Mayor spoke to me in private about the future of our water treatment plant." Our existing plant is twenty years outdated and could fail at any time. A majority of citizens are pressing for Federal money to put a new plant in. Mayor has been stalling, not telling the public why, but he and I are friends and on the same page; in agreement with the WIN group. "Bean, the replacement plant must be up to code, with new advanced technology."

Now Bean remembers a discussion at WIN. "Any power plant built will be computerized, and run remotely out of state, probably by the government." The mayor tells us, Nenana would have zero control over our water even though we are classified as a subsistence community. There are new systems, not computerized, that would allow for local control. We'd have to pay for such a system ourselves, the government would not help. "We cannot afford it."

This past winter the water plant froze during a fifty below zero cold spell. No community water. Because we know our system and it is simple, we were able to piece together parts from our salvage yard and get the water back within a day. If the water was not back on within that time, underground pipes and valves would begin to freeze and break. This could not be fixed until spring, five months away. Who knows what would need digging up and replacing!

Bean nods, "If the plant was controlled from out of state we'd be waiting for a qualified college guy from Washington to fly out here. Good luck with that."

"Yeah, who pays for that?"

"The internet was down for a month. I could picture trying to call the water controller." I laugh, " Press one, press two. 'Please hold, your business is important to us.' Two days later getting someone on line in India who doesn't speak English."

Nenana could disappear and the civilized world could care less. Happens all the time. News of some disaster in some remote place involving a few hundred people - a mudslide, tornado, genocide; whatever. Today's headlines replaced by something more exciting tomorrow, often with no follow up.

Lately word is that Jacksonville, Florida has been a month without water; we do not even know what that's about. *I bet if we were in Jacksonville we would know!* I bet Jacksonville is bigger than Nenana. This is what goes around in my head. A power outage a couple of years ago in some remote country… a year later, still no power to half the country. I forget where, but it could be any of a number of places, including us.

The entire population of our state fits in one city block of New York City. Maybe one building for all I know. The population of our community fits in one room, even in the average citizen's home. Some families are bigger than our entire community. Who are we on a Federal level? No one. Nothing. Not even Fairbanks, our closest community, cares about Nenana. In any large scale emergency, we'd be at the bottom of the list for getting help. Survival means taking care of ourselves.

"Your freezer doing ok, Miles?" Food is a big concern during a power outage. We all have water because the community has an emergency generator for the water plant. Enough power for the basics, not for private phone, internet, household needs, just the water plant and sewer line pumps. We are on two rivers, clean enough water we could fetch in buckets if we had to. I go back to life as it was 50,000 years ago and not only survive, but prosper. I'm unsure how many in my community could be happy. Bean could.

The school also has a generator. We could plug our community generators into the local grid and power our entire community. We'd need a week's time. No money, just a few wires, switches, and local knowledge. Our biggest obstacle... civilization. Fees, permits, insurance, inspections, union workers; stuff like that. We have a bulk fuel plant in our area. We could power our community for years isolated from the rest of the world, if we had to. This knowledge alone is power; has a calming effect on the population.

I reply to Bean concerning the freezer, "So far ok. I used foam sheets long ago to add more insulation to my two chest freezers so they can stay frozen over a week."

"Just you and Iris and you need two freezers?"

"Well, it takes one just to hold my year's moose meat; that's as much as 1,000 pounds of meat!"

"For me it's salmon, Miles!" Yes, we have a good salmon run up the Tanana River and anyone who wishes can put a net out and get all the fish they want. I used to get 2,000 for my sled dog team.

Bean gets his from someone else, I think $2 a salmon. "$20 fills my freezer with a year's worth of eating!" I know. I only want a few salmon. I like moose meat better. Maybe Canada wild geese.

"Then I have the garden to put in the other freezer. Iris blanches a lot, you know... beets, turnips, bags of cabbage..." I trail off and shrug meaning, "all the usual Alaska produce." Bean knows, because his community garden plot is next to mine.

"You do not keep your cabbage in a cool place with your squash?"

"For some reason my cabbage does not keep! Maybe too damp where I keep it. It's not easy! Our answers are not simple, easy, or free. We gave up, and freeze it or make sauerkraut. This year a five gallon bucket. Anyhow, we could live over a year on food we have on hand."

I'm sure Bean can as well. Half our community can... maybe. How to get surplus food distributed will be a topic on the agenda at our next WIN meeting. I like having way more food then I need. In emergencies, food can be worth more than money. Money only works with a functional system in place. I can and do trade food for services and goods I want.

This is not a government program our community is part of. We have no government to speak of. No police, no borough, no state tax. People take care of each other without being told, or made to with threats or rules. We see the need, we take care of it. We put fires out, rescue people, provide our own insurance, like getting together to rebuild lost homes for community members. Bean knows I lived even more primitive for twenty-five years. So this answers, "How long could you live like this if the power never comes back?" It would not bother me to go back to how life was 50,000 years ago.

I'm not sure how many people could. Some cannot live a day without their cell phone. We may have to have discussions at WIN, as to what is necessary, what is a luxury, what can we afford and pay for ourselves. Perhaps ask who we can trust to help us. I think, *Help us what? Live beyond our means? Is that right or fair? Depend on someone else to pays our bills?* I confess, my own personal knowledge is how to go off the grid alone - if there is such a thing. Humans are herd animals. Even out alone I liked matches, propane, soap, stovepipe. I made my own soap. There mostly is not enough time for it. Things like soap, matches, candles, at the right sources cost pennies, but does require civilization. Hmmm.

Topic for another day. I say bye to Bean, guess I'll go home and read a book like Bean until the power comes back on. Like him, I have plenty of water, firewood, food, and a shelter. Within a week everything will be back to normal. I assume.

**Fairbanks, Alaska News Miner:**
Selling almost any item containing ivory is now a crime in the US."

Tusk and I are in the fossil ivory business.

Elephants are seriously endangered. To combat that, all ivory will be banned, including ancient walrus and mammoth, in hopes of discouraging the entire trade. No exemptions for antiques or grandfathered materials.

The article related how poaching supports terrorism which must be stopped. Yes. It all makes sense. Many restrictions make sense when explained. *If the argument is legitimate.* Why wouldn't this argument be legitimate?

People like Tusk, Dodger, Pete, Neil or myself are in the fossil part of the ivory business so know a few facts, study, stay informed. I point this out to Tusk. "Ranking number four in sales of all illegal activity,"

I'm not convinced the world is highly interested in buying poached ivory. This statistic seems very incorrect. If so, what other aspects are potentially bogus; a deliberate altering of facts to suit an agenda.

Lately some news stories claim there is a concern for 'fake news' with little accountability. Part of the concern is people believing it, and calling it 'facts and truth.' I, and those I know, support saving the elephant. I believe in saving the planet! I believe man is the planet's biggest enemy. This belief has a lot to do with why I went off alone to live a simple life in the wilds as a savage! I did something about how I felt! I now believe we all share the blame equally. *Except for me and you of course*! It's those 'other people' we are talking about; the selfish irresponsible idiots.

Tusk is puzzled by the argument for closing the entire ivory trade including mammoth. We all know mammoth ivory is easy to tell from any other ivory based on its growth rings and hashmarks. I add, "Besides, mammoth is 40,000 years old, easy to tell from fresh elephant."

"The argument is the same if we applied it to the meat trade."

I have heard this before. The idea is to stop the sale of all meat so we can control the illegal harvest of wild game, endangered animals, and meat without proper inspections. In fact, once meat is made into burger and sausage it would be hard to name the meat source. There are in fact vegans who will say this is a great idea! The argument will work in the ivory trade because who are we in terms of political clout and money? "Democracy at work Miles, everyone out to hang someone." *Get the public riled up*. For a good cause? Or political reasons?

I prefer to avoid the subject, not dwell on the negative. I do not wish to be known as some radical political activist- protester. Now and then someone like Tusk gets me agreeing, and I put in my two cents worth.

Every single connection I have and big shows I know about that has to do with wildlife is headed downhill fast. Has been for over a decade. No one I know, or heard of, is doing well trapping, hunting, or selling wildlife. There is less space on the planet for wildlife to prosper. I believe the entire planet is getting polluted, with no escape. The air, and the entire ocean is measurably polluted. This, along with global climate changes, could be affecting all wildlife. Meaning, there are no untouched pockets of thriving wildlife to be found and tapped for huge profits, or even wildlife observation. I believe this because I get out into the Alaska wilds where no other human has stepped. Where are the birds, squirrels, frogs, wildlife I used to see when young? Animals not hunted, trapped, or even seen a human appear to be on the decline. There is no abundance of anything! Not even mice! Geese used to darken the sky, and there was a squirrel in every spruce tree. No one I talk to from around the world tells me the wildlife population is healthy, compared to a generation ago.

Worldwide, there are more environmentally conscientious people. Endangered, restricted, illegal animal products are simply not cool anymore. Logic tells me it is not just trappers and hunters killing animals Logic tells me mankind has been hunting as subsistence gatherers for thousands of years. I think, *At what point has there been an animal population issue?* Since modern times, the past 100 years.

I hear, "The population of mankind has increased, so more hunting pressure!" Maybe, but I do not agree. It is easier to go shopping. Food is not a huge high cost.

Hunting is more expensive than shopping. I'm in the woods on the river and there are far fewer hunters where I am than say the 1940s or even the 1970s within my memory. For example, not so long ago everyone had a sled dog team and fished for them. Hundreds of thousands of fish were harvested for decades. Not a huge affect on salmon populations. Fish and Game admits the issue is out on the ocean and may have a lot to do with pollution and increasing water temperatures. Maybe commercial fisherman on the ocean. For sure not subsistence harvesters. Protestors call this my altered bias self-motivated statistics.

"Never mind, it is an ongoing conversation that is an opinion few want to hear again!"

Iris rolls her eyes up, in agreement. Better to keep it to myself. I may have a limited view of the world, living in Nenana, Alaska, population 300. I'm a senior now, possibly not 'with it.'

'Stuck in the old days' some would say. I'm a character. Hopefully, with fire still in my eyes and at least functional, able to take care of myself. *'Adventure' is getting my shoes on in the morning.*

Iris agrees we do ok for ourselves. *So why complain?* Another summer begins, and the crab apple tree is blossoming in bright red flowers. This visual mixes with the smell of lilac fragrance as the garden gets weeded. Iris is more a city girl, who is not as thrilled as I am about the great outdoors and wilderness life. We get along, as she puts up the garden I grow. She cooks from scratch each day. Her long hair is brushed away from her German face, as she rolls her eyes up.

The birch basket guy stops by, Mark. I have dealt with him off and on for many years. A lot of my display baskets come from him. In some ways, he is like another regular supplier of materials, Rooster, or maybe Wes the Mess. Stuck in a lifestyle with no way out, in general, headed downhill fast. There are few artists who make a living. Mark stays the same. He stays even, selling his hand made birch products. At least he has a place to stay, and transportation. I respect him for doing what he wants and being happy. As Mark talks to me from his car, an incident comes to mind.

**Past flash** [1]

Mark is on a selling trip, car loaded down with baskets. He is stopped by the police and arrested for some reason. No. Maybe the car broke down and it was a while before he could get back to it. Or, maybe he did not have it insured or all the proper papers. Meanwhile, the car is towed and impounded, that much I know for sure. The local tow service has Mark's car in custody. They are friends, so I know this much of the story for sure. Mark needs to get at his baskets in the trunk to sell them so he can get his car back! The tow company will not let him have access to get his goods. The car title might be in the girlfriends name, she's in jail. The goods are apparently the property of the tow company. I'm not sure how the law goes on this. I'm no more of a paperwork guy then Mark is. Only, that Mark is in financial trouble. The means to make money is in his confiscated car. The goods are worth far more than the car. *Bummer.*

One event initiates another downhill trend. Wes the Mess goes through the same about this time. Just when such people need a break, they get kicked. I remember, "When rich people get their car towed, if that ever happens, they get their stuff out of the car."

"Oh, I left something in the car, hold on." Followed by, 'No problem.' I'm not puzzled why this is so. I know why. I know why because I was born and raised into the privileged class. It is because this long hair pot smoking hippy freak Mark, is probably going to steal the car back, or rob the tow people. He fits the profile of the poster child of the war on crime, with zero rights. *As it should be!* Thank the Lord I am not like Mark! In general, we all tend to live up to what's expected of us. Mark has a scared face full of pockmarks and lopsided face. I think from burns.

**Past flash ends**

"Not trying to sell more baskets to you, Miles." I had just bought some a week ago, and was in hopes Mark would not give me a song and dance about how he needs the money to bail his wife out of jail, please buy. I already bought more than I need to help him out.

"Miles, I was at the Nenana Culture Center. I have been making dolls and face masks using grouse feathers. They tell me they can't legally sell them! What is that all about?" I know this answer as well.

"As you know, Mark, there is nothing illegal about grouse feathers being sold. The laws concerning birds forbids the sale of migratory birds, songbirds, and scavengers. There are only two wild birds I can think of that are legal to deal with in Alaska - grouse, and ptarmigan." I looked it up. No permit, no restrictions of any kind. I was thinking of bringing a copy of the law to the gift shop.

I explain to Mark what I was told at the shop. A trooper stopped by and told the employees it is illegal to sell any feathers of any kind." I doubt this conversation took place. Mayor agrees. The city owns the museum and shop. I had an issue in the recent past and told Mayor, "The same employee pulled all my mammoth ivory. I straightened that out." I told the employee, 'A trooper or anyone with authority should be talking to a manager or owner, not employees.' I speculated at the time, the trooper read the ivory news article, that Tusk and I read, or had it translated by his boss to the troops. Police are not lawyers or judges.

Local cops are now enforcing game regulations, we are told. "Due to budget cuts." I wonder how much a cop is supposed to know about game laws? Fossil mammoth ivory 40,000 years dead is legal to sell, as long as it comes off land with mineral rights. Fossil hunters tend to get no place pointing this out.

I feel bad for Mark. He depends on his craft to make ends meet. He has little wiggle room financially. He has been using grouse feathers for forty years that I know of. He made a trip to Nenana that cost him time and gas. Now he is told he can't make any money at his lifelong occupation, this is illegal. When it is not! An officer can be mistaken, and that is ok. Lives can get cut off, messed up, without even an apology. *I know the answer why.*

"Miles, we eat grouse! We can use them for bait trapping. No one is killing grouse just to get feathers. So what is going on?"

I have no answer I want to get into. He thinks I may know. I have an opinion. I look at Mark, who is white, and we briefly discuss racial issues between Indian and other races. Mark knows if he were Indian there would be no problems selling almost any animal part. The problem gift shop employee is Indian. I'm not a happy camper.

"Got that right Miles. Hey you want one of my pot muffins?" Mark pulls out a baggie from his pocket.

"No thanks. I just am not into it." I have to apologize, hope he understands and forgives. I am 100 percent certain the officer has no concern whatever for the poor grouse. But I bet he knows what to say.

"You know when you do crafts with grouse feathers, it encourages the killing of endangered birds, by creating an interest and market for feathers!" To get mad and tell an officer he is full of crap, is to only prove his point.

The subject comes up with Bean, who is not subsistence and does not hunt or fish. Bean is an accountant, so among the protected. Not just any ordinary accountant. I keep repeating, "He was once Mick Jaegers accountant, of the Rolling Stones. He prefers a less stressful life here in a small village." Tall Bean argues, this conversation is all about the environment, protecting game, part of the hunting regulations. I often try to make a point by putting something in perspective that affects the person directly I am talking to.

"So Bean, let's say I was in charge. I have concerns for the environment. I contemplate what can be done to help, so I come up with some laws to save our precious world. I notice there is too much car traffic, and studies I fund show this is a major source of pollution. Another study I sponsor shows the public drives a lot in the pursuit of getting food at the store. This is the number one reason for traffic on the road. My studies confirm this!" So I control your getting food by having everyone get a permit. The permit tells you what foods you can buy on what days. You are assigned a number that has to go on the roof of your car. I tell you what kind of car you can use and what route to take getting to the store. $200,000 fine for getting it wrong.

"How would you feel, Bean?" He laughs that it is not even close to being the same situation.

"The resources are limited so we must regulate!" *Micro manage you mean.* Sounds well and good. Bean has a smile and is being sarcastic. We have had similar conversations. I know the basic argument repeated. Where would we be without rules and regulations! There would be no wildlife left! History has shown us! Imagine if we all did what we chose and wanted!" Spoken by civilized people in support of law and order. I believe in law and order too, when it is of the people, by the people, and for the people. I have lived in the wilds away from civilization, among uncivilized villagers. We tend to regulate our lives and each other fine without interference or threats of punishment and fees, or Big Brother's help. It is not usually primitive people damaging the environment. We are. Not big consumers. I've spent years 'plugged in' on every committee, city meetings, heads of anything I could be part of, to do my part to change things from within for the better. For twenty years. Long enough to have a valued opinion on how it works.

I understood what the state expected of those staking homestead land when I first went out, So well I wrote an article for Alaska magazine, 'Would you make a good homesteader?" The state land office reprinted this article in their information on acquiring homestead land.

I GIVE different stories to different people. *The truth will not set us free.* Crazy Lawsen's wife, Hornet, tells me at the Nenana Farmers Market, "It's hard to know what is going on with you, Miles!" She laughs. "I can't tell if you are smart, having different story lines and being full of crap, or if you have a plan!" She is referring to my answers to her question about me going to the bone yard to find fossil mammoth ivory. I pretend to go along with her thinking. The Market is slow. There are five vendors and four customers. Basketmaking Mark does not have it together enough to do a bazaar. I like to give people what they want, so I play along with Hornet. She will not be one to argue with. In a court of law it would help to have witnesses with conflicting stories about what I said." I wink slyly as if we are co-conspirators. She rambles on with stories from the fossil bone yard, and the good old days, when we were all considered outlaws, and that was ok. The public understood. Even Fish and Game looked the other way, but that was the state in the days the state stood up to the Feds. We had oil money. We could afford to be cocky. Alaskan's joked, the Feds would prefer to give us back to Russia as we are so uppity and such trouble makers!" We laughed at what the lower states put up with.

"Try that up here!" Ha! Alaskans tend to have the bumper stickers with pictures of guns on them, and "We don't dial 911!" *If there is a revolution, it will start in Alaska!*

Hornet laughs the loudest. The revolution would start with her. She is a big woman, with star and moon purple dress on. She wants to talk about adventures of long ago.

"Yea, that time the cliff fell at the bone yard, and created a wave that crossed the island where we were camped, and washed us out!" I forget if I was there that time. Hornet continues, "Yes I will never forgot about the beaver you shot, Miles!" I have no idea what she is talking about. She tells others around us, "Miles shot this beaver for our dinner. Crazy Lawsen tells him to kill it on the first shot or it will sink! Miles shoots it and scoops the beaver into the boat with us. It is still alive. "We had a heck of a time with that live beaver in the bilges of the boat! Ha!"

I suppose the beaver flopped around a bit and we had to get out of the way of its teeth. I believe someone had to smack it with a boat paddle. I now vaguely recall. I smile as Hornet goes on, noticing all her stories are about how stupid other people are, how smart she is.

"Miles and the rest of us were sorting mammoth tusks and Pleistocene fossils when the law showed up in a float plane. We quickly hid everything under tarps, sat on the pile, as the officer asked us what we were doing.

"Fishing," we all said, with poles in our hands, bobbing our smiley faces, sharing our coffee with this officer. Ha!" Ken was with us, maybe Neil. We all sold to Tusk back then. There were half a dozen of us together. I hand Neil's book to Hornet.

"I'm still reading it. You'll recognize all the pictures, people, places."

"He wrote a book? I should write one!" Yes, well, all of us are worth a book. "Is it hard Miles?"

Here I am joking around among 'my people' somewhat at ease. I might tell it different to my wife, to the Feds, to the news media. *It's all the truth.*

Hornet notices I am not participating in the conversation. My thoughts are elsewhere. I'm not the people person I once was. "So how are you, Miles?" Said with concern. I know she is referring to my heart attack.

"That's old news, Hornet. Been a month or more now. I'm fine. Got to watch the shaman enter my vein, make his way through to my heart on an overhead screen, put a stent in. It's all magic to me."

She laughs. She believes in magic. That's partly why I put it like that. She'd do well with a crystal ball that would match her purple dress with stars and moons on it. Probably make a good living. She can read the cards with an earnest straight face, filled with sincere words of hope and redemption. She weighs three times what Crazy Lawsen weighs. *But wait! They are not together anymore, she has been gone a few years!* I cannot keep it all straight. The two are still good friends. But she has another last name now. Whatever it is.

Supposedly heart attacks have a lot to do with stress. I like to think I have an easy, stress free life, or it should be. No I am not at ease. I miss having a tribe, a place where I belong. People I know and trust around me. I no longer know what people think. I got turned in to the Feds I am told. Maybe Tusk and Knife were in on being snitches.

I got it made. No real complaints. No bills, own everything I have. Got a nice house, shop to work in, a car, truck, boat, canoe, a four wheeler, ok health, health coverage, respect, good woman, vacation get-away snow bird home all paid for in Arizona. Life is good! I rejoice. I'm happy with the money I make. No financial stress.

Iris and I collect Social Security, but also both work to the extent we want to. Social Security would pay all our basic bills and leave us with not much fun money. I keep up with my art work. My custom knives are desired, and all my art sells. I like to write. It all gives me a sense of self-worth. *Or it is supposed to, or so I tell myself.* I accept not everyone needs to feel self-worth. Other people can be happy without it, but I like to accomplish what I consider, 'worthy things'. This defines my existence in the physical world.

"I am, because," and hold something up in my hands. Hopefully, I do not have to apologize or explain myself. Welfare, food stamps, assistance in a variety of forms is available.

"Just sign here, it's easy Miles!" Watch free stuff arrive in my life. Someone must pay for it. Tax payers I assume. There is a catch twenty-two involved. On the one hand I want to take care of myself as a free man. I also want as few taxes—if any—as possible. Mostly because I do not like what is being done with my tax dollars. I prefer to—and do—voluntarily donate to causes of my choice. I donate more then I owe in taxes, proving it is not about the money. The library, fundraisers for education, and individuals in a bind, like those who had their homes burn, are among my donation recipients. Being in business for myself, I get a lot of tax write offs. My business involves almost everything I do. My life is all about business, therefore all about write offs! I grin to my friends with only half a straight face.

I say seriously and sadly to the IRS. "I'm trying! I truly am! Just any ole time now business will take off, but meanwhile, I am struggling, master! Take pity, thank you for all the free programs to help me get on my feet again!" I feel like… oh… Luke in Star Wars. Out in the solar system or beyond it, adapting to this and that, trying to come up with a part for the space ship at garage sales in another galaxy, without having all the paperwork in order. Not bothering anyone, or a criminal. Just involved in things that get complicated to explain right now to the authorities. In fact, to do good, Skywalker kind of has to bypass those in charge, who are, for the most part, corrupt. Or more, the proper way to behave in this unique situation will not be explained in the book of rules. *The best example is Jesus, who did a lot of good, and look who He was/is, yet the authorities nailed him as a criminal. He took the fall, no plea bargain.*

Hornet catches me again on the street and picks up her conversation where she left off.

"Anyhow, Miles, I need some fossil material!" She runs a business called Mammoth Creations. Last time we spoke she added, "But I don't have any money!" *Yes. Well. It is nice to have a want isn't it?* She may need me to front her some material. Sell it for ten cents on the dollar, or donate it because we are friends. I do not like being conned. She has pulled this before. She has good qualities, cheerful, optimistic, smart, hard working. With some good ideas sometimes. She just gets goofy mostly.

I mention what the news article says about the value of elephant ivory. She only snorts. I do not know who the source of her mammoth ivory is, or if she has any one source. I am trying to be low profile for now. I'm marginally off probation. I tell her, "Well those who charged me, Fish and Wildlife, have no issues over fossils. None were mentioned in charges, none were confiscated. When I asked why, they told me they are not animal products." We both pause and reflect on the meaning of this strangeness. We both wonder why I was not hammered, when the hammering could be done? I press on.

"But my probation officer told me I can have it, just not sell it. I assume while I am on probation." I learn fast, there is only one answer to being able to do much of anything, 'No!' Anyone who has entered the prison system knows that. *It's about learning my restrictions and punishment, not what I can get away with.*

As far as I know, no one else has an issue. I assume Mrs. Probation was concerned I might mingle with the same crowd, get involved with the same unscrupulous criminal people, end up in my old illegal life of animal parts. *Meaning half my village!* That would have been her answer to me. I'm guessing fossils would be okay to phase back into. I'm guessing it would not look good to jump in full tilt, advertising, making big bucks again. *So the plan is to enter the business slow, easy, tentatively, and see what happens.* Nothing on the web site. Show a meager income of a poor person. Let my bank account look like a struggling felon, off the radar.

While I would not mind supplying Hornet, I do not want her having a big mouth telling the world I am happy, doing well, making money, while fooling the government. Or no, she can say all she wants, she just shouldn't be given any proof. I have a few hundred pounds of mammoth ivory. It does not make sense to me, that she has a business depending on this material, but does not have any, nor does she have the money to acquire any. Nor does she appear to have a source. I have all this ivory, and it is just one part of my business. *What is the reality behind her words then?* I have nothing to gain by calling her a humbug. I like her well enough. I repeat that to myself. *Usually in a good mood, full of energy, who makes you smile. I'm told she is highly intelligent with some university degrees. Not a mean person, just living off in space someplace.*

"Where did you get your ivory, Miles?"

I never tell anyone. Not even my wife knows. It's a rude question. I do not appreciate getting put on the spot. She wants to know how much I acquired in Tucson from my Russian connections. I pick up on the cue, and play the part of a poor humble buyer. "It's expensive now to buy, Hornet. I couldn't afford much." And leave it at that. It is in fact more expensive to buy it than to find it. I buy some, because I do not find all the types of material I need for knife handles. I often sell carve-able inner ivory I find, and use the money to buy outer bark I need in my own artwork. It's also nice to have receipts to wave around when needed. Untraceable receipts from another country.

I have unlimited credit with my Russian connection. It has taken two decades to establish that level of trust. All the material I want, with no interest if paid back in a year in one lump cash payment in American dollars. If Hornet had the needy connections as she says, 'I got people asking for tons, Miles!' I could make big things happen. I could come up with hundreds of pounds, and or even half a million dollars' worth of mammoth ivory. Hornet is not in that league. She thinks ten pounds is a lot. I'm probably (*well more like 'could be'*) dealing with Russian mafia. I never asked. Meaning all is well if I pay. No pay, no live. I have no problem with that. Mafia in this situation is about being good to your word, no contract, nothing in writing. I have no problem with that either. Nor do I have an issue with no lawyer, no court, no government, no witnesses. Deals I make have to be serious, with trusted people who understand business. I cannot afford to get stiffed with a song and dance sob story when the question is asked, "So where's my money?!" There is a 99% chance Hornet would say, 'You would not believe what happened!' Someone with a low Jesus factor.

I knew customers in the past who could come up with ten grand cash today, to buy with. That in itself is not a fortune. But it is an amount, if it showed up in a bank account suddenly, I'd get asked how I got it! There would be an investigation. *So? It's legal right?* Yes it is, if you happen to trust everyone involved. It's not exactly keeping my proper place. I am not among the protected. So for that amount of money, both me and the cash could be made to disappear. What if the amount was 100 grand, or more? If it is true or not, I believe it, and act accordingly. *Indulge me, who am I hurting. Some fruitcake nut bush man, paranoid about the government.*

I've already had problems hiding my money getting on planes. It's my money. How much, and where I have it, is none of your business. I don't trust you. Not the government, not the airline, not the bank, and not you. I smile. Iris had some, I had some. In the carry on, in the luggage, in our pockets. No more than ten grand in any one place. Somewhat hilarious when I think about it. Reality is, it takes ten grand cash to acquire a tusk in my business. In general it takes fifty grand to be in the game, at the lowest rung.

That is me, wearing the $7,000 necklace with socks that do not match, Einstein haircut, Salvation Army clothes. Shuffling along. Difficult to put in a category or classify. Safe. That's what I call it. No one you want to hassle. The turnip you are not going to get blood out of. But probably standing out in a crowd. That's the paradox. I have no explanation. *This is me.* Rock stars, eccentric mad scientists, big drug dealers, great artists, Tarzan, all can have certain things in common. I like to think 'us' - 'not like other people'. I hasten to add, not necessarily 'better than,' simply different. Not necessarily a wolf, but definitely not one of the sheep. Another thing we all have in common is, we'd get along with each other in the same room; be allies of sorts; of the same tribe. Many envy us. Few would—or could—walk in our shoes. Many of us have issues in common; similar shared experiences.

*John Lennon was not allowed in the United states; was followed; bothered. Because he was 'odd', and had long hair.* Yes, I know! My unconscious reminds me to cheer me up. Lennon admitted he deliberately detuned his D string when he played, as a way to say hi to his mother. Now that's weird. Still. The best there is at what he does. No higher compliment can be given. It comes with a price. He changed the world in a positive way. A hero to me. On a much smaller scale I too have touched people's lives in a positive way. I've been told I'm one of the best there is at what I do. So says the news media.

When younger, Mountain Men were my heroes. Dime novels were still around. I sigh. Now I can see that most of those heroes were cool all right; strong, wild, free! But devastated the planet. It only took a decade to wipe out 90% of the US beaver population, so the rich could have top hats and Stetson cowboy hats made of beaver felt. The romantic Mountain men were the slaves of people like the Astors and Goldbergs in cities back east who made a fortune off the fur industry.

Extremely few trappers could be referred to as 'free'. They could not afford handmade traps and a horse; that would cost the average trapper more than a year's wages. Most trappers were employees of a fur company—for a while, just one—Hudsons Bay—based out of England. Trappers were fronted money each season, owing it back with interest. In the spring the company put on a big fur rendezvous. The purpose, from the company's viewpoint, was to get most of the profits back by selling drink, women, good times, as well as next year's supplies at profitable inflation, or credit at high interest. The bottom line - the Mountain man trapper who was not free. He was controlled; did not get rich, or even do well. The peak of the Mountain Man era lasted only a decade. No one explained the reality to me as I read these magical awe inspiring stories of great adventure in dime novels, or watched them on the movie screen. Many were like me - robbed of our money - and it's ok, because it is us out there having adventures in the wilds.

"You have money but I have what money cannot buy." Living without money was ok, not ok is being on civilization's hit list.

**Future Flash**

Maggie

It is 2017. She calls me on the phone and leaves a message. I loved her when I was fourteen and fifteen years old. A fork in the road of life. Now a different last name, married, kids, has a good life. Still lives in the same area; never moved. Interested in her family coming to Alaska for vacation to visit me. Memories for me that go back a lot of years, a lifetime ago. Earlier in my life when I knew her, I could have been an artist. Had a more normal life. She is a good woman. Sane. I remember most about her, the ability to smile and enjoy almost any situation. Like dancing in the rain, one of my memories. I could have loved her. Possibly 'us' together for a lifetime. Hard to know. My father decided I should join the Navy during the Viet Nam era. I unexpectedly took off, never to come back. Left Maggie and all those choices behind. Possibly I was not sane enough for her. I recall her getting upset.

I had an art room in school all to myself. Given to me by the teacher. I was working all year on a life-size clay sculpture of a female diver. I finished it, and needed to cast it. I wanted to cast it in bronze! But of course at fourteen years old that is not going to happen. I made a plaster mold and am going to cast it in fiberglass instead. I have no supervision; no knowledge, or money. As I recall, my parents never even saw it. I'm considered a gifted artist. I'm working with fiberglass in a confined space; getting high —and sick. Maggie is concerned! I'd stagger and slur my words.

"Miles ,think of your health! This sculpture is not worth it!" But the sculpture is worth my health and my life. I'm going to do this. I'm going to finish this if it kills me. It does not kill me, and I do not finish the casting I get as far as making the mold. Maggie is the kind who would put her arms around me and gently, kindly, sigh, "What am I going to do with you!"

The school year ended. And I was, off to the military. Maggie expected me to listen to her? Was worried about my health? Would not stand by someone who was so driven? Not right in the head? It is hard to know. We were so young. Could she have come around?

"I see, Miles, you are devoted to your art, and I am devoted to you. I will take care of you, so you can do your art." *That's a rare person*. Become an artist, an art teacher, with somewhat normal life? A woman who made sure I ate right, planned shows, kept me organized. Maggie became a nurse. Taking care of people. Such a giving and kind person. Devoted to her husband, good mother, happy, and say again, 'sane'. Could I have learned to be a good husband, parent, contributor to society with her influence? Among the protected? The road forked, or more like, there was no fork in the road, but I made one with a chainsaw. And is this destiny? I shall never know.

**Future flash ends**

I save letters and emails from people contacting me telling me how I affected their life through my art, stories and help in times of need. I struggle to be true to that. *At least I have heroes who do good deeds even if I can't live up to them.*

This morning the newspaper man was delivering mail just as I came out my door. For the first time in three years we meet! "You must be Miles! Thanks for being so generous with your last year Christmas gift! It means a lot to me. I like to provide good service, but often feel no one appreciates it! You were the only one in Nenana who tipped me." Eight months later, he still remembers. It was $20. Big deal, the price of a meal eating out.

Another example today. I get a phone call.

"Miles, I can't find anyone to cut my wood." I am an elder, cutting firewood is not what I want to do. "But Miles, it's a small amount. Less than a cord. No more than an hour of work for someone. I'll pay $50."

I just happened to have put a new chain on my saw that is running good just now. "$50 is too much. Let's make it $25." Am I a sucker? Good people remember being fair, kind, helping out. In the big picture such people are there for me, or help someone else in return. I notice who helps others. Even if they have not helped me. I may, out of the blue, hand them a salmon, or something from the garden. I simply say, "I notice you do a lot for the community. I notice and this is a thanks." Possibly the same will be returned by them, or by someone else. It's an idea. Who knows? This is not about being some goody two shoes. I see it as survival! We all depend on each other. Let's make it a pleasant experience. I have skills you need. You have skills I need. A community that helps each other, should survive over a dog eat dog outlook.

---

THERE IS MORE than one reason to go on a boat trip. I need to check on the homestead. I review my diary.

> **May 10<sup>th</sup> 2016** First trip of the season to the Kantishna. The objective is to pick up some of my things at the homestead that Foil is wearing out and abusing, and bring them home for fixing. I also want to check out a fossil place for mammoth tusks further up river and enjoy a trip of fishing and getting out with my boat. Weather is cloudy and cool. I would wish for better weather, but sometimes I just have to go for it. As long as there is not a lot of wind. My truck pulling the boat has a battery problem, so needs jumping.

The diary jogs my memory. I recall leaving late. I'm still off by 8:00 am. Fifty gallons of gas on board should give me 400 miles of travel, in a world of no roads, comparable to the middle of Siberia.

Everything on the boat works, which is one part of the mission - to test everything. At Old Minto, I see birds I do not recognize. White. Large. Looking at first like seagulls, but no, they are bigger. I get closer. They look like ducks. Or maybe white domestic geese. Wondering if Minto is now into raising geese for food at their Indian Rehab camp. The Native village is out in the wilds, off the road system, so makes a good alcohol recovery camp. Hard to sneak off or smuggle drink or drugs in. This is an alternative to jail for some people. I keep going, still puzzled. I could have gone through a time warp for all I know. Snow Geese are not supposed to be in the interior of the state.

I catch pike almost every cast at one of my favorite fishing spots. All small so I turn them loose. I have plenty of food on board, and pike is not a highly valued eating fish to me. I get to Tolovana, and had planned to stop, but there are several boats here. I think it might be Griz, the hunting guide, with some spring bear hunting clients. No one waves me over to visit, so I wave and press on. I have places to go, and things to do. This is a change noticed over the years. A decade ago, those of us on the river visited each other, exchanged stories, and bonded as a tribe. Now we are more like disconnected individuals operating on our own. Not as strong. *Divide and conquer.* We may have broken some rule and not know it, not wanting visitors to turn us in. Strangers are now your enemy until proven otherwise. Too many of us got severely punished for being turned in by 'who knows who', for things like using lead shot to hunt, having a campfire without a burn permit, or not having numbers on our boat. Rules from another world imposed upon us.

The wind is up a little, but not serious. Lots of fast moving clouds; a front moving in. Cold. I am dressed warm. Water seems a tad shallow in Caribou Crossing, so I trim the nose down and add throttle feeding water to the prop in water not as deep as the diameter of the propeller. I can tell by the shape of the chop the wind makes on the water what the depth is. When I get through the crossing I trim the nose back up. I burn more gas trimmed down, and travel is not as safe. Out of the corner of my eye I see a bear at the river's edge in a sharp corner. I am around the corner and beyond before the bear even knows I am here, or it registers I see him. I just miss the end of one log, just miss the end of another on the opposite side. All normal travel.

I decide to take a break at a creek I sometimes stop at. I not been here in three years. It's on the other side of the river, and is not on the map. The entrance cannot be seen. I simply know where this is. I slow and idle through the narrow mouth. The creek widens. I coast to a stop, nose to the bank and ground myself in the mud so I do not drift away. There is a porcupine in a spruce tree nearby that notices me, but is not concerned. I step off the boat to stretch my legs, look around. There could be a wood burl, some chaga, a mushroom patch, anything like this to take notice of. This is honestly a business trip. The only thing I notice is a new beaver damn gong in a few hundred yards up that will stop me from going any further. I once went miles up this creek. Life here is like this, creeks close and something else opens.

I suppose this is like city life as well. New roads open, old routes over the years get potholes and not used anymore. Maps can be outdated in a decade. We stay alert, explore new routes and enjoy it. New neighborhoods open up, a housing development, hot areas to shop become slums. We check out the new mall. Likewise sources of goods change, and we shop someplace else. A mom and pop store closes where we once got home-made bread. Same in the wilds. I walk around a little and feel ready for the last leg of the Kantishna river.

I am at the homestead in the predicted six hours in hopes the water level is high enough for me to boat behind the island into the pond where I tie up and switch to a canoe at the beaver dam. Bad luck, cannot get up the slough. This has a big affect on how much I can haul out. There is enough water to drag the canoe through the slough to get supplies to the cabin from the boat docked on the main river, and reverse the process when done. I have two canoes here, and want one to come home for repairs. I am happy to be in the wild, breathing air filled with the smell of poplar trees and sedge grass.

I had hoped Foil would take care of supplies and equipment here since he is using them. The tools need to work for him and me both! The impression I get is, Foil wants me to supply and repair while he uses without paying. It is hard to see how he thinks I would be happy with this arrangement. I get the hidden canoe into the water and to the boat and transfer clothes, food, and other essentials from the boat and and take them over the dam to the homestead. After walking in the water and mud, I am wet and wore out. I'm still recovering from my heart attack! The cabin has not been invaded by mice or squirrels for maybe the first time in my life. Ha! The sleeping bag on the bed has not been made into a nest, or full of mouse turds. *Amazing.* I have a good nap. When I wake up, there are ducks in the creek to watch. A beaver swims by. Sand hill cranes fly overhead as I fool in the yard cleaning up.

I hope to extend a ladder I am building between two close trees to gain height for cell phone service or ham radio contact. All the years I lived here, this was not important. Now I am diabetic with maybe other health issues. Communication is now part of survival. I progress another six feet up, but am tired. Cannot do any more work. I'm slightly bummed out. I have come all this way to do this one project, and am too tired to do it. I try, but no cell or ham contact. Maybe next year." That sounds a bit lame as my years are running out. I am not getting stronger with the passage of time. A year will arrive when it is a chore just to get here at all! Foil has pointed this out saying, 'Miles, we can be partners! I can help you in your old age with projects hard for you to do!' I noticed he built a ladder up a crooked tree using big nails he can climb, but I cannot. He can get phone reception, I cannot.

The cabin is only ok. It does not feel like mine. Foil has rearranged everything to suit his wants. I can't find anything of mine. In the beginning he was very polite and courteous, making sure it was his things put away, and I arrive to how I left it, he as a guest. I now at least have a place to sleep, out of the bugs. I'm not going to complain.

The arrangement of supplies sometimes has a reason Foil is not understanding. He arranges all the food right over the top of the wood stove. When the stove is lit, the food can be fifty degrees hotter than where I had it across the room. *How would he know?* He does not have to know if he simply keeps it arranged as I have it. I explained this to him. Food may not get eaten up quickly. Some dry goods might remain here for a decade. Even pancake mix is not meant to be kept this long. It is therefore important to keep all foods as cool as possible. Temperatures over 100 degrees are most damaging. Not acceptable. My spices loose flavor, vitamins get destroyed. It is in fact a big deal. I'm already tired, so I cannot easily move it all back again.

Next day I planned a trip further upriver. It is colder, almost freezing with some wind, maybe rain due. I take off anyway dressed in insulated coveralls and fur hat. I can stop and make hot soup if I get cold; turn the portable propane heater on. I see a lot of game. A nice cow moose, lots of geese. I may or may not have snaffling a few geese on my mind. It depends who is asking. If I do end up with dead geese or ducks in my boat, I acquired them with my bow and arrow. I repeat many times, I am not allowed to have a gun. I do miss eating wild game. Most of my life I have eaten at least a few ducks and a few geese a year. Most years, I'm happy with one or two. it's not like I desire to abuse the environment for my selfish gain. I understand, animals have more rights than I do.

I check out a couple of favorite fossil spots. One of the largest tusks to ever come out of the interior of the state came from this spot; it was fifteen feet long, and three hundred pounds. I pulled out a couple of smaller ones in the past, but have never seen remains of anything else, like bison or smaller animals. I see nothing. It is possible this is not as good a spot as I dreamed. Possibly the kind of country a lone mammoth passed through on the way to someplace else. Not a hot spot filled with fossil finds like other places.

It is common for me to have dreams, or intuitions, 'knowledge there is no logical way to acquire'. Almost like a connection to a past event, or period in time, or 'called someplace, pulled by something'. I can't explain it any better. Times I stop, back up, poke in the mud, and there is a mammoth tusk. Few others are finding them. Through a fog I see the snow geese again, just like near Minto. Lots of them, as in hundreds. This is not possible. I help myself to the impossible. The geese do not move at the sound of my boat; just stare at me. Even after I kill two, the flock does not take off. How could a flock of hundreds of migrating birds act like they have never in their life seen a human? Such odd behavior. I have seen a lot of odd things, and simply accept them. I eat snow geese and press on.

This visit upriver is not a total loss. It is good to have it known, and prove, I have been here, there, everywhere on the river. I document the variety of places with pictures. If I come home with anything, where did I get it? No one like Foil can pin point an area to grab finds ahead of me.

I hit a few submerged logs hard and cringe. *I hope I have not damaged the engine!* Time is spent checking out the engine lower unit for cracks, bent shaft, or lost lube oil. Everything is fine. This is an advantage of having a huge engine not working very hard. I rarely see anyone on these trips. This means not seeing anyone for a week or more. This as well is my normal.

One of my favorite fishing spots up this way produces a grayling using a huge pike lure for bait. *How odd.* Usually this fish goes after mosquitoes and other small insects. This is a huge fish I decide to have for lunch. The water is clear as glass and I can see the bottom twelve feet down. No one has been fishing here since I was last here a year ago. This is one reason I take such long trips. Alder and willow leaves are reflected on the surface of the water in a mosaic of pattern and color. A few old, tan cattails stand tall, with new yellow marsh marigolds low along the shallow edge. Another monster fish goes under the boat. The skin looks like leather, not scales.

MEANWHILE, 'BACK AT THE RANCH', at the same time I am out boating... "Lawsen! I need to talk to you!" Lawsen freezes in his tracks. When Hornet barks, listeners pay attention. "I got an idea." Lawsen knows what that means—trouble. "I need mammoth ivory. I'm sure Miles is not telling us everything. The only reason he'd not tell us is because he is on to something. Maybe you can follow him for me?"

Like Jack responding to his fat wife, Mrs. Spratt, who could eat no lean. "Hornet, I do not have a good enough boat and engine. In fact no other mammoth hunter I can think of could keep up with Miles. His engine alone is $9,000. Who of us has that kind of money?"

"Then I have another idea. Miles keeps his boat on the Nenana River. You could go down and get his GPS, borrow it for a few hours. I could plug it into a computer, download his waypoints and load it into my own GPS. We could go to his hotspots at our leisure!"

"Hornet, I like Miles and do not want to mess with him, bad Karma anyhow."

"Come on, what's the harm, Miles needs to share. It's not fair is it, that he selfishly keeps information that could help others? Nothing really illegal about borrowing and returning his GPS. If he made a report 'someone borrowed my GPS' you think the cops would care? So what's the risk? With lots to gain!" Lawsen does not reply. There is no arguing with her logic. She adds, "You know, Lawsen, that is your problem. You have no vision. You are stuck in a rut being poor and will never get ahead." Lawsen knows this is the main reason she left him.

Six foot skinny 100 pound Lawsen bows his head as his 300 pound muscular ex glares at him. She finishes up the conversation, "Lawsen, I heard that friend of Miles, Foil would like to tap into the fossil hunting business. He was supposed to be working in partnership with Miles, but Miles cut him out, left him high and dry. I could partner up with him. I know how to run a boat, and I know where to sell the fossils! Foil could use my expertise. We could make a lot of money. I bet Foil has more vision than you do Lawsen."

Back on the river at this exact same time…

Many more snow geese are seen on the way back to the homestead downstream. The water level is low. The main channel is three to four feet deep. The river is a hundred feet across, and very twisty. Spruce and birch hang out over the water, creating a tunnel to run through in some sections. In one shallow spot the water is more clear than usual. The scene is like coming out of a Russell western wildlife painting. Possibly an underwater springs in the river is bubbling up. I am pretty sure I see a huge fish go under the boat. I saw other huge fish further upstream, so what are the odds? I mean huge, as in over seven feet long. Even more odd is the skin looks more like leather than scales, as I noticed before in another area.

I pull over and tie the boat up. I want to stare into this clear water for a little while. In the mud where I tie up is a major footprint. The track of a hoofed beast, but not like the moose I know, more like woodland bison. It supposedly went extinct, I forget how many hundreds of years ago. I did hear the university got hold of an old breed of woodland bison in another part of the world, and has experimentally turned them loose in Alaska. I assume this must be one that wandered far away from where it was turned loose! Wow! Hundreds of miles from where it was turned loose! *How strange!* I spot some plants I do not recognize. I know most plants as an artist, and as someone who studies my environment. I know what to eat, what is a medicine, and what has uses of different kinds.

This is a fern. I only know ferns in the interior of Alaska that grow near hot springs like around Manley, or Tolovana. Possibly I have come across an undiscovered hot springs. I am in a back slough I may have never gone through before. It would not surprise me to see a mammoth step out of this setting to get a drink of water. What an odd sensation and thought to cross my mind! I smile, *I bet this is the exact setting and environment mammoths lived in.* In theory there is no reason for them not to still be here. That of course would be impossible. Still, I daydream, and these are pleasant thoughts. Life back in another time. In truth, who would know if I left time as modern man knows it, and came back again to present time! I get to my homestead and it's been a long day. No fossils this trip.

I'm more tired in the morning than I wished, but get the canoe to the boat on the river, along with generator and other things I want to fix or put someplace else. The weather is clearing as I head for home, but think wind might develop. I advance my throttle more than usual to try to beat the wind later in the day.

**I note in my diary:**
I am home by late afternoon with no incident. Iris tells me Foil and Mad Jay have been making fun of me, saying I foolishly took off in bad weather, and am surely cold and regretting my stupidity. I am not overly bothered by such talk. I am on the river. Where are they? In the peanut gallery, sitting in a bar talking about a dream I am living. A dream they claim is their dream as well. With them it is, 'only if'. In this case, only if the weather cooperated.

Several containers on the boat can be locked. A front deck locks, and several huge totes. I religiously put away valuables, including my GPS. Places I mark as a waypoint are mislabeled. Good fishing spots are labeled 'fossils.' Good fossil spots are labeled 'fish,' and marked exactly two miles away. I would not want Fish and Game to confiscate my GPS and have all my data, maybe give it to the university so my hot spots can be declared archeological sites and claimed by the university in a land grab. I do not worry or focus on negative possibilities. I understand anytime something has value - fossils, land - there will be the potential of treachery. *Knowledge is power!* So while I am considered poor, people want what I know.

I HAVE the bird laws on the computer, confirming the legality of using grouse feathers in art. I decide not to take this to the Nenana Cultural Center gift shop. *Choose my battles.* I'm already somewhat on the outs with the gift shop workers for pointing out the error concerning mammoth ivory. One clerk commented, 'We do not want any trouble, Miles!"

I tell Iris, "I was ordered by the court to stop offering legal advice." The implication is, this would be a crime. The reason for the order was, 'Because you are not a lawyer.' I am in no position to argue, *You are not a lawyer either! Will you get sued for dishing out false legal advice?* I believe the real reason is, the government does not want people to know the rules. Keeping us nervous, unsure, and afraid makes for better control, bigger government. Happy strong confident people have little need to turn to the government for help.

When I was the head of our Chamber of Commerce for several terms, favors were owed to the Lions Club. Favors that screw other people. Decisions have to be made. Someone often gets the short end of the stick and is not happy, while others get special favors. I understand. I can be a team player. The reality is - who you know matters - likewise your color, occupation, income, religion, outlook, and how you behave. *We are not all equal!* I do not get indignant about it. It's the same with wolf packs; most animals.

One road to happiness is to go someplace where who and what you are is accepted, and you have a tribe. Making waves, swimming against the current is not always wise. So Miles, I want to know why you think people want to read about your opinion of the world? What makes you so special? You know about opinions? They are like a*&#, everyone has one, and they all stink! I get the impression this refers to a subject discussed when I am not around like minded people. I nod sagely and reply, "Then my appeal might be to dogs! They like to sniff *&#."

*Yes, for the most part it is good to keep our opinions to ourselves.* People in general ask me questions. People invite me to their parties. *For entertainment.* Pay to have dinner with me, to hear about my most wonderful life. *I think all lives are wonderful! If you are not happy with yours, and your life is not wonderful, why not? How can anyone say they know God, be spiritual, yet be ungrateful for the life given?*

I've met literally thousands of people while doing shows. It is common to get asked, "How do you do that?" When looking at my unique art. "Wow, I could never live like that!" When hearing of my lifestyle. "It must be scary!" "But how do you pay your bills?" "Is that legal?" Others do not appear to get asked these questions.

I answer such questions as best I can. There is no short pat answer. The answer is about a lifestyle, not grasped in a few 'one liner' cute replies. I want to seriously answer the questions. To do so, means to show who I am, how I think, how the world looks through my eyes. Many do not want to hear it! This is understandable and fine. I smile, sincerely wish them a nice day. It is probably true, however, that I like the attention; being the expert, being asked advice, the life of a party. It feels something like respect. It's a good replacement for love and acceptance.

I enjoy hearing about other lifestyles not within my personal knowledge. How does life in the ghetto look, or remote jungle, or under communism? I hope to appeal to likeminded curious people.

If I even hinted at anything 'out of this world' nothing I say would be believed. Worse, I could get locked up. There would be suspicion I am not right in the head. Not right is bad. The not right in the head bad person is probably the one who did the awful deed the community is trying to solve.

"There he is now! Get him! Hang him!" I therefore keep certain things to myself. I do not want to be wacky.

I review one of today's emails

**Hi Miles**

Read the first two books and just about to start the third. I started work at fifteen with no qualifications to my name. I started our own business and got a turnover of 1.5 million so you would think life was great but the truth is we are happiest when we are at our caravan in Matlock spending no money and not having to deal with people, You could say if you want to escape just do it but unfortunately I have had three heart attacks and a five heart bypass all caused through the stress of business and being a diabetic since the age of four, so I will have to admire you for what you have achieved and wish you well for the future. I'm sure you get 1000 of emails but would appreciate any feedback you are able to give All the best **Paul**

I reply, summing up:

Sorry for your health issues! No I suppose you could never be me. No one can. As no one can ever be you. I could never make 1.5 million! Wow! Every choice any of us makes will have a road taken with a reward and a price to pay... I've had what some might call tragedy in my life. People, mostly relatives, who feel sorry for me, or see me as living in a fantasy and denial, full of crap. 1.5 million would probably shut them up. Maybe you can count your blessings?

I ponder this guy's situation for a moment. He mentioned when he started at fifteen with no qualifications, he lied to get his first job. I wonder then if he was not qualified. Part of the Peter Principal- working your way up to your level of incompetence, and staying there. He may have been struggling all along, partly due to the original lie. Maybe he had a dream to succeed and making a million dollars was worth a lot to him! I say, "Great! Go for it!" Of course it has a price! Yet, who among us mortals achieves such a goal? He could set an example of what is possible, starting with nothing! On top of this, overcoming physical issues! I do not get into any deep answer. It is just interesting to me to hear other people's stories. All of us can speculate about someone else's story, "I would have done it different with a better outcome!"

I move to my next email

**Hello Miles,**

I am sending you my heartfelt prayers for healing and lots of love and light! Please take whatever time you need. I would like to have the cap cast on it with the butterfly please. I don't want to have a wire wrap around it. It will have your energy and I will be happy. **Bless you my friend.**

This is a typical customer. I remind Iris, "Remember the eBay customer that got mad at me and wanted to know why I had not checked my email and replied in the past hour?" I tried to kindly explain I am lucky if I check emails once a day, and often the internet is down for days at a time. *This is not Kansas anymore Dorothy.* The customer went ballistic on me. Called me names. I politely suggested they shop elsewhere, I am not interested in their business.

"Then he threatened to sue me, remember? Oh that reminds me Iris, I went to check on the boat at the river and it looks like a lock has been tampered with. I don't think they got in."

"But Miles, what have you got in there anyone would want to steal?"

I agree, *looking for what?* I'm back on track thinking good thoughts about people. I'm thankful I have a customer base of those who know and trust me, who are willing to wait for a product they know they will cherish. Or, I thought I had such a customer base. I admit, after I got arrested and out of the animal parts business, most everyone disappeared.[2] I had to start over. I wonder what it is environmentalists who hate me wish to accomplish?

There are things I could change to make more money. But two of the most valuable things we all have are time and peace. We can't buy it. *But money helps us spend time well and peaceful.* I know people who burn money and deny its usefulness. Like that hero guy McCandless that went in the wild and died. No, not me.

"Speaking of which, Foil is around, Honey!"

I wait for her comment. She says, "I knew he would arrive soon after the ice goes out. Watch him come around and want something!"

I very much regret the day I invited him to Alaska to visit and have a vacation at my homestead. He seemed nice enough when we met. He reminded me we have known each other a long time. I never thought about it one way or the other. I only smiled and nodded.

### Past Flash

I meet Foil as a customer at the big Tucson fossil show in 2001. I also go to Tucson to help, and visit my aging mother. Foil hangs around and enjoys talking about Alaska. It seemed to me, enjoys a break from his hard physical work. Foil buys a lot of fossils and collectables from me. However, he always expects a hefty discount. So he is a customer, but not a great one, when others pay the asking price, or even tip me.

On this day, Foil tells me what a buddy he is, how grateful for my time and stories; what a good artist I am, how he admires and envies me and my lifestyle. Sadly, I eat it up. I invite him to Alaska! "I have a homestead I do not get to often. It seems to me you'd have the time of your life if you were there for a few weeks! Something you can't afford to pay for. I will take you there and pick you up. Just cover my gas costs!" He practically bows to me as if I am a God. He will do anything for me in return.

"It's been on my bucket list Miles!"

All I expect and want is a sincere thank you when it is over as something I could give that he can cherish. I have done this a dozen times before in my life.

Winter ends, spring arrives and Foil shows up as promised for an early trip out into the wilds of Alaska! So excited he is stuttering. Sadly the ice is late gong out. It seems over the past decade that global warming means more of a 'flatline' temperature. Not as cold in winter, not as hot in summer, more of a blend going from one season to the other with no extreme change. Late snows and freezes, late ice break up, yet an over-all higher average temperature.

"Miles, I want to go!"

"I cannot control the temperatures Foil. We have to wait for the ice to go out." I notice many civilized people are not used to being dictated by the weather. In civilization man overcomes and is in control.

Iris and I feed him, help him out in various ways until we can leave. I have to run Foil to Fairbanks several times for supplies, help him out picking things out he needs. Foul had told me in the past he is used to camping, knows what he needs and expects to take care of himself.

Foil loves the boat! So quiet! Loves the expanse - no roads, no beer cans or ties or planes, "Nothing but Nature Miles!"

I nod and we get to the homestead in about eight hours. I show him the canoe and the beaver damn he has to go over and we open up the cabin. Well, it is my small shack now, the cabin burned in a forest fire. I do not have a lot of time. I head home the next day, due to pick Foil up in two three weeks. I supply firewood, propane, fish poles, most of the food, bedding - everything you need to live a month. I have more supplies then I need, no big deal. I'm a little sad I do not have the time I once had to spend the time here.

"At least someone will enjoy it."

I pick Foil up after his month stay at my homestead, and he wants to buy the place.

"Excuse me?" I smile. "This is my dream home, you can go find your own. I can help you find a place like this for sale."

"I want this place! It's perfect!"

I'm not interested in selling, but Foil starts telling me what a great guy he is and how generous. "I have a car in Tucson, you need transportation? It's yours, I'm not using it! Because we are buddies!" Wow! Gosh, golly that sure would help us out in Tucson all right, saves us $500 in rental each trip! "Miles, I know people. I can find you a dream home, cheap, around Tucson, and you can have your own place!" That as well would be fantastic, to get help from a local. "Miles, I help you in Tucson, you help me in Alaska!" It's a long story that evolved in stages. He's been checking on my mother, takes her shopping. Mom tells me what a great guy he is! Even she is puzzled why. She has her boyfriend who drives and looks out for her.

Foil ends up in Alaska most summers. I help him get a cheap place in Nenana. I offered to let him build his dream home on the Kantishna property, there was room for him. I tentatively agreed we could share the property jointly.

I reach a point where I say, "Fine, Foil has a need, Foil can pay for that need." It's not right, I support Foil and take care of his needs, and not much in return.

"Miles! What is it with you! How come friendship is suddenly about money and who owes who?"

**Past flash ends**

Among other things, Foil asks questions, wants advice, but when I offer it, does not believe it or accept it. Then makes fun of the advice, badmouths the advice to others. It's not my way to offer unsolicited advice. If asked, I give it, and if you do not like it, I shrug my shoulders. There are, after all, many ways to skin the cat. That is his right, but by not taking my advice, he has got into a lot of binds that cost him grief and money. I'm not going to be part of these. I do not have them. It's one reason I live in Nenana now, instead of the homestead; I can't afford it. Foil wants his beliefs supported and affirmed. I run into Mad Jay who has a similar personality as Foil. They hang out together.

"Foil, if you wish to be my friend, you have to stop passing personal information about me on to Jay." Foil finds this hilarious, and goes out of his way to talk even more to Mad Jay. Mad brings up the subject of me being a criminal. Then the subject moves to the entire lifestyle of harvesting wild edibles from berries to moose. Living off the land!" As if this is a joke.

Mad and his friends want to know why people like myself are criminals and, "Why not just follow the law like everyone else?" Indeed. What makes me so privileged? Yeah! *Mad is a child molester, so he is just talking nonsense if he wishes to discuss either ethics or laws.* But I know him, and if I bring up the pot calling the kettle black, he has engaged me, can manipulate the conversation and get me to show something negative, like irritation, whereby he gets to smile and look around with a 'See, I told you so!' With nods of agreements all around.

# MILES MARTIN

*My custom art. Pendant cast from old copper water pipes.*

# CHAPTER TWO

## HOMESTEAD PROBLEMS, LAND DISPUTE, GIFT SHOP

The mayor is in his office behind closed doors with the secretary and a couple of city council members. He begins, "I think Miles would be good."

"He's a felon!"

"So is half the community. Miles shows up at meetings, is sober, honest, cares about the community. I'd like him on the city council. Who else is better?" There is no reply. There are not a lot of good choices, fewer who would accept the responsibility.

Someone else adds, "Miles does know our community. He has association with the subsistence Natives, at the same time runs a business. Not sure he represents the bars much!"

"We have others who can do that."

"True."

"Miles is such a sucker, not a good judge of character, so naïve!"

"A person could have worse faults. He still brings a lot of good to the table." No reply, but nods all around. It is hard just to get board members to show up at meetings so there is a quorum. All these people know Miles shows up at meetings regularly.

"Ok, so I'll talk to him."

I just got a box in the mail from this Melissa customer who loves my art and lifestyle who now wants to do something nice for me. It's a box of special foods. One plastic sealed bag is labeled, 'Indian food'! Advertised to be, "Real wild buffalo jerky mixed with nutritious berries." The product looks like White Man factory made Slim Jim jerky, only in small inch squares. Highly processed, with stabilizers, preservatives, processed oils, and trans-fats. I see a lot of salt listed.

I am certain that no Indian a hundred years ago would recognize this as related to what he eats. No Indian would know how to make this, nor have access to the ingredients in the wilds. *If it comes from a plant, eat it, if it is made in a plant, do not eat it.* I am certain if I ate a steady diet of this, I'd die in a short time, as in the 'supersize me' story, about the guy who wanted to see what happens if he ate nothing but fast foods. I am also convinced that modern Indians see this in the market and smile, nod, and agree, 'This is grandfather's food! Yum, this is how our people lived!'

Advertised as 'all natural'. I came to the conclusion long ago, there is no such thing as 'un natural,' as legally defined. It all came from the earth somewhere along the line. That is a meaningless word. Nothing comes from another planet. There is a picture of a rustic Indian on the bag. I do not consider myself a health nut. I'll eat it. It will be good as travel food on the boat. A small amount like this will probably not hurt me.

The issue is, this kind of food is all that is available to a lot of people, maybe most people. I met inner city guys in prison who had never seen a squash or salad. Every meal their entire life came out of a wrapper. They would not eat vegetables because it came out of the filthy earth. I tell Mad Jay and what's his name, "Look, right here in today's headlines. Farmer using sewage as fertilizer loses permit." [1]

Summed up, a large farm operation in our area has been using sludge from sewage lagoon ponds as fertilizer. This is legal. The problem appears to be the size of his operation. I'm wondering why that matters.

Iris comments on the article, "I lived around farm country and it smells. Farmers spread manure over their fields. It's just something you live with being around a farm!"

"Did you see the headlines, Honey? The food we get might not be safe! Oh my gosh!"

*Let's make more rules.* That is what I get out of it. The only safe thing is certified chemicals. Mad Jay does not even glance at the paper I hold up. He is not interested. I believe because his goal is not to get the facts straight, but to cause chaos and discontent.

My well-meaning customer sends this Indian food to me as mountain man food. The sort of thing I am bound to appreciate. She assumes this is what I live on! She thinks this is what I mean when I speak of eating wild moose. This is what she pictures as she reads my books of a lifestyle she envies and admires. This is scary.

Scary because most of those who admire me, support me, are on my side, who wish to defend my lifestyle do not understand what it is they are defending. Reality might shock them.[2] Worse, turn them against me. Who is left who could step out of civilization and make it on their own.

We are all plugged in and totally dependent on those who control us. Run? Hide? Get away? Where? Get back to basics? How? If someone who admires me would vomit contemplating eating a carrot fresh out of a garden because it might have a bug on it, how do we, as people who want freedom, get from 'there' to reality. Because that carrot fresh out of the garden with a bug on it is as good as it gets. It's downhill from there. Imagine that same carrot without refrigeration, several months old, dried and wrinkly, green goo growing on it, and being glad to have it because that is all there is to eat! *I can tell you some bug stories believe me! Just read books one and two.* Freedom is not free.

I open my mouth to tell the world about the great wonders of a magical life of freedom, independence, good health and adventure. I can't even get past the basic concept of our first meal together. It's more than sad. It's heartbreaking. It's lonely. I wave my hand in front of my face to make the images go away.

The equivalent might be a visitor to the city. You discuss food with them and they do not know food can come in a can, and you realize it does not matter, they would not know how to get the can open. They are asking you how to be like Bill Gates, sure that is the life for them.

In my youth, I'd hear women, looked into starry doe eyes. I get romantic, respond, have the hots for them, invite them to a meaningful time together that could be the answer to their dreams, and mine! How many times? Fifty? How long does it last? A weekend? 'You are the man of my dreams, soulmate!' Good for a one night stand? It's more than heartbreaking, it's torture. It's hell.

I'm no longer interested in spell binding crowds with enthusiastic stories from the wilds. Please. Just go away. You are not being helpful. In the end you will have me arrested and in chains in a dungeon. *Over the subject of what I had for dinner.* Yes. I understand. I need to be considerate of others. Follow the laws, my food is precious living wild things that belong to the people. I get it. Living things with feelings. Cute fuzzy cuddly creatures that have rights. Rights greater than mine. I get it. I was warned, then told, and now punished.

I consider the larger cities do not want smaller places like Nenana to survive. The argument being, "Your community is being subsidized!" I even agreed for a long time.

"It is true! We cannot afford our own post office, roads, airport, police, ambulance! Someone else has to pay for it! I agree, that is not right!" My answer was for us to live within our means and not have services we cannot afford. Thus, it could not be said we are sucking off the big city!

Back in the 70s, village life was more self-reliant. Galena, and Manley Hot Springs come to mind. When I arrived in Manley, Karen was a substitute teacher who arrived at school on a dog team. Children arrived by dog team. There was a place to tie up dogs. All grades were in one room, and there was one teacher. It is what the community could afford. Karen was not a certified teacher. Just a concerned parent who needed a job, had good ethics, and reasonable amount of knowledge. The post office was located inside the local grocery store. Heated with a wood stove; the wood donated by locals. Roads were not plowed, or even named. I thought this worked. Everyone was happy, had work, had a home, lived well, did not depend on outsiders. There is still zero form of government in Manley. However, more and more, Manley is forced to live up to certain imposed standards. Follow rules we all have to follow, that cost money.

I GET AROUND to thanking this Melissa customer friend.

I got your package! Thanks! Nice surprise. The condensed dry food is great for travel on my boat when I go out for a week at a time. Iris and I were just commenting that the real and good foods are not very available to civilization. It's 'illegal' even. Fresh fish you caught, might be contaminated. Garden goods might have a bug in it or dirt on it. Canned pickles might not have been properly sterilized. etc., etc. Well got book seven off to be fine-tuned for Amazon and beginning book eight! Busy time of year. Got garden rototilled and flower baskets hung and need to get my canoe ready to patch. It has been raining so want it to be dry. I do not get as much done in a day as I used to. Geez! Anyhow, thanks again for remembering me! Miles

IT WAS my friend Witty from Mental Health who angrily made a valid point concerning subsidized smaller remote places. "Miles, you have been brainwashed! The resources of the country come from remote areas like Nenana. We are the guardians and real owners of those resources. It is on our lands!"

This is true. Our entire state is remote. Without going into details, the remote people feel they own the resources, and they are being stolen. This is the same in the Amazon, Siberia, most remote places. Such places are referred to as, 'The empty quarter' in a well-known book on business and dividing up wealth. Alaska is a place civilized people get high grade resources cheap from savages, homesteaders, and hicks who are not owed anything because we are not strong enough to defend ourselves. We are gullible, not city wise, or civilized, who do not deserve the luxuries of payment. We are inferior in some way.

Big money is made out of Alaska, yet not controlled by Alaska. Some small amount trickles down to us if we humble ourselves. We are treated as if we are beggars, that nothing is actually owed us. Alaska supplies gas to the rest of the country, yet we pay the highest price at the pump of any state in the nation. We even have a refinery here. The winner of the presidential election is announced, while Alaskans have not even voted yet. That's how little the people matter. But yes, our Alaska salmon is desired. That $40 Salmon dinner in New York the fisherman got paid a dollar for.

Much of the fault is our own! I talked to a politician who explained how he tries to fight for our rights. We do not want factories to turn raw into add on value. We sell our timber like it will last forever at rock bottom cheap prices. He said, "Fort Knox roughly melts it's gold into bars, then sends it out to be re-melted and further refined when it should be done right the first time in Alaska, for more money!"

Sarah Palin was reprimanded by the majority for not wishing to accept a Federal grant, saying it is not free, and comes with a price the state should not be willing to pay. I agreed with her. But the majority did not. Sarah resigned, but some insiders say she was forced out for not playing the game. I doubt we will ever know the truth. It is hard to sort out what we hear. Hard to even say a lot of things fit together.

I'm out in the wilds, No camera sees me, no paper trail, electronic trail. Poof, I'm simply gone, and no way to find me. No way to know what I am doing. I was told by the local Fish and Game cop, "We know you are up to no good, Miles! No one goes in the wilds unless they are up to something illegal, and we will find out what that is!"

I got into a discussion on communication available out in the wilds. At the ham radio class Bean and I go to, an expert in radio waves tells me cell phones are deliberately set to shut off when more than twenty miles out of range, when in theory they could have a much longer range. I asked, because I noticed how weird it is, that no matter what direction I go, I get out so far and get cut off! I am shocked by his reply and ask why in the world anyone would do this? He tells me when he asked the experts, 'We do not want people more than twenty miles from a cell tower.' It sounds like wilderness and remote communication is not encouraged or offered, unless we pay big bucks for satellite connection. The wealthy get more privileges.

---

WITTY MENTIONED at one of Nenana's WIN meetings (Wellness in Nenana) that there seems to be a process for forcing people into a downward spiral of having trouble they can never get out of. That has to do with getting into the legal system. Witty's occupation is dealing with addictions and mental health. 'Being in the system is big money!'

Later I want to express another side I see. "In prison, I saw an effort to have prisoners succeed on their own." I felt the message is, "We will give you a crash course on some basic methods. It is up to you to take that, put it to use. It's not up to us. If you do not or cannot grasp it, you can be one of the sheep the system sheers. The choice is yours." It does look good for the system to actually have a few move on, become responsible citizens. The head lady who signed my papers at prison said, "I never heard of you."

I replied "And that's good, right?" Standing out is not what you want to do in prison. I tell Witty, "In any in prison argument both parties go to solitary confinement." At first this seems so unfair! "Is it just to mess with us?" The lesson is, it does not matter who wins a fight. In the big picture, in the real world, you both loose. If there is trouble brewing, walk away, go someplace else. Do not be there when the do-do is in the fan. "There were some bad individuals in the system. Our own boss for the garden for example!" But guess what, I was there on his last day in prison and no one said good bye, wished him well or good luck. I saw that. There were some not so fun things to go through, and the lesson is to deal with it, because real life is like this. Certificates with no meaning, stealing city water, female guards watching us in the bathroom. "Decisions have to be made. It's all a test Witty."

At the WIN meetings we discuss the subject of felons getting jobs, not because we think you will end up a felon! However, if you choose the path with different music in your headset, you risk not being welcome which can equal arrested for being where you shouldn't be.

"It is hard to get a job once you are a felon." I work for myself, but can see how it would be for others. One regular who attends WIN owns a truck-farm and gives her update telling us she has a source of hard working cheap reliable labor. I have known her for twenty-five years. We were actually sort of dating a couple of decades ago. As the meeting progresses, I day dream. We got along all right for a while. She is hard working and loves the outdoors. Reasonably good looking. I'd bring movies to her house and we'd watch them. She'd cook for me back when I was single. I'd sometimes take her out to dinner. What we enjoyed most was going out for long rides on my snow machine.

### Past Flash

I just bought a new snow machine, cash, so am strapped for money for a spell. I cannot afford the gas to go out on a snow machine trip date. I suggest we go 'Dutch' - share gas costs so we can go ride! She does not agree. I feel sharing gas costs is fair, and the right arrangement when dating in such a situation. She believes the man pays for everything if he wants to go on a date.

I can respect this idea, and understand it, but do not agree. This is a common arrangement. When money becomes an issue and you cannot be together due to money, being together is what matters. I believe in equality.

Put another way. She has a car and I do not. When we go to Fairbanks together in her car, I gladly buy the gas and pay for the meals. Neither of us will budge. If I can't afford to take her on a date, then I can't afford to have her as my partner, as the provider. I assume we are not in the same economic class. I stopped seeing her. No hard feelings. I felt used. I have more money than she does, in the big picture. My money comes in 'feast or famine'. We get along and work together sometimes on community issues. We are both on the library board.

**Past flash ends**

So yes, I can understand her outlook here. Hire Criminal X, who cannot get a job. Pay such people less than minimum wage. They are not likely to protest. She says, "Like those who cannot pass a drug test to get a regular good job." It is true! Such people may well starve to death unless you give them peanuts. Such people may well be grateful.

---

SURVIVAL IS IN UNDERSTANDING, and adapting. I do not expect to change things. I tell Josh, my Athabascan Indian friend, "I would have found history fascinating in school if I knew it was in flux and open to interpretation. I would not have been so bored!"

Josh is not someone who would ever attend a meeting or join a group, so our local WIN meetings is not a place he'd show up. He is a respected elder who knows the old ways. As a winner of the Iditarod, and lifetime participant, he understands some aspects of history. We had talked about Indian history and how it is lacking in our schools! The world according to white man!" as Josh says. He's short like me, somewhat feisty, intelligent eyes. History rewritten. Josh reminds me he was around when Alaska statehood was voted on. Voted on three times and failed. He points out what I read, the military was allowed to vote. This is how we were taken against our will as a state. This is not the present news and latest history. The latest is how we all rejoiced when we got to be a state!

Josh changes the subject. "Miles, how is your art coming along?"

He knows I am getting ready for another tourist season. I should get off probation soon. I can change plans when that happens. Meanwhile, I am showing little income with small marginal efforts that looks promising, but looks like the efforts of a slave, dreaming of freedom. I do not give any details. Being a member of a minority race, Josh understands.

"I got a job on the pipeline, Miles. It was easy as a Native, with preference given to us. I was one of the token Indians. I was told to go sit on a bench and stay there with the other Natives. I'd get paid. I refused, and figured out things to do." For all the issues Josh has, he is special.[3] That is one thing that can be said for a certainty. *People who are very good at what they do are often quirky.*

I have to be careful what I say concerning my art. Josh has a son who is an artist, who has problems getting art out and sold. Our Nenana Cultural Center is considered by many to be 'Native Culture' only, and Native owned. Josh believes his Native son is a great artist who deserves to have his art on display, and should be selling well. It is other people's fault the art is not selling. Josh sometimes feels my art sells unfairly because I am white, because his Native son's art is better than mine.

I partly agree. I am a white man in a white dominated environment within a system rigged to favor white people. I understand the system that Josh does not understand as well, even though I myself have issues with the same culture. Josh is lucky to be alive at all considering how he was raised. To be the winner he is, he had much to overcome. He has to have a strong belief in the rightness of himself, what he is doing, and what he stands for. I'd be more inclined to help the son, but he is one who lied to the Feds about me. I heard the phone conversation on tape. Likewise, he has been putting his Indian name on other, white people's art and calling it his. Authentic Indian art fetches twice the price. I do not wish to be associated with this. Maybe it is just too personal and close.

The arrangement with Glitter was similar. Being part of a con job. However, my art is not mislabeled and I was raising the value of other art while mine was getting knocked down, so financially I did not have a lot to gain.

The son's art is ok. He could make a living. The son has talent he was born with, but not doing enough with it. No use telling Josh my opinion. Even though Natives got a raw deal, there are incentives offered to make up for it, such as getting twice the price for the same art. I wish I was Native.

"Well, Josh, I have art doing well in Glitter Gulch at Denali Park." This is a white man tourist attraction Josh has no interest in. Native art does not dominate there.

"Miles, can you sell your fossil mammoth ivory there?"

I hedge a little on my reply. "Some, but I am not sure how my probation officer feels, so not pushing that."

Josh's son spent a life in prison, so Josh should understand how the system is. He says, "The answer is always 'No!'"

I nod this is so. There is a lot we do not need to talk about, we have covered much of this over the past thirty years. I do not think Josh is doing well. He spent his life living 'dogs', raising them, racing them; he has no other life. Josh is now too old to take care of dogs, much less race again. His dog life is over. Nor can he bask in the glory of achievement. His goals were high. Josh won or placed in the top three many times. There was controversy about how he treated his dogs, and Josh wanted to 'clean up racing.' Get rid of drug use he said. Josh had a point, just not a huge point considering he is the pot calling the kettle black. Not drugging dogs, but abusing them. Few took Josh seriously as a result, and now it is too late. Like me, he does not know how to work well with others, especially when he has something he is not happy about.

Since he does not want to talk about it, he is pushing everyone away. I respect all the years between us and help he was in my early years. I had no place to sleep, take a shower, or eat. No money to do anything about it. Josh and his wife fed me, gave me a place to stay, treated me like family, gave me rides to town, fed my sled dogs, gave me advice. I overlook and ignore his attempts at starting an argument that would cause him to order me away. Josh deliberately baits me. I recognize now, Josh has always been abusive, and I have played the part of forgiving him, but that gets old.

"Yeah, that woman of yours bosses you around too much, Miles."

I only smile and nod. *Josh is baiting me.* I say later, "Well, if I expect dinner tonight, I better remember to bring the lily plant home your wife wants us to have. It is better to choose our battles right?" I know Josh does things he does not want to, but knows he better do in order to get along with his wife if he wants to get fed! Ha! Josh is not known for treating his wife well. I tried to defend her years ago. She jumped all over me, joining Josh, so I let it go. It is their relationship therefore none of my business; as long as it works for them.

"Martha and her silly plants. What a waste of time!" Not a diplomatic thing to say in front of her. The mind of someone who focusses on only one thing - that which he is the best in the world at. No one suggests he is good at anything else.

Josh has the thinking of a forager. It grows out wild, go pick it, enjoy it the way God put it down. Why plant trees and grass and go through so much work, when it is already there?" He mumbles, "White man tears the trees down at great cost. For more cost and effort, plants something else that requires money and care. Go figure."

I agree white man spends a lot of time mowing the grass when the forest left alone is beautiful. I'm amused by the point he makes. I smile and say, "Well, your wife could have worse hobbies, Josh!" I wink and elbow him, implying worse habits women can have. He agrees she is all right and his mood changes. I address his wife, "Martha, I rototilled the community garden box you pointed out as yours." I had been tilling my community garden spot when she stopped by. I offered to do hers if I had time, since I have the tiller there. It is a lot of work to do it by hand. I tilled a couple of other people's plots while I was at it. I do not just do everyone's because some do not want this, do not want that, and get mad if someone does it different. Dalia is one, Bearfoot's lady.

"It's sure nice Nenana gives us a place to plant a garden!" Martha is grateful.

We have a small community greenhouse. I am tired of being the only one who fixes it up. I am getting older. Things I do take more time. For twenty-five years I was also the one who cleared the trail across the river. I just do not have the time anymore. I fixed up the greenhouse in the past. More for others than myself. *I have a greenhouse at home!* I mostly plant flowers in the community garden, just so the area looks nice. The tour buses drive by and visitors like to see how we live. If the bus driver says, 'And this is the community garden on your right', I'd like heads to turn and ooh and ahh over the beauty of it. Think well of my community and what a great place this must be to live that has such things in it. *Does your community give you a garden spot and free water for it? Does your neighbor rototill it for you?* Martha and I have the community garden in common, which gives us something to talk about. Only brief talks. Josh does not want much attention going away from him to his wife. Martha runs rough hands through her pretty white hair as she turns away. Josh has indicated she has had enough attention now for a while.

"Josh, are you planning another river trip soon? Looks like the river is coming up!" Josh and I have the river in common.

"No, Miles, too much drift yet." I have not looked at the river today, so did not know it was full of drift. This means there is the potential to flood. We do not think it is likely this time of year, even with all the rain we are getting. Melting glaciers in the mountains usually adds water to the rain for a flood. Glaciers are not melting now.

I tell Josh about the snow geese. "You ever seen them land in the interior before, Josh?"

He used to live on wild game when he was young, but his wife seems not to fix it for some reason. I notice they do not eat any wild game. Josh says it is because his wife does not know how to fix it. But at potlatches, he does not choose wild game either. I dwell on the snow geese question. *Not in his entire life has he ever seen or heard of snow geese landing in the interior.*

The subject of Jade comes up, the young kid Josh sold out to. Jade bought the dog team and rights to the small cabin on the river. Josh does not own the land. He just went and built. No one said anything. Josh has had the small cabin and dog team with river frontage for forty years. Land ownership and title is a loose arrangement in our village. It is common to have nothing in writing or filed with anyone. Locals know you, and accept this is yours.

This spot is where, long ago, Josh suggested I park my houseboat. I ended up pulling my houseboat ashore here and living on it. At the time, the plan was to repair the boat. I never got that done. Instead I changed my approach to how to live and travel. The house boat stayed dry docked for a decade and I finally pulled it with Josh's truck to the Nenana home I acquired. I still park my riverboat in the same spot. Now Jade has taken over the land down at the end of 10$^{th}$ street.

Jade is tall and skinny, reminding me of drawings of Tom Sawyer in the Mark Twain book. He is in his 20s, in his prime, with a lifetime ahead of him. He is the main character in the TV series, 'Life Below Zero.'

Construction of a bridge has begun across the Nenana River just a hundred feet from where I tie my boat. Ownership of the bridge is still in dispute as Nenana waits for a $100 coastguard permit the community was first told we do not need.

This is a boat launch and staging area for gas and oil exploration across the river. Not the quiet place it once was. Lots of weekend warriors park here now in a big gravel parking lot. Military bear hunters park here, so the locals have these outsiders to deal with. Jade takes the brunt of it. They get their trucks stuck, need to use a cell phone, and in general interrupt the subsistence day. *It must be these outsiders trying to break into my boat totes.*

"Yes, Miles, Jade seems to be doing ok. Won a few races with the dogs this past winter." Josh is not happy to have given up the dogs.

"Well, Josh, he is a good kid, perfect for taking over the old guard. He really respects you and your knowledge of racing. He speaks well of you."

"He chases after woman too much and gets distracted, Miles!"

"Have you forgotten what it was like being young, Josh? This is his time in the sun!"

Jade gets paid to be on the show; is a hero, TV star with women falling all over him - a physical stud they want to breed with. There is a Facebook fan club. It might be getting to his head a little bit, but that is the way of the young. *Wasn't I like this?* I had a good hundred women write me, back before the internet. I have my stories of distraction.

"Josh, don't you have some memories from your superstar days of winning races?" I recall Josh telling me our friend Norman Vaughn, the famous arctic explorer, had wanted Josh to marry his daughter, who was a fan. There was also a rich Jewish woman that chased after him for many years. Josh may not have let himself get distracted. Winners often give up certain things in life to focus on being the best at what they do. Yet Josh has a picture of the Jewish woman in a prime viewing spot on the mantle. I sometimes wonder what his wife thinks. I have heard him say, 'I should have married her instead of Martha!' Said in front of Martha. Choices and forks in the road. Josh thinks he picked a nut case when he chose Martha, and tells her so. He blames her for the son going to prison, and tells her that as well.

I often wonder what Martha gets out of it! Being associated with a superstar racer and the attention that brings? Getting attention when people feel sorry for her and it makes her look good as a forgiving kind person? These are my guesses. *Relationships can get complicated.* Other times I see them like two peas in a pod, getting along just fine.

I'm someone that makes an early decision about people. Once I decide you are ok, I am loyal and stick by people. I forgive faults even though I recognize them. Josh has been the only person to head out on the river to see if I am all right, taken the time to teach me so many of the Native ways.

I notice Jade is a long distance foot race runner getting in the front page news sometimes for being a winner. He's just an all-around top athlete.

"Making stupid mistakes, Miles, he does not listen!"

"Josh, did you listen?" I'm trying to put all this in perspective and get Josh to see the bright side, believe everything is fine. The bottom line being, seeing my friend Josh happy.

While I am talking to Josh, across town another conversation is taking place. Hornet meets Foil. "So what do you think of my plan?"

Foil asks, "Why do I need you?" He has no use for a 300 pound person in his underpowered boat. Foil figures he knows a thing or two. "I know buyers I met through Miles before he screwed me. I do not need your connections."

Hornet tries another tactic. "I live here year round and know Miles. You are here a short time. I can keep an eye on things, find out what Miles is up to. Look for a weakness, a hole in his bullshit stories."

"Yeah, I would not mind seeing Miles out of the way, in prison again. It would make it easier to get his land and take over his fossil business. He deserves what he gets, the way he cons people like he does!"

"I like Miles well enough, I do not want to see him in jail! But I need mammoth ivory, and sometimes we have to be adaptable and go with the flow; be aggressive, look out for ourselves in order to make things happen."

---

Meanwhile, back at the ranch.

"Miles, Foil came around and wanted me to take him upriver with a load." I suspected this would happen. Josh and I both know Foil does not have the right boat for what he is doing, so now wants to depend on others to help him out, as a favor.

"What did you tell him, Josh?"

"I do not want anything to do with him. Martha talked to him and told him no."

Martha says, "Because of your experiences, Miles."

Josh is in his late 70's now and really in no condition to take anyone boating in an area he does not know well. I have seen Josh enter a room and forget how he came in. I found him standing at the food line at the senior center not knowing how to put food on his plate. I had to step in and fix his meal before this got embarrassing. Josh had heart surgery and a clot got loose and lodged in his brain. He has good days and not so good days.

Foil did not listen to my advice on what size boat to get. I had one all lined up for him with a great engine at a super deal. He even committed to the deal, then backed out to buy what he decides he needs after I gave the opinion the guy wants too much money, and the set-up is too small. Foil can now pay for his decision. This sounds harsh, but the mistake cannot put Josh or I at great risk, nor financially come out of our pockets, when the cost was easily preventable.

"Let the one who made the judgement error pay the price."

"It would be different, Josh, if Foil had an understanding attitude, saying, 'I think I made a mistake and got the wrong boat and engine, how can I best fix this?' Or, 'I understand mistakes cost money, look at the prices you paid in life, Miles. Guess it is the same for me, part of the learning process.' Instead he said, 'It is you who made the mistakes, Miles, not me, I'm smarter than you. This is going to work out fine.' *Then show me how 'fine' things are and will be Foil!*

Hornet asks Foil, "How come you got such a piece of junk for a boat and motor? I heard you paid too much. How did you fall for that if you are as capable as you claim?"

Foil gets indignant. "It was a good deal, it works. Fast, light, economical."

"Well, Foil, that 'light' part is the problem. When you live remote you rarely travel light. There is always stuff to haul. It's the nature of the lifestyle." She does not want to alienate Foil, so adds, "But I can see why you did not trust Miles and his opinion!"

Foil nods, "You got that right! He had this engine lined up we could not even look at first! No boat, no trailer! What good is that?" Foil is not interested in partnering with Hornet and sharing any finds or profits. But Hornet could be useful. As she says, she is around a lot and could pass on information. Possibly get Miles back in prison, at least help toss a monkey wrench in the gears of his life. "He must be using a gun to hunt with!" Hornet only shrugs. She needs money and ivory, no serious hard feelings against Miles.

THERE HAD BEEN a problem with Josh on the river this spring that I do not know the details of. Martha said Josh hit and got stuck on a sand bar, then needed help getting back home. Josh only said something went wrong with the water pump. Hitting a sandbar can ruin a water pump, sucking up sand. But then Josh was asking me about how to install a gas filter. This would have nothing to do with hitting a sandbar. It is possible Josh can no longer connect the dots. Though it is true, an overheated engine could be gas starved, and not a poor water pump, as is the first thing to suspect. Josh seems embarrassed to admit any mental problems.

"Miles, my dad lived to be over 100!"

I noticed over the years, Josh does not understand white man machinery well. With many Natives, it is 'magic'. The machine works, or does not, and has nothing to do with anything you do to take care of it. It's a spiritual matter. I watched Josh pick up his truck from the shop where the transmission was worked on.

The mechanic told him, "Drive it slow for a little bit to break it in."

Josh nodded that he understood. We took the trip back from Fairbanks, and Josh punches the truck to 100 miles an hour the whole way home, saying, "I want to be sure it's fixed and works before the warranty runs out." He is a man who understands dogs, not machinery. He can read the river, but he is not good with boat engines.

"I should have listened to you about Foil, Miles!" My feelings had been hurt at the time. Josh had sided with Foil and reprimanded me for not helping my friend. Josh was going to help Foil when I would not, saying he is ashamed of me. "You have changed, and become selfish." I did not argue. I believed in time, Josh would understand better what is going on. Still, it would have been nice for a lifetime friend to say to Foil, that I am his friend and he stands by me. Or asked to hear my side of the story before coming to conclusions. But I believe the road to happiness is not in depending on others. When it comes down to the bottom line, we are all alone. So, while disappointed, I was not surprised. Josh had his own issues with Foil when they got into an argument later.

"Josh, someone was on my boat trying to get in my locked containers."

"What have you got that's worth anything?" Yeah, that's what's weird. Break in to steal what? Some rope, an anchor, a sleeping bag? Josh adds, "Probably just kids looking for a fishing pole so they could fish." We have had this situation with the local kids off and on over the years. We agree once again, it has to be one of them.

A few days later Josh tells me Foil took off on the river, presumably headed for the homestead. Foil never said a word to me. I saw him just one time. I was walking home from the senior center. My rich, eccentric neighbor, Atomic, had just bought his very own fire truck. He wants to fix it up and donate it to a community that needs it. Foil was in Atomic's yard admiring it. Atomic came out, wanting to know why a stranger is trespassing in his yard hanging around his prized toy. I suspect Atomic has a surveillance camera on his yard. I arrive while Foil is hobnobbing with my neighbor, discussing the merits of the engine in this puppy.

I simply say, "Hi Foil."

Atomic asks, "Did you get the security camera to work? Was the problem what I thought it was?" Atomic is a genius when it comes to electronics.

I reply, "Yes, it is working. There was a power connection problem with the receiver, now all four cameras work." I was not thrilled Foil now knows I have security cameras out. He may figure out how to sneak around them. I spent $500 on a security system, mostly because of Foil. He has stolen from my neighbor, and taken or 'borrowed' things from me without asking.

Foil's attitude last year was, "What are you going to do about it? You have stuff laying all over the place out in the open that you are not using, do not even know you have! If it is missing or not, you wouldn't even know!" He feels this makes it ok for a friend to borrow it. I have heard this logic before from others.

"If you can't control it, or know you have it, it's not yours." Pointing out that if I have no receipt or proof then I cannot prove I own it anyhow. Followed by, "What right do you have to own such things that other people need? If you do not need it, you should share!" That's the communist view, and has merit. However, our country preaches the virtues of free enterprise.

"You want one too? Get a job, earn the money, go buy one!" Lending implies I have a choice.

"You're all words, Miles, and think you are better than the rest of us!" I have heard this before as well.

One of my faults is that I have issues telling people how I really feel when dealing in the negative. I prefer to say nice things, or be neutral, even if I do not like you. I'm not good at telling people to their face what the problem is. When I do, it is angrily, undiplomatically spoken. Some people feel I am a back stabber. Foil says, 'I at least tell people to their face!' Spoken with the impression this makes him a stand up good guy. Someone telling me to my face they robbed me, does not make them a better person, when, 'And what can you do about it!' is added. The way Foil handles it, it adds insult to injury.

Anyhow, "Nobody's perfect!" Or "Let he who is without sin toss the first egg."

---

MY TRUCK IS PARKED out front with the boat on the trailer. Foil never commented as I walked away. I had said Hi, and been polite. That is the plan, not to make him my enemy, but simply not have much to do with him. Be nice, vague, non-committal till he goes away. I do not tell him what the problem is.

I feel bad for him because I know he could use a lot of help! He needs drinking water, a source of electricity, a refrigerator, tools of all kinds that he used to ask for, borrow, and return. Lending and help is not about me owing. If I say, 'No,' I expect a, 'Well, thanks for all you have done already; no problem.' In most friendships, somewhere along the line the other person gives me a helping hand; lives up to a promise now and then; hands me something they know I need, or does me a huge favor. In some way they show thanks even if it is to simply be grateful, and help me feel good about myself. I do not expect item by item exchange, and keeping track. But when the value of what I contribute is thousands of dollars, the help the other has offered over several years is a hundred dollars, then acts like we are even, or I still owe more, and wants me to feel guilty, then something is wrong with this picture.

Iris took me aside saying, "Just try saying 'No', and asking to be paid for something. You are such an easy mark, Miles! See what kind of reaction you get from Foil! I bet you will see a different kind of person!" I was not as certain as Iris, but thought it is time to see where his head is. Partly Iris is correct. "Miles, you can give away your own money and do with it as you please, but not my money—our money." I have given enough that, financially, Iris is correct, it is her money as well; money we need for the household. I have given Foil three round trips to the homestead for less than it cost me to make these trips. I decide the free trips are over! I had offered to help Foil get started. I had expected him to figure out the lifestyle, take care of himself. Instead he got more dependent on me! He enjoys being taken care of. Foil is beginning to order me around like I am his servant.

"I need to go on this day, and you need to haul this load for me."

I said in return, "Ok, but Foil, this is what I need."

I explained how the trips I have made for him are costing me money. I thought maybe he does not know that, believing the $100 he gives covers the costs. So I broke a trip down into all the costs, because I think Foil will find the costs unbelievable! Hardly anyone believes what boat trips cost! I had trouble facing it myself! Even those who own boats tend not to face the costs! We joke about how a boat is a hole in the water you pour money into! But we do not believe ourselves! Seymour, my survey boss from years ago, had to be able to write off the costs so kept careful records for eighteen years. I did too, and once told Seymour, "I figure about ten dollars an hour for every hour the boat is run." That amount was figured fifteen years ago.

"Yes, Miles, that's about two dollars a mile, for every mile we run."

As Iris and I guessed, Foil goes ballistic! Cusses me, calls me names, will never speak to me again.

"This is outrageous. $300 a trip!" *Welcome to reality buddy.* This is my cost, breaking even. He paid a token $100 a trip to help out, while I chip in the other $200. When I believed Foil was my friend, and trips were fun, believed he was grateful, I did not mind paying the first time or two. I'd be paying on my own anyhow. More like I did not mind helping during his learning stage. *Maybe when he knows more, he can help me out!*

We end up with a written contract, I'm not sure how, but Foil was worried if something happened to me and he built on my land, then what? Well I got the picture.

You have to be civilized for contracts to work. You have to be among the protected. You have to keep records. If a contract is handed to someone in the wilderness and they toss it in the trash, now what? Who will enforce it? How would you even get to the one enforcing it? A whole social structure has to be in place with proof, lawyers, witnesses, times dates, judge, court system.

There are usually too many things I am out of compliance with when it comes to contracts. I am not a rule follower. I do not do directions well. I have little sense of time. I forget names, mix up sequence of events, can't find papers, do not think in terms of 'proof'. Dates are forgotten. Never in my life carried a watch. Exact amounts are not important to me. I do recall the gist of verbal words in the same way I understand the intent of most laws, and can follow that. Most laws, even handshake agreements are covered by things like the ten commandments, or the constitution.

"Treat others the way you would want to be treated." Simple enough. How important then, are witnesses, times, dates, receipts, signatures and details?

I live a loose life of the handshake. Deals that are adaptable with friends who work things out as we go along. Contracts are very constraining. What is important is that we understand and respect each other. I have good deals and working relations with likeminded people. Josh for example is good to his word. Like me though, how could he prove he has any rights to his land? No one wrote anything down, it's just locally understood.

"Prove it!" Is just never going to be an issue.

I get screwed a lot, but hey, that is on them. I do my part, my conscience is clear. Where are these people in life compared to me? I seem to be better off in the big picture than they are so I tend not to sweat it. We reap what we sow. One of my quotable lines on the subject is, "I'd rather be the sucker than the suckee."

Yes, how quaint, how droll, how naive, how open I am to get totally taken advantage of! Ha! You want to sell me the Brooklyn bridge, right?

"No. That is where you are wrong." I simply live by a different set of rules. More along with mafia beliefs. 'This thing of ours.' No lawyer, no phone, no showing up in court, no time wasted, no being a victim, no being sorry. It's like the old way, the bush way, the Indian way. The way it is when you are far from civilization, with no Big Brother's help. The head of the Amazon River, or Australian bushmen. A world where your word is your bond and your good name is the most valuable thing you can have. Friends, descent people, do not study loopholes. Those who deliberately try to get away without paying are dealt with harshly. The thief is the victim, not the one who got robbed.

I was told by a millionaire long ago, "There is no such thing as a contract you can't get out of with enough money and good lawyer." He added, "You either trust someone or you do not. If you trust them, deal on a handshake. If you do not trust them do not do business, even with a contract." Said by one of the top five diamond dealers in the world.

The bottom line being, "You messed with the wrong person, Mr. Foil. No hard feelings! Everyone has the right to make their move as they see fit. Just as everyone has the right to pay the price, see the consequences, learn. Perhaps die trying."

With a smirk on his face, Foil informs me he has a valid contract that is to his advantage and basically I am at his mercy.

"What are you going to do, Miles? You are too old to make this a big deal. Let it go. You are not using the homestead or anything there anyhow. Let someone else get some use out of it!"

I am certainly sorry and embarrassed I invited him to Alaska to be my guest and have a taste of his dream. He thinks he can buy what I have. *Stay tuned to this channel, do not adjust your set, we will be back after this brief break. Will my good buddy, Foil, have his dream, lose it, or die, trying to hang on.*

I NOTICE I am not as involved in my community as I used to be. I have less energy, even less enthusiasm. I keep saying I think the issues is a body chemical problem more than anything. The chemical called "Rah, rah, let's do this!" Partly testosterone mixed with other bodies chemicals, I suspect. I still attend city meetings when I can. I missed the last couple of meetings so am anxious to get caught up.

The mayor takes me aside before the meeting. "Miles, I have been meaning to catch up to you and have not got to it. There is a vacancy on the city council when Soldier quits, to be announced this meeting. We need someone to take his place till elections in October. We'd like to appoint you as a temporary. In October you would probably get voted in."

"I'm a felon, is that an issue?"

"We talked about that, and it's not great, but acceptable."

"If I had been asked ages ago I would have wanted to participate. Now I am cutting back on such responsibilities. Have high blood pressure, need the least stress." I can volunteer to help the community, sometimes be on a committee to resolve a specific problem, but thanks for asking. "Also, the plan for now is for Iris and I to be gone a few months a winter down to Tucson."

"We'd do well to have you, Miles, you'd be an asset to the community."

I appreciate the words, backed by the offer to be on the city council, and decided by the city council. I have Foil and Mad Jay reminding me how many enemies I have, and how many want me back in jail, or to see me hurting. The community in general values me as one of its members. I just have to remind myself sometimes.

I have been attending city meetings for thirty years now. I write in my diary:

**Friday, June 10, 2016 City budget.**
"So we are $400,000 in the red, and we have to submit a balanced budget. There is no more funding being handed out, by either the Feds or the state." Two hours are spent reviewing line by line the community costs and income. There is some revenue sharing from last year of $150,000 expected to come in after the budget is approved. The firefighters say they can survive on $70,000 instead of what they hoped for. Other items are moved around, so it looks like we can be close. For the first time in several years all councilmen were in agreement. This should have been done in a work session, but the public is pleased to hear exactly what the issues are, the item lines costs, and expenses. There has been a lack of trust.

Of interest to me was talk of the Nenana Culture Center gift shop. "It cost $40,000 a year to run, maybe we should shut it down!" I'm thinking, *Are you nuts!* I closely follow what is being said.

The mayor comments, "Well, I have hope for it. Maybe we can break even one day. It does offer local jobs. In fact I think we made a little profit of $150 last year. We are getting a lot of Princess tour buses to stop. An average of 800 people a day. Princess likes the stop and says we have the best gift prices in the state. The only issue is, we need more bathrooms, but we cannot afford them."

I recall speaking up during public comment, and include that in my diary.

"I think you are not looking close enough at the culture center as a serious source of revenue." I go on to explain. There is talk of a lot of theft going on by employees. "An employee was caught stealing shop goods, taking them to Fairbanks and selling them. Nothing was done about it."

"I am an artist, and have worked at gift shops most of my life. I know what can be expected for profits. There is no reason we should not expect $1,000 a day to come through the shop during the tourist season!"
**I end my diary.**

I do not say so, but some of the employees at the cultural center are scary in terms of trust. The gal at the register says the register is old and does not work. She is going to program it. She is a lifetime drunk who was fired from the local store for theft. Such a person is in charge of the register with no accountability? To the tune of a potential hundred grand? Wow! The woman who stole and pawned goods in Fairbanks is back working again. Talk is, the manager is on drugs. Government money keeps pouring in to help the poor Natives. A government subsidized program provides workers at local places like the cultural center. The city takes advantage because almost free subsidized labor saves the city money in wages. But what good is low cost helpers, if 100 grand is stolen? Which seems likely to me.

The mayor takes me aside after the meeting. "No one told me of theft, this is the first I have heard of it."

*Yes, well it is common for the boss to be the last one to know what is going on.* I'm not 100% convinced he is ignorant. *Hopefully he is not taking a cut.* He thanks me for providing copies of the laws concerning the use of mammoth ivory and feathers in crafts to be sold. I decide I need to get involved. I might be arrested for doing so, but I do not appreciate the government not wanting us to know the laws, and intimidating us into going along with some hidden agenda. *Go ahead, arrest me for offering legal advice.* The mayor confirms it was not him who called Fish and Game and told to pull these items. It was one of the problem workers at the cultural center. I also believe there is a reason this employee does not want me there helping, or organizing my own work. I might figure out the details of what is going on.

I realize that one of the Native cultural center employees has a family name I recall from my trapping years twenty years ago. A family I had off and on disputes with over trapping territory. *Is there still lingering anger and resentment over those years?*

"Miles, maybe the city should get some security cameras over there!" I agree this would be wise. He admits he can't keep employees. They come and go. In such a situation, with 800 customers, it is normal to have cameras at the register and elsewhere. In times when the buses stop, I think it would be easy to make sales that go unreported. Even honest mistakes when it's so hectic, as well as customer theft.

So all this 'stuff' going on concerning a city business. The mayor has never run a gift shop and is doing his best.

"Miles, would you run the gift shop for the city?"

Like the offer to be on the city council, I might have been interested ages ago when younger. Even then though, I value my freedom and want to work for myself. It is nice to know I have choices, and jobs offered if I want or need one. I was on the board that helped build and start this cultural museum and shop. I was the first buyer. I have memories, and care about it. Being an artist, I care about our local artists having an outlet. It bothers me the city is going broke, while this gift shop could bring in as much as $400,000 a season... gross, but potentially half as profit. For a community of 300 people, 200 grand has meaning. Another reason to turn this down is, as a general rule, I do not like working for a boss who is broke. Being part of broke is not good Karma. I tend to feel sorry for the boss or broke company and work for much less money.

"The issue, Mayor, is the good community workers already have a job! Also, good workers want good money! Many work for themselves or go for the high paying jobs." Many of our employers talk like money is tight, and they can't afford good workers so offer minimum wage, or as cheap as they can get. Ten dollars an hour is simply not attractive to a good worker who can, and does, get $20 or more an hour happily paid by employers outside the village. In this way, local business drives away good workers who often relocate for better jobs. The subsistence people of the community do not depend on Nenana, but the land, so are not paying it or participating.

Alley brought in 80 grand his last season for the Cultural Center. Yet, he tells me in private, "It could easily be $200,000 and $400,000 would not be an unreasonable expectation." So sure, if he could bring in that kind of return he'd be worth the extra 40 grand he wants. *So where is the other 100 grand or more that is missing, Alley?* I do not believe he stole it outright or benefited directly, if the money even existed at all. I think his relatives and those in need helped themselves with Alley looking the other way. He did not like it or support it, but could not say no. This is my take on the situation.

Making a lot of money is great, but getting the money into the appropriate hands is also important. The mayor and I talk about this a lot, and over a long period of time. Likewise, we discuss the drug crime problems of the community. The mayor had brought up at a city meeting the problems of getting the police to respond. It would cost $150,000 a year to have a policemen, and we can't afford it." We got rid of the one we had ages ago. So we are covered by some Trooper who handles 300 miles of road. He's busy. Since we do not pay, we are not his boss. He'll take care of us when he gets to it. He flat out tells us not to expect a lot.

"In the old days not so long ago, we simply elected someone to be the cop like the wild west sheriff. Pay him 30 to 40 grand a year and we had protection!"

**Past flash**

In book one, way back when young. I was rescued off the Yukon River and dropped off at Eielson Air Force base. It is 50 below zero, pitch dark, winter, and, "You are free to go." I'm hitchhiking from the base to civilization. It is hard to get a ride; it's like 2:00 am. Finally a guy stops and picks me up. I'm explaining how I got here, rambling with him not commenting.

"I got dropped off in the wilds and spent seven months not seeing another human being. My supplies were stolen and I rationed myself. I had to eat a rotten wolverine." I recall something else and say, "I was chasing a mouse crawling on all fours with scissors in my hand and met God! I was starving and decide I have to walk out. I was five days in this 50 below and almost died! A Chinook helicopter rescued me and dropped me off at the base. So here I am! Headed to the Salvation Army to rest up before going back to my trapline!" I had been rescued only two hours earlier. I will never forget what the driver said. Very calmly, "Would you remind me to get a quart of milk when we get to town?"

**Past flash ends**

"Hornet, you think Miles found some ivory on his last trip? Doesn't matter I guess. Did you ever get a copy off his GPS? I can go look myself."

"Don't you mean 'we' can go look!" She glares at Foil. No longer trusts him. She's not working for him as his flunky. If anything, it is she who should be in charge, and him doing her legwork!

"Hornet, if Miles was in jail he'd be out of the picture. We could go look at his hot spots, maybe even tap into his shop where he has it stored! It's not like he can turn us in. Turn us in to who? No cops around!" Hornet just frowns, thinking. Even though Foil is from the big city and thinks Nenana is ripe for the harvest, Hornet has been around, and knows village life.

"Foil, there may be no cops, but that does not mean you can't suddenly and mysteriously end up dead in the river if you seriously mess with a local." Hornet is again reluctant to actually become a serious criminal or wish to harm anyone. "Anyhow Foil, what have you actually got on Miles that you could report as facts to nail him with?"

"Hey, everyone knows how he is! I heard how he killed that guy a while back and got away with it! I heard how he poaches a moose every year! I heard what a shyster he is when he sells stuff! Everyone knows he ran the Native trappers off when he wanted a trapline! Any of that could get him in a lot of trouble! He's a problem person the community needs to get rid of. Just ask around!"

Hornet knows she and Lawsen have been involved in, and lived a similar life as Miles. She feels it is about impossible to operate a small business, especially in a remote area, and stay on top of all the forms, permits, levels of permission, and cuts everyone wants of the take. If Miles could be sent to jail, so could Hornet. She's beginning not to trust Foil, and would not want to be in a position of him having anything on her.

Knowing what kind of boat and how to camp is critical to survival. I live on the boat for a month as I travel.

# CHAPTER THREE

## STUFF NOT TAUGHT IN SCHOOL, MAYOR IDEAS SHARED, I AM THE SOURCE, THE UNPROTECTED

I have heard some stories from outer space before. It is hard to get to the bottom of some people's tales. I believe that is why the police and court system does not want to get involved. That's part of the unspoken rules. I'm going to follow them. I hope to be among the survivors. I want to be on that list of the protected. I mean, I do want to help others. I think it is right to help, but first I have to survive. I did not understand, but now I get it. *Stuff we do not get taught in school.*

The mayor agrees, there is no such thing as equality. We do not have equal rights. It doesn't work, and never has. *So what does work?* I'm still exploring that. Some of the things the fellow mammoth hunter Neil writes about in his book are conclusions I have come to, and agree with. He does not trust the government. I do not feel so alone. One big issue is, how do you get on this unprotected list? How do you know what your status is, like if you are among the protected or not? How might one get off and on the list? Can someone get mistakenly put on such a list?

Most of the people I know and hang with do not share my lifestyle, and have come to different conclusions, based on how they are treated. It's like you have to be 'one of us' to know what we are talking about. Sometimes 'different' gets misunderstood and mislabeled. Perhaps this is the reason people try so hard and care so much about their image, what people think, what the rumors are. Can people who do not like you have you put on such a list without your knowledge? Open season on you, and you do not even know it?

**Reading Neil's' book on mammoth hunting**

"As if the feds do not have anything better to do than to spend a fortune skulking around after a few guys digging in the mud for ivory that would end up ten feet in the bottom of the Yukon anyhow. I told myself not to ever mention it again to anyone. Just try to make a living on the river and you'll end up breaking some law or other."

"Some of us just have our sails trimmed to another tack."

I give the mayor an example that comes to mind about giving the wrong impression and getting mislabeled. I ordered a machete on the internet to cut trap line trail. I started getting ads for pepper spray, how to turn your shotgun into a machine gun, and diagrams of the basement of the capitol."

The mayor smiles and nods that yes, the assumption was, anyone who wants a machete is using it as a weapon to lob people's heads off. "It would not occur to me there is a misunderstanding that needs to be straightened out. This is called profiling, but profile based on what set of collected data? How does one make a correction in this profile?" City people!

I go on, "When I took a basic class at the library on how to use a computer fifteen years ago, one of the lessons was to go on line and get information of interest to us through an internet Google search." I have to add, "This is at a time when search engines were not as user friendly as now!"

The mayor nods.

"I am an artist who collects pictures of wild animals. It might be cool to find such pictures when I need them for my art!" At the time I had accumulated thousands of pictures from Nature, and National Geographic magazines. These pictures have taken a lifetime to acquire and organize for my art reference. "I put in the search engine, 'animal wildlife pictures'. I got pornographic pictures of people having sex with animals!" Yikes! How do I get rid of that? How do I undo that search? How do I let the world know this is not what I meant? Combine this with, 'looking for a machete'. What does my profile look like? What list might I be on?

"We can do research for something, order something, inquire about a subject, and end up maybe among the unprotected, and on the person of interest list!" Likewise, we can choose an occupation, dress different, and not get told, or understand, the consequences. Like loose our rights to be a citizen. "End up with the rights of a slave." It seems to me we should be able to look up our social status and have the right to defend ourselves if we feel a mistake has been made. We are tried, judged, and punished, with no defense, or knowledge.

The mayor mentions my own issue as a felon. He had a wildlife problem at the gift shop. "The gift shop had some items I wondered about the legality of, so I called Fish and Wildlife to get an answer. I got the run around and treated like a criminal. I never did get answers."

I nod I understand.

He says to them, "I'm sure there is a way to do this with a permit, so where do I get one?" The mayor is not stupid, and knows there are legal ways to have things in a Cultural museum and or sold in an Indian village gift shop. Getting an answer of "No! You cannot do that!" is a flat out lie. "Miles, I was treated like a criminal! I certainly did not feel like ever contacting them again if I have a question! They were ready to get a warrant and come out to my home to look for illegal game items."

This conversation with the mayor leads to, in general, small communities and how Nenana wants to run itself.

"Like the days we used to hold car drag races on the airport runway to raise money and have fun, remember?" Not possible any more. The community needs an upgraded water treatment plant.

"Miles, the funding offered is for a modern system run by computer in Washington." Totally under someone else's control. To control Nenana, just press a button, poof, no water. When it needs fixing, no going to our scrap pile and making it happen. Stuck with a bill Washington sends us, using their approved imported union workers. "I do not want that to happen, Miles, and the government is making it impossible to be self-reliant."

I understand, but many citizens do not. Suddenly we need a permit to do something we never needed before, a new fee being applied. Squeezed for money we do not have, putting the community in debt, behind on interest owed. Ultimately in Big Brother's pocket, who owns us now. A goal might be to ensure we do not remain a home rule community as we are now. Outsiders with a vested interest in oil and gas in the area may not want 'home rule'. Such interests may have influence with the government.

"At the city meeting, you said the Coast Guard has had the information about the bridge across the Nenana River." I review in my mind how the bridge project was started. *I was part of that process. About five years ago, the city decided a bridge across the river would open city land over there for farming and homesteading, resulting in city expansion and revenue.* There had been a concern gas and oil across the river would not be a big help to Nenana, based on the history of the companies involved. There was concern, as well, that if the oil company built the bridge, they may claim control and deny access, calling it a private bridge. There might be security issues cited.

If the city owned the bridge, there is the possibility of charging the oil company for use. But for sure we would have access to our own land! We no longer have the money set aside. We have other, some feel, more important bills to pay! The good news is, the mayor was re-elected for about the fifth time, and he did not even run! He won as a write in, without campaigning.

We are a home ruled city, giving us a lot of freedoms and the ability to run our village as we decide, as much as possible. While we are a subsistence village, not everyone lives subsistence. There are quite a few people who live civilized, ordinary lives. They work for a paycheck and depend on stores, services they pay for. They do not trade, have a garden, eat wild game, and may not even have much to do with people who do. They may consider it a class difference. A high percent of the community is involved in the school as teachers or administrators. The school budget is many times greater than the city's. To some people, life is by the book, following rules spelled out, as it is, back where they came from. Balancing the budget is about money and numbers. Black, white, or being red.

There are a few railroad workers, barge employees. A high percent of us are on welfare or some kind of assistance, not directly depending on our communities economy. Many are Native, who have their own tribal government and sources of independent funding. A community of 300 has half a dozen churches. There are alcoholics on the other side of that fence, who are not involved in much beyond their next drink. There are a lot of groups represented. Possibly a majority do not want change.

"A bridge, what a stupid idea! A bridge to the swamps on the other side, geez!" Followed by, "I have been over there, there's no oil! That's not said in the news!" Others say it is simply all hush, hush till the owners have land, gas, laws, and oil secured.

The mayor I think, understands the entire community. Much of our balanced budget has to do with balancing services, trades, favors. We have talked about this a lot as well. I trade and do exchanges in my own life and business. I try more than I did in the past, not to pressure anyone. Like the last farmers market. A local had some seed potatoes I wanted. She halfheartedly mentioned a trade, but really needs the money. She thought maybe I had no money. She would trade since I need the potatoes, just to be nice to me. In the past she had traded for strawberry starter plants from me, but now did not need any. I did not try to talk her into something she does not need. I pulled out money. That works too!

I think poor people learn to trade! The early day hobo would work for you for a meal and place to stay. He did not ask for a receipt so he could do his taxes. You did not ask for ID. The hobo did not try to sue you, and you did not try to turn him in for not having his papers. Around the world there are isolated cultures that do not have easy access to banks or the money system. They are not in the jet age, much less the space age. Some have barely reached the level of the industrial revolution. In such places it is common to trade. Worldwide, there are now 'bit coins' to avoid banks. There are international forms of trading, groups you can join. Trading is not the odd concept it was a few decades ago.

Our community is half Native. Natives understand trading because they had no money before the white man, which was only a generation ago. I find some poor people may not even want money. They do not use banks. Money tends to get stolen. Many trade for drugs. One young local said, "Why would I want money, my parents or friends just steal it from me while I'm sleeping, or beat me up for it." He trades directly for his immediate needs. A meal, enough fuel to give one night of heat.

Iris points out, "We are getting charged five dollars a month by the bank for not keeping over $2,500 in the bank at all times, it's called a service fee." So how many poor people can keep this kind of money in an account when they have trouble coming up with five dollars for today's fuel to heat the house? "No banky? No protect thee."

In my opinion, a community can operate much like a family. The world does not need to know how a small community survives. The world is not kind and understanding. Civilization is well known for butting into other people's business, saying they are right, and everyone else - minorities - have to do it our way, or be a criminal. Long ago there was less communication so small remote places were simply out of reach, off the radar. Few people in civilization understood what was going on, and did not make a big effort to find out. The computer age made us able to see into everyone else's life. With Google Earth, I, or anyone, can look into anyone else's back yard. I think civilization is not responsible enough to deal with the reality of what someone across the planet has in their back yard.

Yes, I think that changes how we live. It's the talk in the remote villages I visit, and among Natives I know who have remote camps. Most no longer like visitors. All of us are not on the same page as it seemed we used to be. Like the tourist at the RV park across the street who would love to go on the river in my boat! I tell the new park owners about the old days.

"I used to hang out at the café and work on my art, drink coffee, talk with locals, maybe order lunch. Tourists would come in now and then. Many were interested in a half hour trip on the mighty Tanana River, maybe catch a fish! I'd take them out for $30. No life jackets, insurance, permits, registered boat. Just you and I, nature, the fish, and sun. All that is given up, because maybe we would have an accident. Heaven forbid.

I am not interested in all I'd need to go through now to be legal. It's a full time experience. Instead of charging $30 to make $20 profit, I'd have to charge $300 for the same $20 profit. Suddenly tourists are not as interested. Neither am I. It's one thing to hang out at the café mingling with likeminded people working on my art as I wait for customers. It is another matter entirely, spending money and time in a classroom getting certified, sitting in a chair waiting to meet with my insurance agent, on hold on the phone, pressing one, pressing two, "Your business is important to us, please hold," waiting on who knows what, my Coast Guard approved lifejackets, fog horn, or running lights. Getting told when done, "Oh no, the rule changed, that lifejacket is no longer approved!"

Such required costs are ongoing. I must have enough customers to cover the high costs. In times past I have no investment. If I get no customers I got some art done in the café, had a good time with friends. No loss if zero customers today. I was quite happy with a customer every couple of days. But even one customer a year, done legally, would cost me ten grand.

The RV park owner agrees. He has seen the changes in village life and subsistence. He retires and wants to run an RV park. "That old life off the land as we knew it is over, Miles."

*So I am not the only one feeling left in the dust.* Much of this has to do with the entire city and how the mayor feels. The village could survive, I think, if we could work issues out ourselves, such as trades, as used to be done within my memory. Someone owes property tax they can't pay with money. So the city has them mow the grass in the city park all summer. It does little good to foreclose, have the citizen move out of Nenana as a result, and the city have an empty house no one wants. The city owes the school, so plows the school parking lot in winter. Everyone may or may not keep track, write it all down. What is it all worth in dollars? What business is it of our Uncle Sam, how Nenana pays its bills, and makes ends meet?

A small community is described as 'like family.' In a family, does each member submit a bill? Is there a discussion of who owes who, what washing the dishes is worth compared to feeding the cat and taking out the garbage? As long as it somehow all gets done and everyone is somewhat happy, isn't that what matters?

Once again I say it is all about lack of trust. If I, or the village, trusted the Federal government to look out for our best interest, we could go along with the program. The mayor repeats what I hear, "I believe the Federal government does not want us to exist." It is easier to keep track of citizens when they are all in one place, with cameras watching. It's not that the community feels defeated, and totally unhappy!

"Got to run, headed to the big city today. It's like Christmas, getting things on my wish list! Anything you need in town?" No, nothing Mr. RV needs. One reason we go in today is the seniors are paying for dinner at the Salmon Bake for all volunteers at the senior center! All the crab, salmon, and prime rib you can eat with all the fixings! We will meet the senior bus there. Iris and I want to drive in, shop, garage sale all day, then meet up at meal time. "What a great community where this happens!"

We stop in to see Knife, one of the few Fairbanks knife makers I know and do business with sometimes. "Knife, did you read Neil's book about mammoth ivory hunting?" We all know each other. Knife is in the ivory business, using it as knife handle material. He gets some of my materials and works with Tusk, who I also do a lot of business with. He's big on the ban of elephant ivory, yet stands up for the fossil mammoth ivory.

"I started Neil's book way back when he wrote it. Did not care for it. Makes all of us look like outlaws!"

Ah yes. Knife seems very careful about his image. He looks over the rim of his glasses. His countertop is filled with blank knife blades, stacks of exotic handle material choices. In the back you can see his work shop with grinders, sanders, torches. This is a nice log cabin- resort looking building off the beaten path at the edge of Fairbanks. Knife admits he has the same issues all of us do—impossible regulations, inconsistent treatment, vague erratic selective enforcements, and the Feds treating us all like criminals and the enemy. Basically no such thing as being nice and having that mean anything.

I agree, "But what else can we do Knife. We have to try. No sense making the Feds our enemy, and try to punch a bully in the nose."

"Miles, anyone can be turned in; be wired, working undercover, and getting recorded."

I take this to mean Knife could be hedging and not expressing anything negative because I might not be trustable. I might be wired. I agree with him, "Neil is a little overboard on his expression of hatred and discussion of the stupidity of everyone making rules. This is how he honestly feels. I see where the feeling comes from, and sympathize. I do not agree these thoughts need to be openly expressed to the extent Neil covers it!"

The book overall tells a good story of a lifestyle and what it is like. His stories are familiar to me. I would not dismiss the book because he hates the Feds. Nor do I agree his hatred reflects on all of us. I think Knife is worried, scared, and wants the heat off our trade, and him personally. I think Knife has a lot to lose, and wants interest directed someplace else. He does not want to be under the microscope. I am not convinced Knife is totally on the up and up, with his talk of being Joe Legal.

Mr. Alaskaland told me Knife and Tusk are business partners in some shady stuff that he has also been part of. I am not interested in details. *A need to know policy is best.* I have no direct knowledge. However, a few personal dealings with Knife did not go well; I felt they were unethical on his part, but not illegal. Knife contacted my customers by phone to mess with my business and get my customers for himself by badmouthing me. He tried to horn in on more than one specific deal. I try to be amiable. Others tell me Knife tries to discover your sources, then take them from you. This is not what I want to be part of.

Knife has said to me, "Miles we need to work together, keep prices higher. You can't undercut me like you do, it hurts all of us!"

What Knife wants is for me to artificially hold my prices high so he has a better chance to make profits. I do not feel it will raise profits for me. I am the source— I live remote, have fewer shop costs or overhead then Knife. I have no employees, so no workman's compensation or required insurance. I sell slightly lower quality at way reduced prices Knife can't afford to offer because of his high overhead. I feel Knife gets more high end customers than I can attract. Because of these differences, we have a different customer base. If I raise my prices no one has a reason to work harder to find and deal with me. People deal with me because I have lower prices, but in return, you do not get to visit my shop, exchange chit chat, or get as fast a service. I might be out fishing.

There are few enough of us in this trade. I respect what Knife knows and like to focus on the good stuff, so enjoy chatting with Knife. He is built a little like me—stocky, taller—of course everyone is taller than me—has long hair, wears thick glasses. Wears the same plaid shirts and jeans our whole tribe chooses to wear as a uniform. I can see how customers might mix us up. He gets very upset when people say to him, "Hi, Miles!" When people think I am Knife, I am only amused.

Knife and I get into specifics about animal part laws as we each interpret them.

"Miles, whale products do not fall under marine mammal protection laws. They fall under OCIA laws. Slightly different. We can deal in whale parts within Alaska. We cannot cross state lines with it. Our tourist customers crossing state lines could be in trouble, but not us. For this reason I'd rather not sell."

I assumed whale material had to be pre 1972, like marine mammal items, involving CITES law. "Doesn't affect me either way, Knife, I can't sell animal parts of any kind. Maybe fossils, but I will stay away from walrus and whale fossils. Too much controversy for me to get involved in." I do not reveal these items were never a main part of my business. I suspect Knife to be one of the competitors who tried to set me up, and may still offer up evidence against me. I'm as careful as he is with my words.

I believe that it is almost impossible for any business to be 100 percent legal and untouchable by the police, especially smaller businesses. Most cops I know have said, 'If I want you, you're had!' None have said, 'If you are honest, you have nothing to worry about!' I look around his shop, and see a lot of questionable things like squiggly long horns from African animals, impala, gazelle, Kudu, and such. I see wild Dall sheep horns, walrus ivory I'd question the age of, and just a variety of animal products I think would be hard to prove the legality of. It's more out at the edge than he admits being involved in. If he was really worried about pleasing the government, he would avoid such marginal items all together. He knows there is money to be made in the grey area. I'm sure he does not knowingly deal in trophies or poached material. Most of us share similar ethics.

One problem for all of us is, if someone we do not know brings us something to sell or trade, we have no practical way to verify their story. Few sellers would tell us what they have is illegal and would we like to buy it? I'm also sure Knife has customers like mine, who cannot provide paperwork because the item originated in the wilderness where these animals live. There is often no paper or pen. Likewise, asking for a receipt is a cultural insult to most primitive and uneducated, unprotected people. We do not discuss such matters. I'd want to say honestly in court, "I have no such knowledge, we never discussed it."

I run into Tusk this same trip to town! His wife has a shop at Alaskaland. I was dropping rocks off to Mr. Alaskaland. Alaskaland is now called Pioneer Park. It looks the same as it did forty years ago. Same historical cabins. No expansion, no changes. People dressed up in Gold Rush era outfits in places like Robert Services cabin. Now a souvenir shop. A lot of memories here. I used to live on my boat at the dock out back in the early 70s. A few of the same shop owners are here! Like Mr. Alaskaland.

Tusk looks older. Maybe because he is. I have not seen him in two years. He dresses like, and is built like Knife and I! His skin does not tan like mine does, and he is bald as I am, but not like Knife. Tusk quotes the latest laws concerning a ban on elephant ivory, and how he thinks it might help, not hurt, our mammoth ivory business. I do not agree. I wonder if he acquired his outlook by hanging around Knife.

"Miles, there is too much money in the mammoth ivory trade for the Feds to succeed in shutting it down!"

There is more money in the elephant ivory trade, and it got shut down. Tusk thinks there are too many rich people with mammoth tusks in their collection. Look at all the pianos that are now illegal with elephant ivory keys.

"Maybe, Tusk, it could be made legal to own, but not sell." I review the potential changing classification of mammoth ivory. The talk is to not call mammoth a fossil, but an artifact, falling under stricter laws.

"Miles, I have been buying a lot of mammoth ivory lately, you should stop by." I'm not sure if he wants me to sell to him, or buy from him. I may in fact stop by. He often has something of interest for me when he needs money, and can give a deal on something. He is not as good at selling as I am. I am in the market for low grade soft rotten ivory that I suspect he has a lot of and would like to dump. I know how to dye, resin impregnate, stabilize, and turn grade D into a valuable usable grade B to A product. No one else I know of is doing this.

"Miles, can you really make that pay?" I laugh. "I'm not sure. A lot has to do with the fact I do not like to see waste. If I can figure out something to do with what is a cool exotic material that is awesome, and just needs some enhancement to be usable, I want to do that."

Tusk has a different strategy. He buys good material, does nothing to it, sells it and doubles his money. He is an artist as I am, but more of a carver, doing large carvings for galleries. He does not retail much, so I rarely see him at shows. Instead, he makes the rounds of the galleries.

"Miles, check out this article on mammoth hunting in National Geographic!" On the cover of the April 2013 issue is a mammoth and one of the dominating headlines is, "Hunting for Mammoth Tusks." A full twenty pages is devoted to the finding, working, selling of mammoth tusks. The story is upbeat and positive, not about a war on crime. The magazine may not be following the government's agenda. There have been previous upbeat articles like 'Raising the Mammoth.' "See, this is not going away!" He happily nods and sounds relieved.

Last time I spoke to Tusk, it was doom and gloom. Needing to get out of this business before we all end up in jail. I forget to ask if he has read Neil's book. Tusk is mentioned in the book. Actually I do not have time to ask. I find it positive Tusk is at least talking to me now. Even interested in doing some business. When I was arrested—and for a year after—he considered I am working for the Feds and setting up my friends. Word has gotten around no one close to me has been arrested. I am not trying to make deals and pushing myself on anyone, asking them to do business with me. Possibly Tusk, like Knife, could have testified against me. I heard a recording that I sometimes refer to. Tusk's voice is on it. Recorded at the state fair. But the recorder could have been planted in my booth, not on Tusk. If Tusk was between a rock and a hard place and the Feds got to him, I forgive him. The Feds are professional, and know how to get to any of us. Threats of all sorts can be applied. Threats to our loved ones, our occupation. I will simply be very careful with everyone. Everyone can still be my friend.

"Iris, I need to stop at Taylor's to see if they want the polished rock slices I brought." I knew the owner slightly, years ago. He specializes in local rocks, cut and set in gold and high end art. When I knew him, he was connected to the stones, and now owns the crystal mine way up north he and I used to talk about two decades ago. I heard he goes in to the mine to get the crystals himself. He has to fly in with a small plane, landing on a mountain top. This connection is unusual in a high end shop.

"Like the fur industry, Iris. Most dealers who sell and buy the final product—the fur coats—walk by those who catch the fur with their nose in the air."

I smile and chuckle as Iris points out, "Miles, I suppose it is true of most industries. Those who eat gourmet foods are not likely to be friends with the muddy farmer. Those who have high end wood furniture, are rarely on equal terms with the lumberjack covered in sap."

When we arrive at Taylor's, I see him in the shop among many other workers. His situation looks changed. This is a high end shop, and looks like he has partners. All employees are dressed in custom made clean matched suits and ties. Barbers style their hair, and the atmosphere speaks 'class'. He looks at my rocks and is not as interested as I hoped.

"I have plenty, Miles."

I'm in business. There is no such thing as too much merchandise and materials. It's all about, 'How much is it?' I only say, 'I have plenty', when I am dealing with someone I am not excited to see. Taylor thinks I could get a higher shine on the rocks.

I answer, "Well, I offer them as raw material, assuming you'd cut and polish them for settings. I tumble polish them."

He looks again, and seems a little impressed I can do this in a tumbler. He hand polishes everything. "You cannot duplicate hand work in a tumbler, Miles." I'm not sure if he is trying to make up excuses, or really believes this.

I explain I take it to a tin oxide polish. "I might do better separating softer stones out. It is possible the harder stones are taking the shine off the softer ones."

The reply is, "You may have to experiment with the polishes, some work better on certain stones."

I know I am not trying to reach the clients he is. You can see your reflection in the shine on my stones. That is good enough for me. He acts like he might like to simply chat about finding rocks, and the people in the business. I'm sorry about your problems with the Feds. Such a shame..." He trails off as one of his business partners shows up. With no apology to me, "We have customers, if you are not buying, there is work to do."

I take that to mean I am not a potential customer or seller. I need to leave. The tone is curt, stern, firm. Given as more of an order, given by someone who considers it beneath them to acknowledge, and directly address me. I am guessing my presence is upsetting the balance of the shop. I am scaring the customers. Everyone is high end, rich, looking at very expensive jewelry. I come across as a street person. The police will be called if I do not leave. I have clean clothes on, and am neat, but it is a pair of jeans, and plaid shirt. I have a $5,000 necklace on, but it is fossils—teeth, sort of 'folk art' looking. Rustic. Not for suit and tie people.

I am not looking for the market Taylor is. Knife is a little like Taylor's as well. Wants that high end customer. I had thought about this over the years. *Who is my customer?* I do not wish to appeal to someone I cannot relate to, where I have to pretend I am someone I am not. If it were summed up in one word, when a customer looks at me, I want them to think "Source!" I am the source. I want them to imagine, "Indiana Jones!" The man with the story of how the rock was found and where. I sell magic, memories, energy, hope, excitement, adventure, provenance.

Taylor's used to find their own stones, but have moved up in class going for bigger bucks with higher end customers. *When you find it yourself it may not be the best, it is what it is.* More important to me is local, story, affordable. If what Taylor's wants people to think were summed up in one word it would be 'class'. "Buy something from me and be somebody important. You have arrived." I overhear a customer, "Oh dear, I hope I can pick this up by two o'clock, I take off for Europe this evening, Ta, ta then!" It would be so 'uncouth' in this setting, to find a rock yourself, and tell of it. You might have mud on your shoes.

I know now, Taylor's has changed, and not to be a customer buying from me.

Iris comments, "Yes, Miles, their work comes with a plaque from a certified trained person who has a degree in a frame on the wall who graduated from the Latin school of La de dah!"

"Hoity!" I correct her.

"Graduated from Hoity? Never heard of it."

"Hoity-Toity."

"Ah, yes."

I chuckle at the joke and slight distain. I'm not jealous of these high end guys, more disappointed that our worlds do not mesh better. I am not envious, nor wish I was like them. I'm proud of what I do and what I offer and to whom. I explain to Iris, "Taylor seemed impressed what I could accomplish with stones in a tumbler. He examined the stones under a magnifier when I told him." *I'd never think to examine my work under magnification.* "His work is all done by hand, and I agree it is impressive!" I'm processing fifty stones for each one he turns out. I believe I can accomplish ninety percent of what he can do by hand, in my tumbler. "To gain two percent improvement, he increases his time and prices by a factor of ten." I pause, "So, the question for the customer is, are you willing to spend ten times as much money for a two percent improvement?"

There are those who want the very best and can afford it, who reply, 'Yes!' Good for them. I smile and say, "Next!"

I feel this kind of high end customer, and those who cater to this customer, are involved in a great deal of waste. They want only gem grade mammoth ivory for example. Bluest of the blue. Everything else goes in the garbage, or ninety-eight percent of all raw material is not usable. I do not mind so much being the garbage man. *Cleaning up your mess. Salvaging from your garbage heap.* It is thus understandable, Taylor's does not want the garbage man coming in the front door, discussing business in front of customers. The garbage man belongs discretely unseen, out back being quiet, after hours. Customers are not to meet the garbage man. I simply misunderstood the situation. No hard feelings. Taylor's used to be one of the garbage men. Well not sure 'garbage' is a correct term! More like, I take ordinary bypassed raw materials, and make them beautiful. I take the ordinary and get us to stop and focus on it, appreciate what is common, because it gets handled in a unique way. *It was all made by God, thus beautiful. My job, is to remind you.*

When I first showed up at the Tucson show, I was interviewed with a story put in Gem Magazine. The owners of the magazine would not let me pay for an ad in their magazine. Because I'm not qualified. It might hurt their reputation. If they allowed me in, my gosh where would it end? Just anyone would be advertising, and that messes with their exclusive discriminating customer market. It's like the black jazz player is welcome to entertain you in your nightclub. He cannot go out with your daughter. It's all about knowing your place and staying there. Go out of the social bounds, and go to jail.

I understand why, and the reason is acceptable. I'm not totally opposed. It's about being safe. Safe is your tribe. That involves your class; those who are like you, understand you, and will defend you. Other tribes equal, 'threat'. That is discrimination, but also human nature, animal nature, all nature. Necessary. I had not gone into Taylor's to make a point, exercise any rights, cause a problem. I simply made a mistake, remembering a different time a few decades ago. I'm guessing too, those were different, pipeline days! Rough and tumble dirty booted guys, had big bucks, would walk into a place like Taylor's, and buy high end jewelry. Back in those days the state was rugged. Few of us in the interior wore suits and ties. Bank tellers and jewelers wore jeans and work clothes. Few women wore dresses except on special occasions. I smile at the memory. It was common to see trappers and miners walking down the street with guns much like the wild west in the movies. In those days such people had a few grand in their pocket and willingly put it on the counter, "Show me something nice; got to get a present for the lady here!" People who had no education, zero class, could find gold nuggets the size of eggs, or be a janitor on the pipeline for 100 grand a year.

Sure, I miss those days. It is normal for elders to speak of the good old days! Ha! The good old days were the 40s before statehood! No, the roaring 20s before the depression. No, the Mountain Man years of the 1850s, and the free trapper! Because of this, I smile, take change in stride, try hard to see the good in the 'now'. I could never have gotten my books in print in the good old days. Maybe the good days are yet to come! *Going to Mars! Talk about adventure and space.*

I see a movie and there is a line, "If you stop a motivated person from doing what they love, they will go insane."

"Speaking of adventures, Josh told me Foil is around asking for help." I'm not keeping track of what Foil is doing or not doing. I refuse to focus on the negative. But, now I think about it, since the subject came up.

"I wonder where he is getting his water, electric, laundry done, meals, and such? Who does he borrow tools from?" I have not heard. All that used to be donated by us. But after three years he needs to have his own solutions. Foil has been spreading the word he works for free, and loves to help the seniors. I hear, "Did you see the wonderful job he did on a sidewalk for Kathy?"

"He is tarring Robert's roof!" Mowing lawns using the senior mower. Hanging out with the seniors. Hobnobbing, getting free coffee and I wonder what else. Valerie thinks he is the cat's meow because he cares about her seniors so much, volunteering like a good citizen should. She gave him firewood, the key to her house, and the senior shed, so he can get tools to help others.

I'm sincerely puzzled, as it is he who needs help. He should not have time to go around volunteering free labor, while, at the same time, not taking care of himself. This issue is not about borrowing from Peter to pay Paul. This is borrowing from Peter, insulting Peter, being nice to Paul, who is Peter's friend, in order to avoid paying Peter. Foil should have his hands full getting supplies to the homestead he says he wants. Valarie says, "Well he speaks highly of you, Miles. It's worth paying attention to who badmouths who." I take that as a hint to mind my own business, say nothing, or say only good things.

I hear from one of the seniors, "Miles, where is your buddy Foil? How come you do not see him much these days?" Another of the seniors knows it was me who invited Foil to Alaska, and introduced him to Nenana. It seems strange not be on speaking terms now.

"Surely things can be worked out!" That is the attitude I get around the seniors. *It's me who is so hard to get along with, look at how kind and helpful Foil is! Everyone gets along with generous Foil.*

The impression I have is, Foil wants to have the community on his side so if I 'retaliate' in any way, Foil can say, "See how Miles is!" And have me rejected by my community. He feels I would not want this to happen, so will go along with what he wants rather than risk being excommunicated by my community. I feel, in the big picture this is not going to work. Bottom line being, Foil is not from around here. There is an advantage to being from around here for over forty years. "Miles is one of us—who is this Foil guy? " Hopefully figure out on their own, *"Some kind of con artist!"*

What does a person do in this situation?? :( :(

I answer, "I began helping Foil when he first arrived. I'd like to continue to help, but just cannot afford to anymore." I leave it at that, and am not asked to elaborate. I said that politely and truthfully.

There has been no talk of Foil's failed trip to the Kantishna homestead. I heard him reply when asked at one of our senior meals, "Yes. I made a trip out and back."

Iris mumbled loudly, "Yeah right, he never made it!"

I suggested she keep quiet. "He said he made a trip out and back, not necessarily to the homestead. He is probably embarrassed." I decide once again, it is not making me look good to point out that Foil never made it to the homestead and needed rescuing for about the fourth time! Iris and I both wonder why he is getting discounted senior meals and hanging out here. He is not a senior, so does not qualify for $2.00 meals. It's not our place to point that out.

I so much want to take Foil aside and scream, "You F*&ker!" Except I know this is what he expects, and even wants. Part of his game strategy. If I do this, he can prove something is wrong with me, while he is such a nice person! This would go against me in court.

"I was trying to be reasonable and Miles has gone off the deep end on me!" Get witnesses to prove it. Everything in writing supports him. All our talk and verbal agreement has no witnesses. He knows I can't go to the law. He thinks he knows me and has me under a rock, stuck in the mud I created. I will use this false belief to my advantage.

Foil does not know that someone who helped him on the river called Iris and I. Foil gave him our number just in case I would run out to check on him, *for the fourth time, for free.*

We were told, "Foil was about twenty miles up the Kantishna River having trouble with his engine. It kept slipping out of gear." Any machine being overworked is likely to have gear slippage problems. This is part of what I told him he can expect with the wrong engine and boat for what he is doing. "He unloaded a lot of weight and got back downstream to the Tolovana Road House." The caller says, "I spent two hours trying to help him with the engine. He was way overloaded and underpowered. When I first saw him I was in hopes it was he who could help me haul some of my load!"

The caller left his boat at Tolovana, called a friend, and had the friend fly in with a float plane to take him back to Fairbanks because he was going too slow and running out of time. I'm guessing that two hours helping Foil set him back, and may have cost him finishing his own trip. The caller says Foil had a spare engine as well as his cell phone. Foil had several choices - float to Manley with his load or come back to Nenana empty with the small spare engine, or call someone. There is no emergency I need to be involved in, unless he was a friend, which he is not. I'm not feeling that good anyhow, weak, dizzy.

All I know is, here is Foil. I do not know how he got back. I doubt he has told anyone. The news will not come from me. His engine is off the boat, so assume in Fairbanks being fixed. He is laughing, talking, like he has no care in the world, helping others. Hanging out in Nenana for weeks, not at the homestead he claims to want to be at.

"Miles, been meaning to talk to you!" Hornet's loud voice calls me over. She shuffles, and is having a hard time speaking as I wait. "Miles, you be careful of Foil. He is not your friend!" She volunteers no more. Spoken as if she has some direct knowledge.

"Anything specific I should know about, Hornet?"

She answers fast with a giggly laugh. "Well I just do not trust him myself!" I suspect he tried to get her involved in some shenanigans of some kind.

"Well, Hornet, I appreciate the heads up, thanks." I do not want her to feel uncomfortable or sorry she gave me such news. I'm not going to interrogate her, nor confront Foil.

Not long after, Hornet is gone, disappears. No longer in charge of the Farmers Market. Crazy Lawsen is vague, saying she had an emergency back home in

Montana. May not be back, ever. I hear around she had been bragging about her accomplishments. I heard more, but may as well leave it at that. Just talk.

A MONTH GOES BY; Foil has not made one trip, or spent one day, at the homestead. He still owes me for three rescues, plus the yearly $5,000 land payment joint ownership agreement. He is only halfway through paying the full amount. Iris points out it was not smart of him to be building on land he does not own yet, and then getting controlling and bossy about it. It seems clear to me it will be impossible to share the homestead as discussed and as part of our written agreement. I had suggested several times we should change the agreement! "Amend it, update it, or something to reflect reality, Foil!" Foil is happy with the agreement how it is. *Well sure! It favors him!* When written, we agreed we leave it flexible so we can adjust it later as needed. One reason I was not very concerned with the exact details.

My lawyer friend from last year explains, "If Foil is halfway done paying and defaults on payments, you owe him for improvements, the way the contract you wrote up reads, or that is how a judge would probably see it. The problem you would have is, Foil could name some outrageous value of his cabin work, more than the land is worth. It might be a set up, and he knows what he is doing. I have seen these land scams before."

Nenana is still a place where we help each other. Rooster is passing through hitchhiking to Fairbanks. I watch a local give him five dollars so he can join them and eat. He is maybe homeless, for sure broke. There is a teenage girl trying to raise money to take an educational trip. I overhear locals passing the word to buy her cookies to help. The message gets to our WIN meeting; many in the group have donated money to help her. She has raised several thousand dollars.

After the farmers market, I have lots of rhubarb left over. I joke, "I was the only one who had any the past two weekends and sold out. I had an early crop. Now several other vendors are carrying it and rhubarb is all over town! Now I have sold just one bundle and I have ten pounds left over, dang!"

"I'm going to make pie with mine, Miles." Iris does not like rhubarb in a pie or any other way.

I see a woman in the parking lot I know mostly as the wife of Jim. Jim is way older than her and has health issues. She took leave from her job for a year just to take care of him. So this year they are living frugally! Since I know this, I hand her a couple of bundles of rhubarb. I tell her, "I do not know what I am going to do with it all!" She is excited to be able to make her husband a pie.

"Miles, your art is so exquisite, so individual! Where did you learn your trade?"

"I graduated from Riff!" I proudly say. She looks puzzled.

"Never heard of that university, where is it?"

"Riff Raff! There is a branch in every state!" *Better than Hoity-Toity.*

She nods; she understands. "I guess your art is kind of high end folk art!" This is how she describes it.

She and her husband love the outdoor life, and are used to commercial fishing; being on the river. Jim never learned to read or write! He is a Native senior who simply worked hard since he was a kid, and never went to school. He has lived a life a book could be written about.

"How are your books selling, Miles?" I tell her I am doing only ok. I sold a lot more when there was just one book, and I was busy doing shows and hustling. Now I am slower, so the books have to sell themselves more. Lots of competition. I rarely get local comments on my writing and the book series.

I usually get, "Him? Miles? Write? Hard to believe!" Or, "Good idea, I should write a book, tell my life story!" I smile, like it is a good idea. I encourage people to do what inspires them. It is rare that a first step results in being up and running right off. No one has asked me for any tips. No one has suggested I do a book signing, reading, or work with the kids who might like a writing class after school. The library has a local book section and has had guest speakers to tell of their books. I have not been asked. There is a book reading club with a variety of books that get picked over the years. No one suggests reading something by a local author. It would be nice to get local input, exchange ideas, offer insight into why a certain style got chosen. It is rare at a book club to have the one who wrote it being present to answer questions. So this is like a one man show without a lot of fan fair or encouragement. The drive to get it done is totally coming from within. If anything, I am discouraged by others.

"Who is on the phone, Miles?" Iris is upset. We do not like our quiet TV time interrupted. We lose track of what is going on with this 'who done it' murder.

"Rooster," I tell her covering the phone. She nods like it figures. I understand people get phone calls! That is part of life.

I agree with Iris, I have no reason to talk to Rooster. He sounds drunk. I tell him, "We saw you the other day hitchhiking from Fairbanks. I was half-asleep. Iris told me and we were already past you. We might have turned around but had a really loaded car, no room and needed to get home." I thought Rooster would have seen us and been upset we did not pick him up. Iris said he looked drunk when we drove by. Giving drunken people rides is something we do not do. *Drunks can be so unpredictable.* Rooster told me he crapped in someone's car when he was drunk then threatened them when they got mad.

"They should understand, Miles, stuff happens! That's life!" That may define his ordinary life. I am not so sure how many people crap in someone else's car, and call that normal. What can I say is my own normal is the outer limits.

"Miles, I need that gold wire I sold you!" First, I'm puzzled, *what gold wire?* "On the small spool, you know, the really thin stuff!"

"That is brass, Rooster. I never got real gold from you." I would remember something valuable like that. I got rocks, and he tossed in some junk he did not want to carry—used brass wire was one such item. Why want it back? It's worth fifty cents. This long distance phone call asking about it cost more than the wire is worth. It's not his, he sold it. I'm puzzled.

"Miles, I got a friend who can use it." I'm not going to give it away if it is in fact gold.

"What have you got for trade, Rooster?" *That's worth fifty cents, and half a day of hitchhiking to come get, are you nuttso Rooster?* I have no reason to say this out loud.

"I have a little money now, Miles, got a $50 down payment on my canoe."

"You had a canoe Rooster?" He should know I am always looking for camping outdoor type stuff. I'd buy a canoe. Turns out he is selling an 18 ft. freighter in installments for $150.

"Are you crazy Rooster? You can get $400 for it easy!" It's about $4,000 new. He says it is in good shape.

"Well, I can't remember everything Miles! I can only remember in sections. One section of information at a time!" He forgot to tell me. I feel for him, to sell a nice canoe so cheap when he is obviously so broke. I do not feel so much kindness that I want to get involved. I think he would bring me down faster than I lift him up. Rooster tends to space things out. It is common for him to tell me his valuables got stolen, misplaced, or he forgot where he left something. Someone else finds it. Finders keepers. I think the drinking fried Rooster's brain. It's too late to fix that, he's almost a senior now.

"Yes, Rooster, I'll be home tomorrow. I got a guy coming from Fairbanks to look at knives at noon. Don't forget, Iris and I are at the seniors having coffee until ten o'clock!" He said something about heading out early hitchhiking. I do not want him wandering around my property with me not here. I do not treat Rooster well, but suspect I treat him better than most do. How sad. "

"Oh, Miles, I left some art at Wal-Mart, and one of the galleries in town, and the Nenana Culture Center wants more, so that is one reason I am coming down your way!" He says something about having a kerosene lantern for me and five gallons of good kerosene. I usually can't afford that, or did not in the past, so use diesel fuel in my lanterns. That works ok, not quite as bright and dirtier. Especially on the Aladdin style lamps. I assume I am mistaken and I have some real gold wire, which would be worth a lot. *How did I miss that?*

After I hang up I explain to Iris. She says, "It is hard to believe he got anything selling in Wal-Mart because we tried, and their buyer is in the lower states."

"Maybe he meant he sold in the parking lot, or to one of the employees." I go check on the gold wire when in the shop later doing other things. Just as I thought, it is brass. Rooster must have told someone he could get some gold wire for them. Well, I can give him a little money for a lantern and real kerosene.

I am building a small fish camp, trapper shed, get away place, just a half hour boat ride away. Iris keeps asking about this place. There is an eagle's nest with young. Many baby ducks, geese, wolf tracks, songbirds of all kinds and a great creek with good views. There is a lake not so far away, sandbars with nice rocks and drift wood. An outdoor person would be in heaven. I think Iris would love the view and wildlife. There are not enough hours in the day to do everything. I'm always busy with this project or some other. I believed at first, an old shed along a remote creek I found was private land. I cannot find the owner. I'm told some guy who used to work on the barge line owns this. He left twenty years ago and has never been back. I see survey markers, but no survey plat recorded. I then wonder if this is state land. Possibility state monuments, not a private survey. I am only seeing the bearing tree markers, no monument.

There are no doors left on the shed, and no windows were ever put in. A flood long ago left debris on the floor no one has cleaned up. Animals come and go with no doors. However, the walls are ok and the ceiling is the best, still intact tin with no leaks. The plan is to put doors and a window in.

I got a pile of windows four feet big, just because they were cheap. I had no immediate use for them, but you cannot go wrong owning windows. Double pane in aluminum frames for seven dollars each. My neighbor had them, the father died, wife is moving and plans are all changed. There is stuff laying all over the yard disorganized, getting stolen and broken. This is where Foil stole the boat gas tanks.

So I look forward to a nice window in the shed. I will leave a note for the owner to find, if the owner ever returns. I can just walk away and hopefully get a thanks for fixing it up. At worse told to leave. I see nothing on any maps or records. I can put in a table, bed, stove from my surplus stuff. This kind of work is an old habit. It is how I spent most of my life! Building, fixing up old places, moving in stuff I buy cheap whenever I find things at garage sales. I have not given that up, so buy sleeping bags, small beds, cheap wood stoves, lanterns, and such. Possibly out of habits that die hard. This is a place I can hunt for moose, catch fish, even get out and take the laptop computer and work on my book in the quiet of the wilds. Like old times! So, I am a little excited. *Not the stuff great adventures are made of, but it still beats working for a living.*

I have already dropped off an old canoe. The one I took from the Kantishna and fixed up. The one Foil said, "You are not using any of this stuff, so you may as well give it all to me; what else are you going to do with it?" *Well good buddy Foil, this is what I will do!* He thinks I am too old to do anything. I had been having trouble justifying spending $500 to go to the Kantishna homestead just to fish or relax. It is a six-hour run. I no longer live that lifestyle where I need that remoteness. I can now get out for a few dollars of gas, half an hour time, and enjoy the same benefits. As I get even older, I should be able to manage a half hour boat trip for more years then a six-hour trip. If income drops and I can't afford supporting 115 horses, or all that gas to the Kantishna, I can get a little boat and motor, and make it out to my sugar shack on an amount of money Social Security supports. *Thinking ahead, leaving myself options, that's what survival is.*

I'd still like to visit the Kantishna land now and then if I have my way. That is part of the agreement with Foil, and why I sold for almost half of what I might otherwise get. He promised I'd be welcome to share, and keep my small cabin. I'd hunt moose maybe. But if not, I can move all my things out, and let Foil build his dream cabin and pay for the land. I think of Karen and the kids who visit sometimes and have memories of the place. I think of the people over the years I have dropped off there to enjoy the place and thank me. I did not realize how much that means, to be able to offer this, that was to be part of the agreement with Foil, now being ignored.

---

"MILES, what are you going to do today?" Iris and I are at the senior center doing our usual routine. Iris is getting paid. I am visiting her, drinking coffee, reading the paper. She's doing the same thing, but being paid to answer the phone if it rings. It might ring a few times all day. Our time to be together. She often wants to know my plans for the day. I had outdoor plans when I first woke up, but we woke up to rain.

"I'm not sure yet. Something indoors." I could work on my writing. On the walk home I decide it might be a good day to get the shop wood stove going and clean it up, sort a few things I have simply set down inside the door meaning to get to later. Like now. I never get to this in winter because it takes too much to heat this main section of the shop that has half a dozen rooms. I just store stuff here in winter. I can come in long enough to find something, then take it to a warm room to deal with. In summer, I end up too busy to ever get the sorting done I plan to do. Today is a nice enough day; a small wood stove fire will take the dampness out of the air and so here, I am! While I am in the shop, I run the rock saw so I can manually reset the next cut when I hear the saw shut off.

"Hello. Anyone here?" I recognize Rooster's husky voice. I was not sure he'd hitchhike in the rain.

"Yes, I checked the wire; it is not gold, just brass!" He needs it anyhow. His smile shines through a bushy beard and full head of hair. He looks like a muscle bound boxer. Leaning over, arms forming a circle like an ape. Maybe muscle gone to flab now.

"I couldn't bring the kerosene. No one would want it in their car." Two very old beat up kerosene lanterns is all he brings. "I have already been to the cultural center and they do not have enough cash to give me."

"Rooster, they are getting 800 people a day in the shop, a good $1,000 a day. I doubt they are low on cash."

"Yeah, I know." He accepts they simply do not feel like paying him today. He is spending his entire day coming here to not be paid. He's a drunk. If the cops see him, they'll run him out of town. The community has its hands full supporting local indigents. Rooster is an outsider. The community is open to those in need, but not chronic sycophants. So while we'll help him as he passes through, there is no encouragement to have him move here.

"Miles, I need some amethyst crystals!" It sounds like his plans are changing. I had not seen him in a few years, then heard he was gone. Rooster shows up this past spring and tells me his home is someplace else now! He is only here to wrap up business, gather up his things and head back home out to glorious Cisco, California! This is why he was showing me all this stuff he is selling. Now he needs more money and wants some of his tools back and is acquiring things like crystals. He wire wraps them for add on value and sells them as pendants. These are crystals I break off a cluster with a hammer. My cost might be ten cents each. Rooster takes ten for the two lanterns. I understand Rooster saying he might get ten dollars each for these amethysts once wire wrapped.

"Miles, I'll make $100 off these when I sell them!" He is all excited, but seems to not have a grip on reality. He will hitchhike all day to go to a shop to sell one then reward himself by getting lunch, which he deserves. Lunch that causes him to break even financially. Hand to mouth, no real grasp of finances. *No wonder his girlfriend finally kicked him out. He sounds so full of promises of success when you first meet him.*

"Rooster, do you still have your wire wrap pliers?" I'm not sure he kept anything when he said he needed money. I specifically asked him if he really wanted to let go of the tools of his trade! He said he is giving it up, selling out, going out of state and doing something else.

"Yeah, Miles, I'm finding stuff I forgot I had. I also see some of my stuff up at the lodge! So that proves those neighbor kids stole it and sold it to the lodge!"

We both know there is no use telling the police. Rooster has no rights. He has to accept there is open season on him and his goods. He lives in an abandoned railroad car he does not own, on land that is not his. The general public is annoyed, wants him arrested, locked up, so he does not rob them or create a scene of some kind. I do not like to see him around, beyond a hello, exchange of smiles and the odd small trades now and then.

"Rooster, did you stop by earlier?" I checked my security camera earlier and saw footage of Rooster going up to the shop door maybe trying to get in.

"Oh, yeah, Miles, I forgot." He pauses. "I saw your shop door open and closed it for you!" I do not let on I know this is not true because I have video footage of him. It is in fact nice to have camera security. What I see is, those who can afford this $500 camera package get protection. Those without the money often get robbed blind.

I recognize an issue we face. *What does society do with folks like Rooster?* I cannot help but think we, the people, are part of the problem; we help create and maintain it. We made money off this sap selling him poison. We treat him like a criminal, not a sick person. I'm told he had a choice! He chose to go on this path! I have known Rooster for thirty years and there is more to his story than that he simply chose to be a destitute drunk. He had an attitude all those years ago I thought would not work in terms of winning friends, getting ahead, surviving.

He had a girlfriend for years. She had it together better than him; made more money. He treated her all right. She was pretty, younger, sober, smart. Back then, Rooster was handsome, easy going, fun to be around. He dreamed of doing better with his craft. Thirty years ago Rooster was a cute teddy bear that women could adore and cuddle. She allowed him a room in her house devoted to his crafts that would one day pay the bills. I did not see him seriously analyze his business, work towards making more money, changing to meet his market, look at advertising. He enjoyed exactly what he did, and was not very responsible nor adaptable. He used to appreciate it when I offered selling tips on displaying his work. He saw results and thanked me later. This tells me he is capable of responding to an education. I'm guessing no one took the time to teach him life skills. I'm not going to take him on as a project.

I never thought he was artistically talented. He was capable of low end crafts in the way a child is. Find a rock and say it looks like a face. Glue a simple cap on a tooth or small item and call it a necklace. Nothing that required equipment. Good for the tourist trade, not galleries. Not the $20 an hour he says he is worth. However, if he understood what he offered, he could make a good living. The tourist industry badly needs locally made items under $5 and up to about $50 high end.

The girlfriend finally got tired of supporting him. Partly he did not see it that way; did not appreciate what he had. I'd say the entire situation was marginal for success. It could have gone either way with just a little help or luck. But his luck tipped to bad luck. Long ago I recall him saying, "I don't need her, Miles, I'm headed to Fairbanks to get drunk!"

*Getting drunk on her dime.* His answer, many people's answer to what ails them. I see the outlook promoted in many movies. Something goes wrong, and the common movie line is, 'Come on buddy, let's go get drunk together, it's on me!' That is what Rooster's heroes did. *What other way is there?* No one showed him. I can picture a relationship deteriorating when his answer is to go get drunk whenever she wanted to talk serious. He never did get back on his feet or get ahead after she left. Possibly he is not someone who can make it on his own. That in itself is not a crime. He is not mean. He has simple inexpensive needs, kind of an easy keeper. He'd even be an ok stay at home father, back when he had his lady. He's happy go lucky. Another woman might have understood the plus and minus, shrugged and taken him on.

It is possible I deal with the likes of Rooster so I can look good, better, compare, and be glad my life is better than this! I say this because there is a group of people in the art business doing financially better than I. I do not hang out with them. Most own their own shops - like Taylor. Knife has a shop; Mike has a gold shop and half a dozen artists I know. All have offered to either hire me, work with me, help me out, or offer friendship. I have been stand offish. They tell me it is not financially worth doing farmer markets, bazaars, or fairs. I look at them and their operation. They are almost all running $300,000 a year gross operations. Presently out of my league. I'm at the bottom of their financial food chain. They all buy new top of the line equipment, pay for permits, taxes, insurance, employees, workman's comp, etc. They are participating members of society; among the protected. When they get robbed, it makes front page news. The police are right on it. Insurance covers them. They may well admire my adventurous and free life, but would not pay the price to trade places. I admire the money they can gross, but would not wish to be in their shoes with all the 'hoop-la' that goes with that. Their personal take home pay is not so much greater than mine.

Interestingly I get insight when Rooster says the same about me. "I love being free, Miles! All I own is in a backpack!" Like the country song, 'King of the Road'. Happy or not, Rooster is making problems for other people, being a burden, and this is where I do not agree with him. He is not drinking with his own money, nor smashing up just his car when he runs off the road.

While high end artists with shops recognize my talents, and believe I could be 'one of them', they also put out a hand to help me get there. But if it's not what I want, they understand they need to smile and wave good bye. Being too close to me could put their status of protection at risk. My father had to make up a new life; got fake papers in order to rise to a different class just as my son needs to not know me if he, as well, wishes to be a professor.

I earn what I feel is a lot of money, but 'a lot', I keep saying, is a relative term. *Compared to what and who?* Compared to Rooster or Wess the Mess, I'm a millionaire. But Taylor probably makes in a month what I make in a year. Do I want, and am I happy being, a big fish in a little pond? Where I am one of the best artist in a neighborhood of 800 citizens and two dozen artists? I smile, because this is what I wondered when I went to the big Tucson show for the first time. They are going to kick my butt. The country bumpkin gets his just desert.

I was told this, I considered it a possibility. I had never done a bigger show in my life than Fairbanks! I only broke even for my costs the first year in Tucson. Still, I was featured in Gem Magazine. I was not laughed at. I held my own as at least another 'one of us' professionals - 'world class'. Even if on the bottom rung. I met a quarter of the venders who were in trouble, lost big time, going home in debt, some not even able to pay their way home. I also noticed when I pay $1,000 to set up and sell, the police consider me among the protected. We are buddies. Contributing to their wages. *We are not all equal.* The world is viewed entirely different. I'm blending in with the civilized. I could not keep that up. It's just a two week show. If we are not all equal, then there is discrimination. Perhaps there should be. Discrimination is the definition of individuality. Our tribe feels superior to other tribes, even as we allow them to walk under the same sun. We protect our own. We do not protect all people equally. How could we? I understand protecting Rooster now would be a full time job for a dozen people, and even then, he'd figure out how to screw it up. He border-line needs to be institutionalized. Rooster contributes very little to anyone but himself. What slows me up helping Rooster, is the fact he steals from me. He seems happy enough with his lot in life. No serious motivation to change. "I like being free Miles!" I smile at Rooster, nod, and wave him on his way.

I mention to the mayor that I ran into Rooster. Mayor is ultimately responsible for buying inventory for the cultural center. "Yes, he stopped by here a week or two ago. I bought a few of his things just to help him out." Mayor tells me the product looked used. Rooster told the mayor, as he told me that all his belongings got wet because he had them in a pile under a tarp for a year. He had no place to stay, or keep his stuff. Tourist do not want to buy wet photographs. Now Rooster thinks he found a gold mine and is here regularly looking to be supported. No, it was a onetime courtesy because Rooster needed help.

This puts in perspective, my own younger years. I used to bring art to sell to civilization. "Customers loved my stories, Mayor! How I made the art without electricity! How I hauled the art to civilization, and it took a week to get here in an open boat, or by dog team in winter! But oh no! The tourists did not want to have the art work look like it had been on such a journey; covered with river silt, damp, bent display box, tarnished." The mayor smiles and nods he understands. My feelings would be hurt. I'd fight resentment. Not accepting the reality of the stories they are eager to hear.

"They want to know I slaved away at their ten dollar necklace for a week creating it. Telling me how talented I am. At the same time, smiling at the deal, understanding they are paying me five cents an hour for my time." I had to figure out on my own, there has to be a better way! Rooster never figured it out. *Or knows, but is somehow trapped by his addictions.*

*Subsistence fish wheel in Tanana river.*

# CHAPTER FOUR

## TUCSON COP, ANILCA AGREEMENT, SUBSISTENCE LAWS

I see a pile of birch baskets from Mark on the mayor's desk.

"They sell well, Miles, and we can hardly keep up!" I smile. Mark has his business figured out. A little like Rooster, with the big difference being, he knows his market and targets it. Mark has little money, but can get bark off trees free! Folds up a square into a simple basket and it's a $7 tourist item he gets $3 for. Not a lot of money, but he can make one in under ten minutes, so that's $12 an hour, maybe clears $10 an hour, higher than minimum wage.

"I see grouse feather art is back in the culture center gift shop. I was there checking on my own art, and saw the feathers. What's her name, Dim's lady is kind of in charge. She thanked me for bringing in the paper with the copy of the law that says they are legal." I add, "Not to make an argument! Just to politely show, and kindly say, this is where we read the policy, so if we have misunderstood, please show us in writing!" Dim's woman says the cops tend to have a personal grudge for some reason, either against the community, or an individual they want to mess with. Nenana is possibly viewed as a subsistence low life community, not contributing much to civilization. Not among the protected.

"Only 'maybe', Miles, it's speculation." I press on as if she hadn't said anything. Loudly from my soapbox Maybe police resentment; as a community we voted not to pay for a cop, can't afford it.

"No pay, no protection!" Thus, open season, vulnerable. Maybe even in the way some people enjoy pulling the legs off spiders. In the way some cops get into this line of work so they can bully people. Bullies target the weak and unprotected. Not all, but it only takes one to be a pain.

The mayor explained to me his take on what he sees. "Criminals stay in touch with each other, Miles. Word is out that Nenana has no cops and no one cares. So naturally a criminal element moves in out of the reach of the law, and just an hour drive from Fairbanks where there is law and order!" Some sort of good news is, the criminals do not necessarily target us, the Nenana poor. It's just a home base for them, no hard feelings against us. Also the criminal element is not the sort who mug people on the street. We do not have a lot of violent crime against the general population. Violence is a way of life among some segments of the community who prey on relatives and in-laws. Ordinary citizens are off limits. So you are safe out walking your dog, Children are safe out alone, as well as women. Criminals may have a code they live by. To mess with us is to end up with a hired cop. Certain things the community simply puts up with. Just leave me alone, and you go do your thing." I sometimes have that view. But more and more I feel what happens to some affects us all. Mayor also points out, "Fish and game issues like the feathers has to do with Nenana and its attitude I bet." I see his point.

"The past Edmund days." He nods. A Native elder character used to illegally sell fish and eggs, and engage in high profile illegal wildlife activity under the guise of Indian subsistence. He got way with it, but it upset Fish and Game. Nenana backed Edmund, almost as a local hero, so the entire community got on the outs with the law. We may have acquired the reputation of an out of control community that needs its chain yanked. No one wins against the Feds.

Mrs. Dim used to drink a lot. Sometimes she was even a serious problem. She and the husband beat each other, called the police on each other, sold each other's goods to raise money to drink. When sober, two peas in a pod. She loves the trapline, skinning fur, running snow machines, river boats, cutting fish and can be a hardworking, knowledgeable helper. She got fired from more than one job for drinking, stealing, and such. Here she is, operating the cash register, and being manager. Word is, she has not had a drink in several months or more. She comes across as intelligent, competent, dressed well, shows up on time and a genuine interest in doing a good job and being responsible. Possibly an example of someone who was given yet one more chance after a lifetime of chances, and something clicks, there is a change, and a life is saved and straightened out. I'm in favor of giving her a chance from what I see.

The mayor said, "Miles, I am having trouble keeping workers there and no one in charge. No one wants to work!"

Optimistically, I do not agree. I think it is more like good workers already have a job and are in demand. Those who do not have jobs, claiming to be looking for work, are last in line to get hired. Possibly the answer is to offer enough money that a good worker will leave where they are, to take this job. The mayor says we cannot afford that, the cultural center is not making enough money.

One reason we are so popular with the tourist, is that we buy and sell a lot of Native crafts. We are the only authentic Native population on the road system the tour buses hit. We are the only major gift shop featuring local handmade, not imported normal tourist 'crap'. We offer the lowest prices for gifts in the state. The mayor is firm on taking only twenty percent from the artists. Normal is sixty percent. Good artists are coming to us to sell. For the first time we have not had to hunt around for inventory, or make buying trips.

"Miles, Princess tours wants us to pay to have buses stop. They want a kick back." Most tourist traps pay to have buses stop. "I told them no. They can stop because it is a good place to stop where the tourists have good things to say so it is worthwhile to you. Otherwise, do not stop." They keep coming. Up to nine buses at a time and now, all day long. Not even the Denali Park gift shops get this many buses stopping. Mr Glitter told me he only gets fifty to 100 visitors a day at Denali Park. He has high overhead and still makes money. This is why I cannot believe the city takes this lightly, and thinks breaking even is a good goal.

Helm asks how my book is coming along. I give him an update.

"I do not know, Miles, if this is going to be interesting reading. It is normal daily life no one cares about!" My long time German friend is only concerned for me and making money from the books, not meant as a put down. I study him before replying. Helm is big, looks like a boxer, square face, filled with determination.

"Maybe, Helm. Maybe I am setting the groundwork for something more interesting, like murder. By setting the stage, getting into the dynamics of how a small community operates, with all the politics, grudges, hidden agendas and how the money moves." Helm only laughs.

The police are not doing anything, so many citizens are getting fed up. There is, in fact, open season on a few characters. It is expected for them to turn up missing, or for us to hear gun shots followed by, "Do not know a thing!" Those who are the target may not fill their accepted role and might shoot first. Stay tuned to this station! Do not adjust your set!" Village shoot outs make dandy civilized news! Stories right out of 'Deliverance!'

"Anyhow Helm, it might be interesting to hear how small communities operate. Especially if readers contemplate moving to one, or are interested in where they will get supplies if homesteading or being off the grid. What is it they can expect? There is the dull and boring, along with the good, the bad, and the ugly. I used to trade with some druggie people. At first they brought me antlers and carcasses of animals they trapped, or found lying around in the woods, maybe at their fish camps." This worked out well for a year. I only began because of talk at the community Wellness meetings, about how the community needs to be responsible for all its citizens, even the poor. The poor came with a hand out looking for free money. Instead, I suggested ways they could earn it! I'd pay them! Give them their respect back.

After a while this element needed more money, but had already sold me all their antlers. So they began to steal stuff and bring it to me. I explain to Helm, "I started getting phone calls, asking if so and so brought me such and such, because it was theirs." I did not enjoy being the first on the list to call if something was stolen! I'm out the money and the product.

"There goes any good reputation you had, Miles!"

I tell Helm, as I have said to others, "I even told the police I was developing some good connections with the local underworld, and would be willing to help clean up the town, help stop crime, be a good citizen and all."

"So what happened Miles?"

"It's in one of the books, I gave the details, didn't you read it?" I just do not know if the entire system, from the top down is corrupt.[1] I remember being told if I had anything to do with these low life drug people, I would not be among the protected. Flat out told this by the police more than once.

Mrs. Dim says, "It's in the news almost every day. Another cop, another official, another trusted person in power, caught taking a bribe, raping a secretary, doing drugs. Basically the same as common citizens like us, only we get arrested!" I chuckle that I notice that as well. She tells me what a shame it was, what happened to me, a felon now and all. The implication in her voice is, as bad a person as she is being accused of, she has not done jail time and is not a felon. My crimes are greater than hers, and it is me who needs her understanding, more than she who needs mine. She is the better citizen, as judged by society.

However. However I notice the news is about selling papers, sensationalism. Ninety percent of that appears to be bad news! All I hear is how so and so got beat up by the cops, or nothing was done about their missing stuff. The police must help people. *I mean otherwise why would the public hire them?* They must be protecting someone, right? What's reported is how the cops showed up to clean up the mess afterwards, mop up the blood as the war on crime continues. Mrs. Dim adds, "Yeah, where is the news so and so had their things returned with no one getting shot? Does that ever happen? "

I'm puzzled, as I know more of a variety of people then average. I do not have enough information to even form an opinion. I'm simply puzzled. Why would a trooper come to our village, tell us we cannot sell grouse feathers? Why not straighten out the mistake, or apologize for the error? Why not be accountable for their error in the same way I am accountable for knowing the law. I'm puzzled. I'd think the police would want good relations with the public, indirectly, their boss.

I tell Mrs. Dim about a Tucson incident with the police, in contrast to our conversation.

**Past Flash**

Iris and I are at the gas station near my mother's in Tucson. We go there each morning for coffee and chat with the clerks who are becoming friends. We noticed over time, that several homeless people hang out here. It is interesting that they are not run off. The owner and clerks are friendly to them, talk to them, help them. Even to the extent, "Hey could you watch the store, I have to go to my car a minute." They are trusted! They enjoy being homeless as a way of life, or accept it and do not appear angry. They do ok pan handling.

One guy tells us he used to be a long distance trucker. "Did so for a lot of years." He got into a truck accident and 'damaged' his brain. He is not the same; cannot hold a job; cannot concentrate. His mind drift, he cannot tell time. Apparently there was no compensation, or if so, he spaced it out. He is now a street person. His friends are on drugs, but do not steal. They all live in drainage culverts and have no bank account, ID, or belongings other than what fits in a back pack.

One day a cop comes in. Iris and I watch him buy a bunch of burgers. We joke that he must be hungry! He only smiles. We walk outside to see him handing out the burgers to the homeless he knows by name. He trusts them and they trust him. The homeless group tells the cop that they are looking for a bike reported missing by a gas station employee a few blocks away. These homeless are the ears of the cop, not dirty low life snitches. But treated like humans. I'm shocked. This is what I envision things ought to be like, can be like, and in my view should work! Here is an example right in front of me at the gas station.

Then, there is Andre, who helped me at the show for several years; recommended to me long ago by a friend as someone I could trust. He has a wife and child who live in another state. Andre asked if I might find him a little work or do some trading when he hitchhiked to the show. I can't 'hire' him as such, due to the laws. He helps me and I give him pretty rocks. Tall skinny, always has a smile, carries a skateboard like teens do, only he must be thirty, so suspect he is a little 'off'.

Andre was constantly beaten by the police for being a vagrant here in the same area. He now has brain damage, I believe, from the beatings. I think so because I saw him change after being beaten up and not seeing a doctor. He had blood coming out his ears after one beating. He was working for me, and this is when I suggested he sleep in my booth as security at night. We had to clear it with the hired security.

I asked him what he said or did to get beat up. Surely the cops had their version of the story! Maybe he left out the part where he was robbing a store or raping someone.

However, after several years of Andre working for me, I find him to be kind, honest, hardworking. He seems to have a concentration problem or attention issues. He's only good for menial jobs, but that is ok, and he is ok with this - a very happy go lucky sort. He says, "I was sound asleep under some bushes in a quiet place away from public view. I woke up to cops dragging me out of my sleeping bag and clubbing me." After that Andre started giggling, humming, having a zoned outlook, and acting even more like a child. Even a year later. So are there cops with different views on different beats?

**Past flash ends.**

All we need is just one good cop in the area we can trust and work with. Who might look out for us and protect us. Who in turn we can help. This is my vision of what I hope will happen. Not in making all cops our enemy. But also not trusting every cop for no other reason than because he wears a uniform and carries a gun and says he's a cop.

"In most cases you want to run!" Mrs. Dim says with a chuckle, and I agree. *These are dangerous, even ending times we live in. Who can we trust?*

I tell Mrs. Dim about the local cop who helped me get my driver's license, otherwise, I'd still not have one. Nice guy. "Yeah, Miles, because you are white, the chosen race. Try being Native or Black and see how that cop treats you! He beat the heck out of Dim! Put him in the hospital!"

The community got rid of this cop. There were too may accusations of harassing pretty women and beating up minorities. I never believed it at the time. I'm now more inclined to think there was some truth to the talk. I had heard 'the other side' a couple of times. How individuals taunted the cop, baited, threatened, cussed. I thought at the time, the cop should have training in how to handle difficult people and situation without taking it personal. After all, he is a professional and is the one who is armed. While I did not condone his behavior, as a human I can sympathize with wanting to clean someone's clock when they have an attitude, spit on me, and insult me. Chinese proverb, 'Poke finger in someone's chest, feel nose get broke.' *I made that one up.*

I have known Dim for over thirty years. I respect him and trust him. I lent him money so he could buy gas and go trapping one year after he recovered from one of his drinking binges. He paid me back. I do not know him as the sort who would yell, cuss, or threaten. I view him as a likeable happy drunk. Dim is one of the many Natives who has given me written permission to look for mammoth fossils on his Indian land. In his prime, ran 300 miles of trapline. I respect what it takes to do that. If he got beat up by the poe-lease? I take my stand at Dim's side. I know Dim. I do not know the police. I try to know the police. The police rarely want to know me or be my friend.

"We do not have to be friends, we carry the big stick." *Good luck then! See how far you get without allies!*

I tell my dubious friend, "This story, these details, concerning the relation with the laws, police, government and interpretation of the laws is crucial to understanding my lifestyle, Helm."

Helm is German and discusses issues in his country.

"Helm, who is 'us'? Who is 'them'? Who is the enemy?" What about equality? What about laws applying to us all? What about discrimination? You want people to know about the wilderness Helm? Then hear about wilderness people! Hear what concerns us!"

"Miles, people want to be entertained, want the dream. They do not want to hear about problems and politics! This is what it is they wish to escape when reading!"

"What about Animal Farm, Diary of Anne Frank, Utopia, 1894, The Jungle?" These are well read classics that are not about escaping reality. "There are a lot of books, Helm, fictional books about Mountain Men written in second person by city slickers interviewing a few people, looking up a few facts to support a fictional story. How many such books are done in first person? How many could get into the delicate details? Why do people tune in and drop out, Helm? Why did I come back to civilization after working so hard to leave?" I weave the thread that defines the web. Owned by the spider, that catches the flies. Helm is stubborn, and there is no changing his mind. One aspect of this trait I like is, once such a German is your friend, he is your friend for life. He stands by you. It seems all the women in my life that I stayed with are German. Imagine!

"Miles, that is easy to understand, it is because Germans are the best looking women. Everyone knows this!" We both laugh.

I head over to the city office to chat with the mayor. The mayor is curious about the laws concerning subsistence, as it relates to buying, selling, trading or even displaying various craft items using animal parts. We have already reviewed specific laws about bird feathers and fossils.

"Miles, I tried to Google the subject and no actual laws in the law books show up. Nothing I can quote to the artists, tourists, who ask when buying, or troopers who tell us we are doing something illegal!" He says, "I heard Nenana is a subsistence village. So does that mean all of us who live here qualify? I hear talk subsistence allows for some amount of barter, and even money exchange. Yet, the newspaper has articles saying someone is under arrest for doing so. The game regulations hand out say 'No'. How can the game books be wrong?"

This brings up the issue of what laws are, who makes them, who can change or enforce whose laws? I'm not going to go there. In fact the game regulation book has a disclaimer in fine print on the last page, "This is not a legal document, please refer to the actual regulations." There is no information on where these actual laws can be found. So the book handed to us is only a suggestion? What is it exactly, if the suggestion is not the law?

I recall a front page story the mayor refers to, of a subsistence person trading extra moose meat for firewood and getting arrested. There was such a protest from the public the charges were withdrawn. This did not settle the argument; if this is legal behavior or not. It is common behavior for village people to trade wild game and other natural resources, but is it legal? I ask the mayor if he is familiar with the ANILCA laws?

"I could not find any reference on the internet, Miles, but heard something about this agreement between the government and the Natives that also effects all subsistence people in Alaska."

I tell him I cut and pasted the relevant sections of the agreement in a diary, and I'd print it for him. This affects him, myself, and the community. Specifically, concerning items for sale at our gift shop.

Anyone interested in living remote needs to consider options for paying bills in an area there are so few regular jobs. What is it we can actually depend on? Ourselves! Selling raw, or doing crafts using 'things off the land', should be considered. This includes rocks, wood, furs, flowers, anything we can gather that might be something a civilized person would not have access to and might be willing to pay for. Laws concerning what, where, how and when would help keep us out of trouble with civilized dealings.

Why aren't the laws concerning this, available to the community and easy to find? *Why is it we can Google for anything under the sun and get information, except laws.* A judge, well actually just a magistrate, told me the requirement is for citizens to know the law, but the legal system is not required to provide easy access to or be helpful in finding these required laws.

The impression I get from this is, very few people, even in the system, know the law. Judges have admitted to me they do not know the law, until a lawyer points it out in writing and the judge verifies, "Yup, there it is all right!"

This is so important to me and my own lifestyle, that I review the laws, and read them again carefully. I look them up again to verify nothing has changed.

A summation of Subsistence laws relevant to issues at hand

### Federal

Under ANILCA, rural Alaska residents are eligible for the subsistence priority. Rural residents make up about twenty percent of the state's population. Rural residents are defined as all Alaskans except those living in and around Anchorage, Fairbanks, Juneau, Ketchikan, Adak, Valdez, Wasilla, Palmer, Homer, Kenai and Soldotna.

Note: Nenana where I live is specifically named as a subsistence community.

## Subsistence Economies in Rural Alaska

Some subsistence products are bartered and exchanged through customary trade networks via small scale transactions involving modest amounts of money. Federal and state laws prohibit the sale of subsistence products at commercially significant volumes. The subsistence socioeconomic system in rural areas is most properly understood to be a mixed, subsistence/cash system. Thus, subsistence and cash sectors are interdependent and mutually supportive.

The rural preference in federal law provides a tool for fish and game managers. Subsistence harvests by communities classified as `rural' can be recognized as distinct from the recreational harvests and commercial uses.

Subsistence users are enabled to pursue cultural patterns without conflict with regulations from the federal and state governments. When wild resource populations cannot support all uses, customary and traditional subsistence uses are restricted last, after commercial and recreational uses. In this manner, Alaska communities with the greatest dependencies on fish and game are provided an opportunity to continue ways of life built on mixed, subsistence/cash economies.

5 AAC 92.200. **Purchase and sale of game** (following is a list of rules restrictions )

Sec. 16.05.930. Exempted activities

e) This chapter does not prevent the traditional barter of fish and game taken by subsistence hunting or fishing,

Sec. 16.05.935. Restrictions on cooperation with Federal Government.

The power to control the management of fish and game within the boundaries of the state is an incident of state sovereignty, and that the Federal Government cannot commandeer the lawmaking processes of the state

Subsistence in Alaska

Customary and Traditional Use Determination

In Alaska, customary and traditional uses of fish and game populations are protected by state law, and the Board of Fisheries and Board of Game must provide for those uses first before providing for commercial or recreational uses.

Sec. 16.05.940. Definitions.

In AS 16.05 - AS 16.40,

(2) "barter" means the exchange or trade of fish or game, or their parts, taken for subsistence uses

**(B)** for other food or for non-edible items other than money if the exchange is of a limited and noncommercial nature;

(7) "customary and traditional" means the noncommercial, long-term, and consistent taking of, use of, and reliance upon fish or game in a specific area and the use patterns of that fish or game that have been established over a reasonable period of time taking into consideration the availability of the fish or game;

(8) "customary trade" means the limited noncommercial exchange, for minimal

amounts of cash, as restricted by the appropriate board, of fish or game resources; the terms of this paragraph do not restrict money sales of furs and fur bearers;

(33) "subsistence uses" means the noncommercial, customary and traditional uses of wild, renewable resources by a resident domiciled in a rural area of the state for direct personal or family consumption as food, shelter, fuel, clothing, tools, or transportation, *for the making and selling of handicraft articles out of non-edible by-products of fish and wildlife resources taken for personal or family consumption, and for the customary trade, barter*, or sharing for personal or family consumption; in this paragraph, "family" means persons related by blood, marriage, or adoption, and a person living in the household on a permanent basis;

Sec. 16.05.930. Exempted activities.

(a) This chapter does *not prevent the collection or exportation* of fish and game, a part of fish or game or a nest or egg of a bird for scientific *or educational purposes*, or for propagation or *exhibition purposes* under a permit that the department may issue and prescribe the terms thereof.

Below another source of info. with links for more information.

Barter complements the environmental movement that has gained traction in the late 20th and early 21st centuries. Consumer and small business websites such as Barter-Quest.com promote bartering as a green alternative to buying and selling.[8]

Modern trade and barter has developed into a sophisticated tool that can sometimes help businesses increase their efficiencies by monetizing their unused capacities and excess inventories. The worldwide organized barter exchange and trade industry has grown to an $8 billion a year industry and is used by thousands of businesses and individuals. The advent of *the Internet and sophisticated relational database software programs has made it easier to conduct these activities and has further advanced the barter industry's growth.* Organized barter has grown globally to the point where virtually every country now has a formalized barter and trade network of some kind. Bartering benefits companies and countries that see a mutual benefit in exchanging goods and services rather than cash, and it also enables *those who are lacking hard currency to obtain goods and services.* To make up for a lack of hard currency…

I get interrupted in my review of the laws. I admit I am not clear on the difference between a statue, a law, a regulation or a declaration. It's all 'stuff you are told you have to abide by', 'or else'.

Iris ran into Rooster and asks me in irritation why I even say 'hi' to him! Yes, a good question.

"Well." I have to pause and tilt my head with a smile. "Rooster was not always as he is now." I explain that the nice diamond rock working station I have now, "I got from Rooster!" For $300! It's worth new about ten grand! It has seven stations. When it runs you cannot see the shaft move or hear it. I'm not even sure something this good is available today at any price. Huge solid roller bearings, easy to replace. That was a Rooster deal - the last deal, "Iris, there was a five gallon bucket of grade A Nowitna agates I have no reliable source for." Some of the best I have ever seen, made up now into $100 belt buckles. That's $100 a slice, or an ounce. When I paid $100 for the whole 30 pound bucket. "Sometimes Rooster comes up with cool stuff!"

"He's just a bum, Miles!"

"Perhaps. Maybe not always. He used to work a regular job in Denali park—gold panning with tourists." He was once quite an outdoorsman, getting out to remote places, with knowledge of how to live off the land. I always respected that part of him. I usually enjoyed hearing his adventures. We had a love of the outdoors in common. I tend to believe in loyalty. I usually do not desert people I know in their time of need. I do not dump them because now they can't help me anymore, or have changed social status.

Iris shrugs her shoulders, not sure she understands, saying, "I was just wondering. I remind you of your problem with saying 'no'!"

I do seem to empathize with underdogs—those with baggage. Because I am considered such a person? Even Iris, who is supposed to know me best, would admit I am a problem person. Certainly she has a very long list of things I should change. Authority issues for example, the very subject at hand. Who can tell me what to do! How to do it! I do not want to think about it…

I get back to the laws I looked up. I now review what the state has to say on the subject.

In reviewing the subsistence laws, I note some repetition between Federal and state, and different sections within. One aspect is unclear and might disqualify me. Barter, trade and cash is legal *only if it is small amounts, noncommercial.* Maybe I do not qualify because of the amount I was dealing?

I am told, "Come on, Miles, you cannot have it both ways! You cannot tell me you make a lot of money, and at the same time claim to depend on the land as a poor person!" True. I walk a fine line. I know and admit this. I wish I had some guidance, direction, a government I could trust in such matters.

I have said, I tell different people different truths or stories. I can say, "A lot of money!" But that is a relative term. I can also add, "Remember, I lived a couple of decades on $2,000 a year. Double that might be considered a lot, but still below poverty level." Remember too, I do make 'wads of money' but it is feast and famine, not consistent.

Getting all wrapped up in causes. It's a problem when it might be the wrong cause. It's an issue when you feel suckered, brainwashed, used. Believing in something false, due to wrong information is not exciting. I'd rather find out there has been some mistake or misunderstanding, that can get corrected. So here I am. I calm down once again. I hear arguments against subsistence, against various methods of defining subsistence and reasons white people should be exempt. I say, "The intention of having a group of people designated 'subsistence' is to protect a way of life which is based on a dependence and special relationship with the land." Witty and I discuss our community and what getting judged means.

I mention the word 'peers'. "Judged by my peers." Define peers? Defined on the internet on Google search: A person who is equal to another in abilities, qualifications, age, background, and social status.

Another older definition is 'companion'. I have said, "Someone who knows me, who I respect, understand, and trust, 'of my tribe'. From a legal viewpoint, 'group of people whose rules I am expected to know, understand, and follow.' This is who I am to be judged by. I believe I understand what my community expects and will accept. I pay attention. I ask, discuss this with other subsistence people.

I discuss my leaving to sell in Tucson, or selling the internet. Natives do the same. I point out, "Indians go to Tucson to sell. There is, in fact, a special Native section of the show. American Natives set up at the Flamingo. Native villages and corporations have web sites to help Natives sell subsistence crafts. The law states subsistence people should be innovative, including taking advantage of modern technology, ways and places to market, to survive. Trading with other races or foreigners is traditional and has gone on for thousands of years." I have another argument I have repeated till I'm blue in the face. I do not recall getting an answer.

"If I do not qualify for subsistence, why was I issued a subsistence license?" Why didn't anyone explain my permit is not as valid as the same permit issued to an Indian? Show me someplace it says this. Neither my lawyer or the judge had ever even heard the word 'subsistence', and did not know what it meant. I got a puzzled look.

"What does that have to do with anything?" *It's my 'get out of jail free' card. It's the law. This is scary, to be judged by people who never heard of subsistence, when it is so much a part of my life and the life of my peers.*

Imagine being civilized, getting judged by hicks who never heard the words: contract, lawyer, bank, loan or interest while facing a financial bank discrepancy charge. Likewise, you do not understand the words of those judging you. They use words like 'pilgrim', or 'cheechako'. Phrases like 'educated idiot'. You are puzzled being asked which way your stick floats. Judged in a place called 'Minchumina' that you can't even pronounce. No one is in a suit and tie which is your tribes uniform.

"Hi Mom! Wanted to let you know I am ok. I got in to see the doctor finally and they took an obstruction out of a main artery and put a stent in."

"Are you ok, Miles!?"

"Yes, in some ways feeling better than I have in over a year. My blood pressure is almost normal now, with half the pill dosage. I'm expecting to recover, even heal and be better than I have been in years!"

'Did you have a heart attack, Miles!?"

*Rats. I hoped she would not ask that.* "Well, I'd say no. But, yes, the doctor said I did. It didn't feel like a heart attack. It was just the usual soreness in my chest I have had off and on for a year. I did not feel any worse than other times this happened."

I hope Mom understands and does not panic or get excited. She had been saying on my visit with her, "Looks like I'm going to live longer than you!" Maybe she is worried because I am supposed to be executor of her will. So it is important I outlive her. Otherwise, things get complicated and she is not up to making a lot of changes. She wants to be in her declining years with all in order and taken care of. If it's true or not, I want her to relax and believe all is well in her world. I mean, none of us can promise the future. How secure is anyone's life? No use stressing over it. We are not promised who will outlive who.

I recall Mom's words concerning her sister who lived next door. "I have to outlive my sister! It would not be fair if she lived longer; she smokes!" This was a driving motivation to live, press on with a goal to outlive her sister! I heard this every time I visited.

The word 'heart attack' is synonymous with, "You lived this time and by the grace of God, might have a year left if you are lucky, before the big fatal one comes." I suppose in the not so long ago past not much could be done when the heart went bonkers. With the technology today, I am in the hospital two days, and home. A tube was stuck in me and run up an artery to the problem area. I was conscious the whole time. Not long ago it was major open heart surgery!

"Miles, I'm glad you called and are all right. You sound well!" It is hard to know what Mom really thinks and how she's handling it. But I have done the best I can here, and let it go at that. Iris overhears the conversation. She as well, wants the details of what happened in the hospital.

"Well, you know me, Honey, not great with numbers, times and dates. I already lost track of what day this is!"

"Miles, the VA doctor will want records of what happened, so maybe we should review what was going on, what got done and when. It might help with knowing how to handle the future!" So we review the past events on the time line. I sometimes have to let time pass to process events for a while before I talk about it.

**Past Flash**

I tell Iris...

"My first memory it hits me that something serious was wrong is about a year ago. It was when visiting Mom! We are on a walk to the coffee place we go to each morning. Do you remember?" She does not recall. "I am in the middle of talking, and lose consciousness for a part of a second. I am hit instantly, like a baseball bat, 'blam'. It did not hurt. I was simply 'gone', with no warning or dizziness. I feel weak after, and my legs are wobbly." Since I cannot explain what happened to Iris, and she did not even notice, I keep my thoughts to myself. "Maybe just indigestion. I burped!" I make a joke of it, so she does not now recall.

In truth I am worried. If I had been crossing a road and dropped, or been driving, doing anything important, then what? What if it happens again? Going instantly unconscious is not a good thing. In hindsight, this may have been when plaque or a clot broke loose in a vein and lodged itself for the first time. Or a mini stroke. Visiting Mom is stressful. Who understands?

I resent my probation officer saying, "Now isn't this a wonderful thing, helping out your poor mother! How thoughtful, yes we will allow this because it is healthy for you! Being close to a loving family!" I smile, nod, go along with the program. I like Mom. She's ok. We enjoy being together, mostly. But I would not call it love. I think I remind her of my father who I look like, the one who left her. For me, she is the mother who abandoned my sister and I, so can't be totally trusted. She simply lacks mother instincts. Spending a life without a reliable mother is stressful. There are ways we do not agree, and have conflicts. I think she did the best she could and is a good wife, good person, had her own childhood issues. Not every good woman is a good mother.

The trip seeing Mom is a big change in eating habits, amount of exercise, opposite climate, life in the big city, which I find stressful even though I look at the bright side. I'm breathing pollution, dealing with honking cars, interactions with strangers that I am not used to. It might be good to experience change, but still, it is at least a slight adjustment and pressure.

I can't eat Mom's cooking. She insists on cooking with lard, serving a lot of cheap noodles, starches, white sugar. I explain I am diabetic and can't eat like this.

"It's just a short time, Miles!" and "Look at me, in my 80s and healthy as a horse! What's wrong with my cooking!" Very indignant and insulted. It's true, no matter what, you can't insult your mother's cooking. *Is it better to die?* So Iris and I go out to eat a lot. Restaurant food is not as good as our own home cooking of garden foods and fresh wild game. One proof is my blood pressure and sugar numbers look better at home on a diet we are more in control of.

In Alaska I have access to my blood pressure cuff and blood sugar gauge and can get in to my own doctor if I have questions; can call and make an appointment if there are issues. In Tucson I am a stranger in a strange land far away from my doctor. Tucson feels like a big crowded place with a lot less personal service than what I am used to back home. So when I do have issues in Tucson, my tendency is to tough it out until I get home and can see my own doctor. If I was at home and had an incident of lost consciousness, I might have called my doctor and got an opinion. What am I going to do in Tucson? I do not even drive. I do not know where the VA is. If I called, what is it I'd say? They are busy. Hypochondriacs call every day saying they do not feel well. Even back home.

"Well, let's see, we have an opening in two months. Do you prefer Tuesday or Friday?" In two months we get to discuss something that happened ages ago that seems silly now.

We are on our way to have our morning coffee and Iris and I do not want the routine to be interrupted. Something negative and unpleasant to forget. This has been my outlook all my life. Focus on the positive, forget the negative. No one wants to hear complaining.

Both my blood sugar and pressure have been all over the chart. I had concerns. I know these numbers will not kill me immediately, but this puts stress on my body, and in the long run are harmful. I'm visiting my eighty-five year old mother who has her own health issues! She wants me to have morning grapefruit with her.

"Mom, the doctor specifically named grapefruit as something not to eat with the pills you and I both take!"

"Miles, that is ridiculous. I have been eating grapefruit every day for years and I'm fine!"

I am not so sure Mom is just fine. Her legs hurt. Her feet are blue. She sleeps almost all day. She feels awful. She denies being diabetic. At the same time she says she is healthy as a horse, she says she is going in for blood work, and it sounds like having her blood sugar taken. One of her pills I recognize as what I take for diabetes. My opinion is, the doctor gave up on her. If she does not care, refuses to change her diet, tells the doctor what she tells me, I'd tell her the same thing. It's your life, I'll keep quiet about it then and not bug you." There is long list of cooperative patients who care, listen, and the doctor can help. Anyhow, what right have I got to speak of healthy choices? Me in my early 60s on my way out, talking to someone in their 80s in better health! Grandma on her side lived to be almost a hundred.

"You get your bad health from your father's side of the family!"

God is somehow involved in this, having to do with the wild and evil ways of the Miles clan. So my mother would say, and implies.

I'm guessing she is too old to have her feet amputated and survive. The doctors have better things to do than change the mind of an 85 year old. What's it to them if she dies? As long as they do not get sued.

"We gave her advice, she wouldn't take it." Case closed.

The situation is different for me. I am 'young' yet—in my 60s. I could and should have a long productive life ahead of me. I'm still active, still working. I'm busy, with a lot to live for, and a lot going on. I'm not ready to call it quits yet. Due to this, I have been willing to make major changes in my life to ensure a longer, healthier life. I have been willing to have a much stricter diet and change my eating habits for example. Mom does not agree with me.

"That's ridiculous! No one is going to tell me what I can eat and not eat! What good is life if you can't enjoy eating!"

Dad said exactly the same thing and died at seventy-two, in my opinion, at least a decade before he had to. "Mom, there is another way to look at it. There are a great many foods in the world, with many choices. There is a long list of food that tastes good and is also healthy!" I go on to tell her my changes. When I need a snack I eat a carrot. At first I had trouble accepting this, and called it rabbit food! But in the choice between eating like a rabbit and dying, I chose to eat like a rabbit, and learned to like it. I can eat all the vegetables I want, mom!" She tells me she will fix me vegetables and I can eat with her. How can I not eat with my mother? She refuses to let Iris cook in her kitchen! Mom is not interested in eating how Iris and I like to eat. We are visiting her and in her home, we eat what she likes, how she likes it! This makes sense when you are healthy enough. It is a short enough time, so oblige the old woman.

Morning grapefruit is an important time for us. She arranged to have plenty of good grapefruit from her neighbor's trees way in advance of me arriving, so we could share them. We sit and talk together. Just her and I in a meaningful way. I do not wish to deny her that, and I too, value the conversation. But eating grapefruit is known for interacting negatively with at least one of my medicines. She serves vegetables and I help myself! Then she happily tells me she cooked them in four sticks of butter, on canned salty vegetables.

I can make up for it when I go home. *But can I?* I am getting discouraged. We try hard to eat better while the diet gets more strict. I am not sure the results are helpful. It's easy to give up. I'm amazed how many people around me say, "Oh, just this once break the diet, geez, Miles!" At least once a day. I'm amazed how common sweets and bad food are. On Valentine's Day everyone gives everyone else a box of sweets. It's common to have a box of doughnuts put in front of me with, "Here have one." I feel like an alcoholic being asked, "Sure you do not want a beer, just this one time?" I'm dealing with Iris eating potato chips in front of me, having sweet rolls with her coffee every day, leaving snack treats on the table. It is not easy to ignore all that and press on with a strict diet no one else but me believes in. Slowly, over five years, Iris comes around. She admits for example, she used to live on normal people food, junk food at work. Frozen meals, fast foods, etc. Now she feels better, is losing weight, has not had to see a doctor in five years, which is unheard of in her life! She has to admit, good food seems to be a huge factor! It is just hard and slow to change. Especially, when with her, it is not life threatening in the immediate future.

Back in Alaska.

I tell my doctor after we return from Tucson, at my next general health appointment and checkup, "I've been getting chest pains that concern me." I forget what the exact reply is. I assume he asked me some questions about it. Angina it is called. There is acute and not acute, depending how often, how long, and how severe and when. I probably said after exercise or work, and not that often. But he diagnosed a heart condition because he issued me a vial of nitroglycerin. If it gets worse take one of these." Is all I recall, and the subject was never brought up again. No one asked on any follow ups if the pain was more often. No one suggested there was reason for concern. It may not even be in my records. My doctor is not concerned, neither am I.

"It's just part of getting old, I guess. You are going to have aches and pains."

Iris's father died right after retiring; she was taking care of him. He was diabetic and did not want to change. It was not a good experience for Iris, so she may not want to look at the possibilities. I'm fine, going to be fine, nothing more to discuss. So tend not to talk about how I am feeling. I have chest pains more often.

"Yeah, just the usual, I'll be fine after some rest." I feel weak, out of sorts, not myself. There had been times when others asked for, expected, demanded I 'work' or do things for them, or expected me to commit to something and I knew enough to refuse.

This is when Foil had wanted me to launch my boat, haul supplies, join him on adventures that would involve physical labor, being remote, and dangerous. I did not wish to discuss why I do not want to. He said I owe him, am obligated. I will do whatever it is at my pace, when I feel healthy enough. No one is going to make me do what I feel I should not be doing. Foil had a habit of pushing, testing, challenging. When I called him on this he says.

"It's how I was raised, Miles. Me and my brothers were always pushing each other like that. It's normal!" Normal for him. I'm not even sure it is true. When I met his brother, the brother did not seem happy with Foil. I did not see friendly banter and fun challenges going on. I recognize my health is not great, without knowing why. I wait for good days and get as much done as I can. I rest on bad days.

Now and then, more lately, I can't sleep, and the pain is painful. I can't breathe. I'm planning on telling the VA doctor when I have my next health visit.

"When is that, Iris?"

"In another month, Miles." *Dang, I can't wait that long. I can't make it another month.* Iris is used to the routine. I take a nap every day. I think this is serious now. The doctor asks questions and thinks yes it's serious enough to come in and let's have a look.

"I can see you in a week, maybe, if I have an opening. Call that morning to verify." I make the appointment. I hope I can hang on until then.

Tripod weekend is here. This is the big spring celebration when we put the tripod on the ice and have a party. I am concerned about lifting and work setting up my craft table. I'm not feeling up to it. But I am not going to tell Iris and get her alarmed that I am not up to setting up! It looks like I am being lazy for one thing. It's one of the big events of the year and we expect to make some money. It's a fun time we look forward to. *At least I do.* I load up the snow machine sled with all I need Friday night. I'm not feeling horrible. Even so, something is not right and I know it. Even though I have a tentative appointment for Tuesday, this seems a long ways from Saturday. I run into someone I know who is head of the Nenana EMT medical team. I engage in conversation.

"I have not felt well in a while and it is getting worse. Chest pains. I think it's serious. Can I stop by the fire hall and get looked over?" I make an appointment for Sunday morning. He is nice enough to come by just to check me out. I'm hooked up to the heart machine. We both watch the graph forming.

"Looks good and strong, Miles! You look healthy as a horse!" Another monitor and, "98 oxygen, Miles, very good for this bad air in here!" I wished we had found something obviously wrong. Something is wrong and it's not showing up. Dang and rats! All I can say is thanks, and walk away. I even feel good enough later in the day to dance one dance with Iris when one of the entertainment bands plays at our event. We make almost $800 over the weekend! A nice financial shot in the arm when my Social Security is $600 a month.

I give my usual line, "Beats working for a living!" I joke with the other vendors and all is well. When packing up I laugh with other venders. "This trunk I'm pulling is heavy! It's the money! Just the 100 dollar bills!" Tuesday I'm going to see the doctor and find out once and for all what is wrong. If it is not my heart, as implied by my read-out Sunday, maybe it is a reaction to one of my medications? I can review all the possibilities.

One thought is a pinched nerve. I maybe am getting all worried about nothing! Just a pinched nerve! Ha! How embarrassing. No emergency! Because I moved my head and the pain increased. I have a neck nerve injury from long ago that could send pain waves down my arm. That whole area hurts. Around my neck, the back of my head, somewhere in my throat. However, if asked to describe it I'd say, "It feels like my arteries!" I know where they are in the body, and it feels like someone cut off the arteries on both sides of my neck. Like I'm going to pass out from having them cut off. Like I want to loosen my shirt collar but there is nothing squeezing these arteries. But surely if it was a heart issue the monitor would show problems? It feels like my heart is big and cannot fit in the cage. But the EMT guy says probably this would show up in one of the readings he is showing me. Oh, ok, eliminate that then. Anyhow, I will find out Tuesday. One more day! Monday I survive ok. Just weak, I manage to go out and get one load of firewood cut. I have not felt like getting firewood. We need wood! I do not feel like it, but need wood.

Riding the snow machine is ok. After I cut two trees down I have to sit on the snow machine to rest before cutting them up. This is not like me. I cut them up and rest again before loading them in the sled. I get the wood in the sled and home ok. I feel a little better, but that is all relative. Just less weak than I did an hour ago. I have not felt really good in a year. I assume I am just old now. *Get used to it.*

The offer Foil gave of being my buddy, standing by me, looking out for me sounded good. I teach him and he pays me back by helping me in my old age.

The awaited day Tuesday comes! I am to call at 7:30 am. I call, and no one answers, so have to leave a message. "I have my ride to town, will be available at 11:00 as requested" Asking him to return the call.

Our cell phone rings and it is the doctor returning my call. He says, "Not all the doctors are in today, so it is busy. Can't make that appointment. I have another opening in a month " There is nothing to say. He did not ask if my pain has increased or what is going on. Because the heart monitor I had done said my heart looked fine. What is it I am going to say? I have a regular doctor appointment soon. I forget when. I'm not really with it and not thinking well. Dizzy all the time. Everything is far away. *That must be normal for being a senior.* Others are worse off than I am! I'm not going to complain. When I finally get to see a doctor we can figure it out… maybe. Maybe there is simply nothing that can be done. No one knows what is wrong with me. Never will. Age is the answer. Oh well. Iris is not concerned. So I am not either. Life goes on. So does our routine.

**Past flash ends**

# MILES MARTIN

*Another time zone up ahead in the fog.*

# CHAPTER FIVE

## HEART ATTACK

At some point Iris tells me my regular doctor appointment is not until the end of the month. Two weeks away. I honestly think I'm not going to make it that long, but no use saying anything. I've been saying for a year something is wrong! What is wrong, and what has been done? Nothing. So I shut up. There is nothing worse than some old senior going around saying something is wrong and no one knows what! When something breaks, we will know what is wrong, and hopefully it can be fixed. So that is what we will wait for. Something is going to break, and I know it.

I have seen the doctors off and on. I had blood work done. My A1C was bad. It had been ok earlier. I got a new medicine for blood sugar. I wonder if this medicine is at fault because as soon as I took it my blood sugar dropped 100 points in fifteen minutes as advertised. I felt it like a roller coaster drop. Wow! I begin using it but it is almost scary. My numbers can go up and down so darn fast with this! I monitor the blood sugar three times a day now. Keep a chart for the doctor. I am not happy with it bouncing all over. Between seventy and 200 in the same day. I do not think this is good, but I am faithfully recording so it can be examined in a month. My blood pressure as well has been way high. One thing we learn in prion is not complain and do what you are told. *It's almost like brain washing.*

Wednesday morning, Iris and I do our usual walk to the seniors. Sometimes when we walk I get short of breath and have to stop and rest, get dizzy, feel bad. I make it to the seniors ok. Iris asks, "Do you want me to make you a grilled cheese sandwich, Honey?" I have to think about if I should have bread or not. On my diet, bread is not great. Cheese is better.

"I guess it will be ok." I am eating this with the usual morning coffee and have pretty normal chest pains. Pretty normal arm going numb. It's all the same. Headache. Trouble breathing.

"Go home and lie down, Honey." I know I can't make it home. Maybe after I rest. I kind of wish I could lie down right here. How absurd. That would alarm Iris. How inappropriate! There is nowhere to lie down except on the floor! I feel I might pass out. I really want to lie down. Iris is annoyed with me that I am not just going home! Since this has happened before, and no one considered it an emergency, I'm not going to panic. It will pass as it always does. I have no idea how I am going to make it a month before I see the doctor. But, one day at a time. This is lasting annoyingly longer than usual, but I had one incident where it was like this almost all night and I could not sleep. That hurt even more than this. I'm aware of trying to raise my right hand to signal I am not ok. I might be trying to wave good bye.

"Should I take your plate?" I can't talk. But soon. Once the pain goes away.

I hear the door open and know it is our friend, Valerie. As soon as she comes through the door she says, "Miles is having a heart attack, call 911!"

I do not agree, but think it is a good idea to find out what is wrong with me. If someone would just drive me to the doctor. But why? I already know the doctor does not have time. I'm seeing my doctor in two weeks or maybe it is three weeks. This is not a heart attack or an emergency because this is the same old thing that has been going on for a year. It was never an emergency before, why would it be now? I'm feeling a little better by the time the ambulance arrives. *Maybe we can finally find out what is going on. Too bad we have to zoom in under emergency mode with an ambulance.* We are going to find out, as in right now! That's a good thing.

A major issue is cost. If it is me saying I have an emergency, get ambulanced in, and doctors decide it is marsh gas, not an emergency? I owe the value of my house. VA does not cover false alarms! Where as, if a medic, EMT type calls it in, I'm not liable.

The doctor did not even ask any questions. He looked at me and knew what it was! A heart issue! Why is it so obvious now but not before? No tests! I'm on the operating table. The VA is going to pay. Who is going to pay needs to be made clear before anyone touches me. Everything is on hold until it is determined who is paying. Not, "How do you feel?" but, "Who is paying?" I give consent for this and that, told the risks. The operation could kill me, but better than the odds of doing nothing. I understand, let's do this! I'm conscious the whole time. A vein is cut in my leg and a tube run all the way up to my heart. I hear everyone talking. I'm watching an x-ray of my heart pumping. Pretty interesting. I hear on the loud speaker all the noises and closing and opening of heart valves.

A stent is put in. I'm wheeled into a recovery room. I fill out forms and have lunch.

The doctor called it a heart attack. I did not believe him. But the doctor said there was damage to my heart, so it must have been a heart attack. I would not have lived much longer. Maybe not another hour. My artery was completely blocked. Good news though is, I'm coherent with no visible brain damage or impairments. "About a golf ball size of the heart is damaged from lack of oxygen, but you have a strong heart, so even damaged, it is within normal function range." He adds, "It is good you do not smoke, drink, are in good health, and eating well. Otherwise, you'd be dead now." *Talk about a high Jesus factor!* "Arteries in general look strong and healthy. No cholesterol or plague. Just that one area, a lesion of some kind. Could be hereditary. A clot formed. We fixed that."

I SEE Valerie for the first time since I got back from my hospital stay. "If you hadn't come in when you did, I'd be dead now the doctor says. One more hour is all I had. So glad you came in."

I think Valerie does not want to face what she understands as the truth. She knows I am right.

She stuffs her face with chocolate as a main diet. Has a hip problem, maybe due to being overweight. She hobbles along smiling, doing for others. All her money, all her time, goes to others. "Don't worry about me!" She cheerfully says.

"When are you going to get that hip fixed, Valerie?"

"When I have time! No time, too busy!"

If she fell over dead most seniors would angrily say, "Hey! My lunch didn't get delivered yet!" I'm her friend, and care. I try to thank her and do things for her, to show she is noticed. I doubt very much she would be one to holler, "Call 911! I think I'm having a heart attack!"

I understand how my grandfather died. He and his wife were side by side reading. He put his book down, leaned his head back, and died. His wife never knew. Well, maybe when we love people, we are in denial. Who knows? We can't accept we may lose them.

"Valerie, I was out cutting firewood the day before! Imagine if I was out on the river and had this happen!"

"Didn't you have your cell phone, Miles?"

"No." The reason is complicated. I lost a cell phone once. I've had dreams of the situation, laying on the floor dying, with a phone in my hand, and can't think of anyone to call. Just setting the phone down. Like the song, "Here's a quarter, call someone who cares." It is different when young. I never called anyone for anything, or asked for help of any kind. I walked fifty miles once in an emergency. I'd do that before bothering anyone. It's just how I am. I took a cell phone with me on the river long ago. During an emergency the phone fell out of my pocket into the river. Five years later Iris remembers, and has never forgiven me.

"Do you know what that phone cost?" Yes, $20 to replace. Ever since, I'm going to lose the phone, get it dirty, damage it. That's fine. I do not touch the precious phone.

Iris was upset when I got home from the hospital. No one called to tell her I got out of the hospital. She was waiting to hear I got released. Sat around all day waiting for a call. I called her the night before saying how it was. I think I called early in the morning saying I'm supposed to get released, but the doctor had not showed up, and Josh's wife was coming in to run errands and pick me up at the hospital. I hate phones. I assumed Martha was in touch with Iris. Iris could have come in with Martha and been in the know. She said she could wait. Ok. Then it's all taken care of.

Martha has shopping to do, so I need to wait an hour. On the way home Dalia keeps calling Martha, who has to pull over to the side of the road as they text back and forth about every five minutes. I wondered if Dalia knew Martha is driving? Why not call when she gets home, or stop by? They both live in Nenana! It is dangerous to be on the side of the road in a curve or on a hill. It didn't sound like an emergency. Dalia is such a busy body. I assume she wants all the latest about Miles's heart attack. Why else text back and forth instead of talking, if it is not a secret? So she can go tell all her friends. I just want to get home. I'm not interested in making any calls and having Martha listening in so she can contact Dalia with the latest. Getting home took a few hours longer than expected because picking me up was not the main purpose of Martha's town trip.

---

"MILES, you kept telling me you wanted to go home! But said you couldn't walk from the seniors! I suggested calling the ambulance and you said no!" I have no memory of any of this. It must be so. I recall seeing a sign at the hospital I asked about. "Who poached the moose?" Clearly written on a white board in magic marker. I asked why that was written, as it seems odd in a hospital to have that written on a public board. No such sign was up I was told. I imagined it. Yet it was so clear to me, it is hard to think it was my imagination. The heart attack involved things I do not remember or remember incorrectly. I smile. Possibly the last thing I'd see before dying is, 'Who poached the moose?' *Was that St. Peter at the Pearly Gate, asking me from the other side of the river of life?* I have to grin.

Iris and I are coming back from the A frame, the Sunday coffee shop on the two lane highway a couple of blocks from home. A voice comes from off to the side, "Want to hear us sing?" I turn and look. Two young pre-teen girls are in front of their house on the side of the road with a can on the ground in front of them, along with two children's bikes. I am not sure what these children mean. Iris keeps on walking. *Don't get involved, Miles!* I stop and say, "Sure!"

"That will be a dollar!"

*A scam!* I tell them, "That's too much!"

"How about a quarter then?"

I fish out a quarter and drop it in the can. The two have rehearsed this. Iris and I are their first audience. One starts the beat while the other watches to come in together. One, two, three, four..." One of the girls begins an imitation of modern dance, doing pretty darn good. The other sings what sounds like a song they wrote themselves about Nenana, walking a dog, going for ice cream, meeting with friends. Mostly putting on a performance without being very nervous, staying in tune. Very good experience for them.

I see a time in life when it is important to build confidence. Towards the end, the singer forgets some of the words and hesitates, so I think the song is over.

"Wow! Very good!" The two start babbling excitedly about making a quarter, screaming to the house to their parents.

"We made a quarter! I told you so!" As Iris and I walk away, we can still hear them babbling excitedly about making their first quarter.

Iris wants to know why I bothered to stop. I repeat my outlook, "It's a little time and effort on my part that I can afford and means little, that I can donate. This means a great deal to someone else." Iris actually gets along with people better than I do. She knows better how to engage in conversation, inquire about a baby or children, and remembers names. She can talk about normal everyday things. There are not a lot of normal everyday things I either do or understand. I do not care who you are, what your name is, or anything about your kids. *I just want to help, do something for you, offer you a positive experience. What difference does it make who you are?* I've said before, after maybe five years I might recall your name. In twenty years, maybe we are friends. Individuals are not part of the joy in my life.

---

OVER A YEAR after my heart attack.. I'm not good about following a time line.

### From my diary Tuesday, June 28, 2016 Fossil hunt

Writing this on the laptop on the Cosna River ivory area. I have been out three days, and feel strong and healthy. I had been concerned due to feeling so weak on my past two long trips!

A lot of rain, but this has been all right. Not cold. A larger new tent is nice to stand up in, as well as to sit and read. However, I find I spend most of my waking hours outside the tent, so perhaps this tent is not normally needed. I like setting the smaller tent up on the front deck, and thus able to live on the boat. I cook, eat, sleep, and no supplies leave the boat.

I talk about this often - how camp is set up, the routine we choose - has a lot to do with my success and happiness. Others who travel the river say the same. "Making camp right is important!"

My diary just covers some basic highlights of my trip. The entries help me remember the trip and the details.

I find two mammoth tusks in a cliff my first quick look-over. I ignore them, since they are impossible to reach. If I can find enough fossils without these two tusks being in the inventory, I will leave the tusks, as I have done in the past. I find some fossils, but only ten pounds of ivory. *A little ivory makes up for 100 pounds of other fossils!* Ivory is the priority. I find a nice mammoth lower jaw with two teeth intact. This should fetch a good amount of money at the fossil show. More exciting is this is a section of an old homestead I have permission to be on, but have never found anything at! I suspected all along, this is a productive section, and in theory should find fossils. This find confirms my opinion. Being mammoth parts, there should be tusks if I keep looking, even if it is other trips. I now have a new hot spot.

I come to the end of the private land parcel, but a little further along the cut-bank I come to another old homestead not marked on the map, that may have been returned to the state, but potentially legal to look for fossils. I make a mental note to inquire as to the status of the land. Most important is, not Federal land.

I keep looking up at the tusks I first spotted whenever I make another pass. *No I better not try.* I spot a very decayed partial tusk further along the same cut bank and just twenty feet up the bluff. I think the incline is an angle I can scramble up and hang onto without going backwards into the boat or the river. There is a prehistoric ice age log to tie up to. High above me - about 200 feet - is an overhang of mud and ice. This is so far bent over with a curl that the trees on top have tops pointing down at the river. This looks somewhat like the curl of a perfect surfer wave, only black. I study this only briefly to verify it is stable and not moving or about to fall, but I never really know. *Fossils are not free.*

I have a backpack over my shoulder with ice pic, folding shovel, rope, ice cleats and other things I might need when I get to the tusk. I need the point of the ice pick to stab into the mud and use to help stop me from sliding. The mud is a foot thick and moving. I am for sure going to be very muddy when this is done. *I love mud, don't you?* I have a big smile, and answer myself, "Yeah!"

I wiggle through the mud and get up ten feet and stop to rest. I hear a rumble from above and know what this means. A hunk of mud the size of a house is letting loose above me. There is no time to look up to see. I have less than a second to decide what to do; the falling mud will land where I am. I flatten myself in the mud as close to the cliff as I can. There is a 'woosh' and like being in a vacuum. Many tons of mud falling from 200 feet brushes my back, and lands in the mud, missing the boat.

It hits with the sound of a cannon going off. The impact—like that of a comet—sprays chunks of frozen river silt all over me and into the boat. It is enough to rock the boat and it takes on some water. I'm a little further away, so less affected. I freeze in position for a full minute in case more is falling, and letting everything settle. Crazy Lawsen once said, "If the cliff above gives way, do not try to get away. Cling to the cliff as close as you can. Falling debris in most cases, will travel two to three feet away from the cliff face." *Like riding the inside of the surf wave tunnel. The falling outside curl is a few feet away if you hug the inside.* Instinct took over, memories, data, in my unconscious, took over to make an instant correct decision. I'd be dead now if I had tried to dodge this.

I have to wipe ice and mud out of my eyes to see. The falling mud clipped the tip of the tusk and ripped it from the ice. It is laying in the mud on the beach at the water's edge below me. The mud is so deep, and like quicksand, that I have to lay down and move like a seal to get to the tusk and my boat. How do I answer, "And what did you do today?"

---

I FINALLY DECIDE to try climbing the cliff. There is a ravine 300 yards downstream with access to the top I think I can negotiate. I try to pinpoint this spot by looking up at the tusks, imagining what this spot should look like from on top looking down! *Where the big willows end and there is a low patch of grass.* I assume I can find this. *I am only moving 300 yards once I get on top!* I haul a backpack with climbing rope, harness, and all the gear I think I might need. I take my time. I have had little to no experience with climbing gear. The hope is to dig the main tusk loose, and tie strong thin parachute cord to it. The other tusk is barely showing, and may be too deep in ice to get. It might be worth tying this off as well. If so, I will need a lot of rope. Stronger rope weighs more than I can carry.

I will be adaptable as to how it plays out. I might lower the tusk and retrace my steps over the top and back around to the boat. Climbing down with tusks might be over 200 feet, and not be at the boat. Before I take off, I check the option to walk the muddy waterline as far as I can at the bottom if I choose to come down the cliff with the tusk. There is fifty feet I cannot walk. I might try to swim if I choose this option. The water here is slack. I check the depth at eight feet, so not likely any undertow. Possibly I could toss a rope and hook to the back of the boat if I can see it only fifty feet away. This way, I'd have something to pull me over to the boat while I am in the water. Just an option that requires I bring this hook to the top. The river here is, in general, not just a little risky, but very dangerous. Logs get sucked underwater, never to come back to the surface. I focus on the tusks.

I might loop either or both tusks with the lighter line, and toss the line down the cliff to within reach from below if I boat around and retrieve the tusk from below. This option allows me to tie the heavy rope to the light line, and pull that heavy rope up and around the tusk, with a noose. I have done this before, and been able to use the boat to yank on the tusk to pull it free if I cannot dig far enough around the frozen mud to set it completely free. *For sure I am taking my time!* I do not want to get over winded due to my heart surgery! *My doctor would have a fit if he knew what I am up to.* Thinking ahead like this allows me to pack materials for the various options, like a roll of heavy fishing line just to loop around the tusks if I choose. Possibly I could do this from the top without having to come down to the tusks at all.

I overexerted on my last boat trip and it took over a week to recover. My last successful fossil trip was last season, before my heart attack. Hmmm. *It would be nice to think I can heal and be stronger, but the trip to the homestead this spring wore me out.* Normally, if I rest an hour I am back up to speed. This climb is going to be taxing! I am concerned if I am up to it. No use falling over dead trying to get at a tusk or two or using all my energy to get halfway there, unable to go up or down, with heart pains. This is one reason the best option might be to simply walk away, and accept this ivory is not available to me. I wonder, 'what if'. *Could I acquire the ivory and make a ton of money?* In the past I simply did not feel up to it. I knew in my heart, I could not do it. For the first time in a few years my body feels like I could possibly be up to it! I'm going to go for it, and see how it goes! I promise myself that at any time if I feel I can't do this, I'll back off and walk away.

I rest. While resting I review a long ago dream.

### Dream

A female helper. Smile, and sigh. 🩶

A long ago dream. A young thing that knows how to climb and repel; athletic, healthy; loves the outdoors. Perhaps a college student who has summer off, but no money. Looking over her options. Likes to travel, seeks an adventure, but no way to travel or pay for it, broke, like many young people. I offer what she seeks. Something just for the summer. Call it an 'arrangement'.

An open minded woman who accepts the ways of the world. Life is not always easy. Getting what we want often requires compromise, sacrifice. Our own happiness can involve pleasing others around us.

"I'm not even saying it has to be sex!" A father-daughter type relation. A student - mentor. Teach someone something, offer a vacation, a good time, an adventure of a lifetime in return for friendship, appreciation, respect. The ability to get tusks I cannot.

At the end of a summer, "Thank you, Miles, I had the greatest time of my life and will remember this forever. I have learned so much from you! What a great vacation!"

"It has been good for me as well. I never could have gotten those tusks without you. We had some good conversations and good times together. Good luck with your career!" It ends there. No regrets. Not trying to make more of it than it is. Maybe, probably, never hear from each other again. Repeat the process again next summer, if I wish, with someone else. Or… who knows, she may want to return next summer.

The dream has been long term, but never made into reality. Best left as a dream! Reality requires conditions that seem not to have ever been present. I want someone young enough to handle the physical feats required, yet mature and wise enough to understand and be accepting of the arrangement. Good character, honest, kind, intelligent. Most potentials I have met in life are up to something with a hidden agenda of their own, not up front about the arrangement. Or, feel it is me up to something.

"Men, you are all alike!" Are women all alike as well?

One girl lied about her age and activities. Others were whacked out on drugs they depended on. Several from other countries, needed green cards were ready to lie to get what they want. Some were mentally unwired and not sane.

When I was young, I was not ready for one good woman I met. The time was wrong. Some had good qualities, but could not get away for the summer, or even a few weeks. Others had their own idea of what life as I describe it would be like. Either they or we had someone else in our lives, or obligations.

The concept of exchange of services is not a common thought line. Reality is, too many factors have to come together for the dream to be realistic. This did not stop me from day dreaming and wishing. How nice it would be to say, "Well this is the situation kid, there is 20 grand up that cliff and you need to scramble up there with this rope. I can't."

A youthful hug of exuberance, "That was so cool, Miles!" A 200 pound, twelve ft. long, 40,000 years old mammoth tusk in the boat. Sharing that. Changing someone's view of life forever. Ah well… sigh.

**Dream ends.**

I'm alone.

I have a life partner now, a good choice, a compromise of all the things I want and can offer. *Not an outdoor athletic type. It turns out, I mostly like to be by myself. In this way, the trip can be all my way. No matter who I end up traveling with, there is always advice offered. I'm not looking for advice. I am rested now and need to act.*

I begin the long climb up the nearby ravine that gives me access to the top. Back pack with all I think I need in tow. The boat is still within sight below, bare aluminum where paint wore off glitters in kaleidoscopic flashes and Morse code dashes, spelling out 'help', as the sun moves behind trees and mud pinnacles at the top of the cliff. I begin the cross over towards the left where I need to be. Always going up, facing the dappled light of the sun through the leaves. The first part is the hardest, with nothing to hang on to. I use an ice pick I found at a garage sale. It may be a rock pick. I do not know the difference. *So much for specialty tools and their use.* I know the tool digs in the mud and gives me something to pull myself up with. Crazy Lawsen had only a rope. Pictures in Neil's book show him with just such a rope.

After the first 200 feet, I get to some small bunches of grass I can use to help pull me further up. From here on, there are small diameter willow trees to brace against and pull myself up. I stop three times to rest. In half an hour I am at the top, but now unsure where the trees stop. The low grass place I saw from the bottom is not identifiable. *It all looks different from the top.* I wish now that I had created a method of marking the top by using a dye gun or something. Forgot.

I think I am near the right place, so walk to the cliff edge to peer down. *Maybe I will be able to see the tusk or a reference point.* I get vertigo seeing the river so far below, and say to myself. *No way I am going off this cliff!* I feel I do not know enough. I have zero experience. I have bought all this climbing gear and have not gotten it to work. I cannot even climb to the top of my house roof or let myself down. I was unsure how to hook up to climb down a cliff in the past. There had been a forest fire, and all the trees to tie to were burning. I decided it was not a good time to go off a cliff tied to a burning tree. Even then, I was unsure how to hook up all the gear. *I think all I need is a weight on my rope end. This is my only problem!* I have not found an expert climber who can tell me how to set this equipment up for this specific use. They all say you need two people. And two ropes. *Yeah right.* I will be lucky to carry one rope! It's heavy stuff! Likewise, no one suggests I try with the diameter rope I chose, the minimum required. All advisors I meet learned by the book, teach the accepted ways with no innovations. I want to know what is wrong with my light parachute cord which has a holding strength of 1,000 pounds! Me and my gear should be a mere 200 pounds! But no climbing gear I can find, fits such small cord, so 'oh well' to that idea.

Likewise, most climbing is done on mountains of rock, or faces of glaciers. I agree that rocks or ice might cut such small rope. Who knows about mud and permafrost climbing? No rocks anyplace. Who has knowledge of plucking tusks from cliffs? Then coming down with the added weight! Most climbing involves 'getting to the top' as the objective, a long ways up. Not 200 feet, then back down with an object. Few climbs are about making money, where the number one priority is not to be safe. Money and safe are not commonly used in the same sentence.

As I rest, I explore my options, and let my imagination come up with future solutions. I go back to my daydream as I get my breath back.

My dream is interrupted by a flit of what looks like a girl's head and leather dress, but is gone in the thick brush. I frown. *Did I see that or imagine it?* Logic tells me I had to imagine it. No one could be here. There is no boat. It is hard to know the difference sometimes between imagination and reality. Or no. I do not suppose I have a problem telling the difference. I just say this to myself as the only logical explanation. I heard nothing, but was sure I saw a face, hair, leather dress. Looking like a cavewoman of some kind. That's not possible, therefore it did not happen.

---

I HAD SEEN a show where professional tree harvesters use some sort of electric winch to go up and down tall trees. I have such a winch, used on a truck with remote control. *Press a button... zzzzip... up I go. Press reverse... zzzip... down I come. I mentioned this before? I forget.* This still requires I get a line up and secure, even a pulley, and have this all set up.

---

As I CATCH my breath and my heartbeat returns to normal again, I enjoy the coolness of the permafrost ravine. An extremely rich micro environment with its own kind of mixed sensations, grows in areas like this. Cool, damp, with very little wind and well fertilized soil. The smell of black, wet, rich soil and the pungent scent of thick greenery and flowers is heavy in the air. I listen to the sounds of exotic birds and spot a Wilson's warbler flitting around. There are tiny bright yellow warblers that remind me of canaries. I realize one is acting like it wants me gone, is nervous, possibly a female with a nest nearby. Not a common bird in this section of interior Alaska. She comes within two feet of me, yelling in her twitter language, warbling her wings in anxiety. I spend plenty of time resting, enjoying this scene as much as any other part of my adventure. I feel ready for the next step.

Here is the face again, of a cave woman. Then the whole body. She is more of a child, about ten years old, dressed in smoke tanned leather. No adornments. She could be one of the local Athabascan Indians from a nearby village. *I do not think so.* This is no place for a village, or any ten year old to be out wandering around. *I agree.* These are dangerous bluffs and cliffs made of frozen mud. She looks worried, afraid, but beckons me to come, to follow her, hurry! I know she can't speak English and she knows I am not going to understand her words.

I get up and follow. I'm slow, not as fast as she is. Around the corner is a spear leaning against a tree. She grabs this and hands it to me for a walking stick. I gratefully put my hand out. I look much like one of the hobbits, ambling off into a fog in a strange looking world of thick foliage. I'm a bent over old man, holding a seven foot shaft with an eight inch flint blade in it, three inches wide. I know a little bit about artifacts, and this looks like the kind of spear and point used to kill big game like mammoths. The point is held in place by sinew. I do not take this in because there is urgency in the task ahead. Cave Girl comes to the edge of a crevice and looks down, expecting me to follow and look.

A man is down twenty feet below us, stuck in a narrow place where the frozen mud has cracked. Water has made a hollow, much like a sinkhole. This looks like it might be the girl's father. Like her, he is dressed in a leather robe. These two look like they might be primitive Siberian hunters. Their features are dark and Mongolian, far north China, Russia. looking like pictures I have seen on Nova. Or more like people in my dreams. I may have even seen these two! *That's it, this is a dream!* I thought I could tell a dream from reality. Since my heart attack I am not so sure. I see the problem; cavegirl needs help getting her father out of this hole! I peer down and he seems unhurt. I get the impression she told him to wait, and she would go get help. Her desire to find help was strong. *Could such a desire bridge time?*

I have had out of body experiences before. There was that time long ago. [1] A winter arctic memory of crawling in the snow at fifty below zero and knowing I am going to die unless I get help. I left my body and found a pilot. The pilot met me later after the rescue and told me he had to fly that day and took the jet. He did not know why he had to do this until the stewardess told him a passenger spotted a distress signal down on the Yukon River ice. I did not say much at the time. I smiled. I nodded. I was not about to talk about out of body experiences, time travel, leaving the planet and stuff like that. I know how that would look! I simply 'did it' and kept my mouth shut!

So I am not terribly shocked being in this situation. *Someone needs help. I'm going to help. There is nothing here to be afraid of.* I have my climbing gear with me. I set my pack down and look over what I have brought that would be most useful. More than just a rope, I have to find my small rope block and tackle. This item is advertised to lift deer up into a tree to skin and store. The pulley multiplies my strength. I am not strong enough to pull this man up! I find it odd Cave Girl does not know what this rope is. There is a tree twenty feet away, which is further than I wish, but this area has been sliding. Closer trees have slid into the crevice. I do not recognize this tree, beyond it being in the spruce family. There are ferns and other plants I have not seen before.

I get the end of a short rope around the tree and hook my pulley rope to it, then toss the long rope into the muddy ravine. I take the other end of the long rope and run it through the pulley wheels. Caveman knows to grab the rope. Possibly he could climb the rope on his own. I never thought of this. Most civilized people I know would not be strong enough to climb a rope. I begin to crank on the winch, pulling the rope up. Caveman knows enough to wrap the rope around himself. One leg appears to be sore, but he uses his feet to help scramble and push as I crank. Cave Girl is almost in tears out of the way, squatting down holding on to a doll.

Caveman hands come into view above the rim. Tan and leathery as a mummy hand. I am still cranking twenty feet away. His daughter drops the doll and comes running to grab her father's hands. Once he is up, Caveman wishes to build a fire. I can tell he is cold from being trapped in the underground ice. I do not know how long he may have been down there, but possibly for days! He waves his hand and the daughter brings him what looks like a mountain man possible bag, as the only thing I can compare it to. He speaks to her, but I have no idea what language this is. I have traveled a lot, so can recognize many languages by sound. Latin based languages will have a word here and there I can understand. I know when someone is speaking French, German, Spanish, Russian, Chinese. These guttural sounds do not remind me of any language except perhaps Gwitchin, an interior Alaskan Athabascan Indian language.

They glance up to see if I am following her words, but they do not expect me to understand. Both are avoiding me as if I am an apparition or something conjured up… maybe to be feared. I understand some things because of past dreams and a kind of telepathy. In the same way I understand what is going on with most living things, animals, even plants! All my life I have studied and imitated animal sounds, got them to interact with me. I talk to ravens, and when hunting, figure out what is on my preys mind. To defend myself I study predators habits.

I watch closely and decide these cave people mean me no harm; they are just afraid. Afraid because magic was involved in the rescue. I understand magic is powerful. Anytime we deal with power, there is a need to be nervous. We never know what God we may be dealing with, and what agreement or price there is to pay. Was this heaven sent, or did they just sell their soul to the devil! How do I communicate? As with any living thing, talk slow, no fast motions. Keep the hands visible, keep my profile below theirs.

While thinking this, I see Cave Girl's doll near my winch. As I put my equipment away, I pick up her doll. *I have seen dolls like this in museums.* My friend Dodger buys them up north from people who dug them up in ice caves. He sells them as artifacts. Only this one does not look old. I notice the doll is missing an eye. *Now I understand.*

I packed something I did not understand or know why. I had this notion I needed to put a glass eye in my pocket! I use these eyes in my artwork. I often simply do what my unconscious bids without question. I have a very good relationship with my unconscious self. So here is an eye, with small amount of instant glue in my pocket! Now I know why! I glue the eye on the doll. In ten seconds the glue is dry and the doll is fixed.

I get Cave Girl's attention by gently touching her on her bare shoulder. She does not pull away in fear, but is slightly wary. I speak soft and gentle.

"Here, here is your doll. I gave her a new eye. She is not broken now." The cave child's face lights up when she sees the new eye! Breaks out in a smile and hugs the doll. Shows it to her father and speaks. He is amazed, in a good way. How can I be with the devil if I fix a little girls doll? I am obviously not here to hurt them, or extract a heavy price for a lifesaving rescue! There could have been no better way to put them at ease.

Caveman gets out his flint, steel, and pouch of fire starting fuzz. I have lived primitive, so know what all this is, and how it works. I get my own pile of lichen off the tree branch overhead and loose dry bark and leaves. Much bigger material than Caveman is getting. I have a few twigs and larger sticks nearby. My Bic lighter comes out of my pocket. I see a stop in their conversation as they watch me with curiosity.

I pretend I am doing magic. "Hocus pocus let us focus!" I chant in a commanding strong voice. As if I were in charge of the God of fire. Which indeed I am.

"Let there be light!" I spin the Bic wheel and instant fire. I play the part of calm acceptance, no big deal, as my companions have seen a miracle! I am, after all, an actor, seller of goods and stories. I move the flame to the pile and 'poof' instant fire. No mess, no fuss, no waiting for flint sparks and blowing carefully on perfect tinder dust. I hear babbling, very excited. I know what it means.

"Wow! Did you see that? I wonder what other magic he has! Amazing!" I have their full attention. I hand Caveman the lighter and he studies it, turns it over. Of course, he has no idea what it is for, beyond 'pretty', and made of material he has never seen before. I slowly wink and smile. I move my hands, open them, "Nothing in my hands- Nothing up my sleeve." I reach in my pack and pull out my Stanley thermos with hot tea in it. I'm sure Caveman is cold and thirsty. I pour a steaming cup of drink and hand it to Caveman. He is not afraid, but is not sure how to deal with this situation. I make motions that yes, this is for him, it is something good to drink! I hear, "Daddy, what is it? It looks hot! Is it good? Is it really hot?"

"Yes, My Child." How can this be hot when it came out of a carry bag? How could the bag not burn up? How did he heat it?" Cave Girl lets go of her doll with one hand, takes the cup of tea and takes a sip.

"Oh! This is too good Papa!" I take the cup and fill it again. I take a sip and pass it around again. As we relax I look around. Eating together is a bonding thing. All living things have known this for a million years.

I had not paid a lot of attention when I was focused on the rescue. I now see a tall glacier in the distance, less than half a mile away looking down the length of this ravine. My unconscious understands things I do not. I have always known this. The world of instincts, past genetic memories; everything I ever saw, heard, tasted, read is stored someplace where my unconscious can access the data. Most of us initially have an ability to access this information. Many of us chose not to… or learned not to. Some of us know this in an emergency. Suddenly, we do things without having time to think. We bail ourselves out of bad situations, or decide not to have anything to do with someone. 'Premonition' it is called sometimes. Or simply 'instinct'. Well described and studied in the modern book, 'Blink'.

I know I am in the same area I was when I arrived in my boat. Only in a different time. I recognize the rock bluff. I know because this takes thousands of years to seriously change. I notice the mountain is not as tall. The river below has not had centuries to cut deep into the land. I understand that long ago there were glaciers when the mammoths lived here, whose tusks I seek. I am here during this time period. Arriving here takes great energy and strong desire. Cave Girl needed someone to help very badly. I was nearby and I also had an extreme longing to leave where I am in time. I'm not especially puzzled or shocked. This is almost 'normal'. It took no machine, no magic potion.

I saw the snow geese, the huge fish. I have often felt I was 'someplace else' other than where ordinary people can go. I go alone. Partly because I have 'abilities' and do not want to let on. I do not wish to be a freak, locked up, or exploited. There had been that time years ago I was helping Will, one of my best friends. We sold cars! Four, five times or more! He'd fix an old car, I'd tell him where to take it, and it would instantly sell. Until Will got excited by my ability, and wanted us to go into business together full time! I can call animals to me, speak to them. 'Power' is not something I seek. Yet in times of need I can reach out and pay attention to things I was not taught in school. Trust what I know and believe. Not depend on the world man created. It's not exactly 'special'. Most of us could do this if we wished… and tried… or had a reason. I'm not much for the world of human beings, so I focused on other worlds. I often joke, "It's a Zen thing!" In truth, how would I know. I never formally studied Zen, or been an apprentice under anyone who taught as a master.

**Future Flash**

A special on TV about the brain. It has been proven by experiments the size of our brain and how it works is directly related to outside events. Nutrition, exercise, but also how we use it. Einstein's brain, when analyzed, had way more of something than the average person. Musicians and artists have a different brain development than other people. The changed brain was not there first. The change came with practice. People who practice hearing, and must, have a higher developed hearing section of the brain. Same with sight and with those who work with their hands, those who read a lot, those who are spiritual.

A taxi cab driver's brain was said to be different from everyone else's. The indication is, our brain can develop to suit our individuality, needs, and occupation. As with muscles - 'use it or lose it'. We exercise certain muscles and they get stronger, more coordinated, bigger. There can be a huge difference between those who develop the brain, just as between a weightlifter and sedentary person. An office person struggles to lift 10 pounds while someone who lifts weights could lift close to 1,000 pounds. A hundredfold difference, starting with the same body. The brain can be the same, a hundred-fold difference with practice.

I have lived a different life. Twenty five years alone in the wilds is bound to develop a different kind of brain than a civilized one. Exercising my brain, training it to notice subtle changes in the weather, animal behavior, playing with time, remembering bends in the river, and the various survival skills in my lifestyle is going to strengthen different parts of my brain. But at the expense of not developing parts of the brain civilized people develop. This makes sense then, that I might be expected to have skills based on my experiences and practice, that are different than the ordinary person. I keep saying, "Not better then, just different." I never believed I was smarter... only that I developed the same brain as yours, along a different path. Here is scientific proof backing up what I have believed for a long time.

So I can image a trained weight lifter who can lift 1,000 pounds being unable to imitate the moves of a baby. In an experiment, an entire football team was told to imitate the baby up front. None could. Few football players are geek Einstein types. Few Clark Kent types are Conan.

With this line of thought, a great many things become possible. There is a man, Jim Brown, I think, who is a world class tracker I read about. He finds lost people. Tiny bent blades of grass, a twig out of place, a sense of smell, highly developed sight, geared towards the passing of a body through the forest. It's unbelievable to the rest of us. The proof is, he finds lost people in the wilderness. How? Is it magic? That which we do not understand, we call magic. Magic could be explainable if we studied it long enough and made it our life passion. All this may help explain what I do. Or it may not! Just tossing out theories here.

**Future flash ends.**

There is much I wish I could communicate to these two! What is it they know or believe? What part did they play in our meeting? Do they have abilities like mine? Are there more of these people? Do these two have some kind of magic and are they medicine people among their kind? I make the sign I am puzzled. I get a nod these two understand and how may we help. I have their full attention. I suppose I can see if I can tell them my name. What is my name? For years I was 'Wild Miles'. However, for even longer I have been 'Sunshine'. More people call me Sunshine. I may as well go by this. I point to myself, "Sunshine." I point to them to ask if they can say this word?

They mumble "Soonshine." Good enough.

I point to the sun. "Soonshine" They look puzzled. I said my name is Sunshine, so how can this ball of fire also be Sunshine?

Cave Girl speaks excitedly to Papa. I think she is explaining. "Papa, he showed us his fire! The sun is fire! He gets his magic from the sun. He is a small piece of the sun! He is named after the Sun God!"

I suppose this will work. After all, why did I get this name? I try to bring warmth, light. It's a noble wish I can't attain as I dream, for I am mere mortal flesh. Still, I can be named for my goal and what I strive for. These two may understand me better than civilized man! What a thought!

Cave Girl touches my arm gently to get my attention. Points to herself and says something I do not understand. I know, trying to tell me her name. How she is called among her people. She looks around. There is a freeze dried flower nearby and she pulls it out of the ground, points to it, points to herself. I get it! I say it in English, "Flower!" I point to the flower and point to her.

She repeats and can say, "Flower". We know each other's name now.

I point to Caveman and say, "Papa", he smiles and repeats, "Papa" and knows this is his name in my language. I sense the doll is sort of an important medicine tool. I assumed more of a toy, but no, the doll has powers.

Flower looks around and spots a plant, digs up a root and another root. I understand this is food. I dig in my pack and pull out an aluminum pot that holds half a quart. I find a crevice that has ice in it and pull out my knife and chip ice. Flower and Papa chatter, and I understand. They have not seen such a knife! Or such a pot! After I have enough ice and put the pot on the fire to melt the ice I hand the knife to Papa. I move my thumb along the razor edge to show him. He tests it and is impressed. I already know flint is actually sharper than steel! However, the edge is not continuous, nor straight for the length of the blade. Papa tries the knife on his spear. He nods and goes to work giving his spear a better shape. Flower understands she is the cook, and I have given her water and a pot so she may cook the root. Because of all my past experience in the wild, I know this root is starch and will taste better cooked.

Some meat would be good in this soup. I had weapons in my boat, but not with me. I do not know if I can get to my boat. Maybe never! If not, I'd have to take stock of what I have with me and wonder if it is enough to survive with! I'm not worried. If these two can survive, so can I. I have used slingshots. I see my elastic strap on my pack. I find a forked stick. As the water is warming, I fashion a slingshot. It's very crude, but I notice little pika looking creatures nearby going in and out of the cracks in the ground looking for grass roots. There is only a fist size amount of meat on one of these, but it's flavoring for what looks like a potato type soup. I grab up a pebble and test it in this slingshot. Papa has never seen a slingshot, so when he sees the pebble fly he is curious. I get another pebble and make the motion to be quiet. We get up together and silently walk a short distance to where the pika are moving about. Papa holds up his hand, he has spotted one.

He has hunted in a group before. He gets the pika's attention and I understand what he is doing. I can sneak up to within ten feet from the side, unnoticed. I kill two pika with this slingshot. Papa knows how to skin, so I am sure this is familiar food. I spot some wild mint. Not the best flavor for this stew, but better than no spice at all. I know mint goes with meat. I had seen field mint growing here in the future! When? Maybe 10,000 years from now? So some things do not change a lot.

Flower knows this mint, and what to do with it. She is only a child in my eyes, but am guessing these people do not live as long as the civilized. By ten she is a woman. By thirty she will be old, by forty dead of old age. I'm over sixty-five, so a very old man by their standards. *Old but wise!* I smile at this. Back in civilization I am a nothing and a nobody. My skills are not important, rejected, punished by my tribe. I have trouble with the telephone; do not drive. I am mentally challenged, not so far from getting sent to assisted living. Here? I know how to build a fire, find game, make tools, identify edible plants. *In just a short time, look how useful I am! Stuff I get locked up for in my world!*

However, reality sets in. I accomplished my mission; saved Flower's Papa. Fixed her doll. It is time to see if I can get back to the place I came from. I have my boat, my civilized life, a wife, things expected of me. I'm not sure, at my age, I could live like a cave man! I make a sad gesture and point the way we came. They understand it is time for me to go. I tell Papa he can have the knife. I have no way to tell him I am a custom knife maker by trade! Flower hands Papa her doll and puts her arms around me, giving me a warm child's hug. I rock her back and forth. She is real. It is hard to think this is a dream! She lets go, I sigh and step into the fog I arrived in.

ALL LOOKS as I have left it. The plants and trees are the same as when I was sitting here in modern times. I feel rested. I may have just been dreaming, ok, time to try to get these tusks.

I HATE to get back to the top only to realize I forgot something, or realize I have to hook myself up another way. That is where experience would be good. So here I am at the top for the second time, in hopes I have everything I need and figure out how to connect the dots. I am looking down this mud cliff 300 feet, *maybe only 200 ft.* thinking - this is nuts! I think I am at a high elevation. I went too far beyond the estimated 200 ft I had guessed from below. Where I need to be looks like half this height! If so, maybe it will look less scary. I get to where I hope I need to be, but am unsure. I move around to see down below. I go to one spot and can see the boat. I adjust, move to look again, and can see the tusks, but far off, closer to the higher elevation I was previously. But I was not quite sure when there, exactly where to come down! *Dang*! Due to being unsure, tired now, my lack of experience, and my being scared, I give up. I will climb back down. Let these tusks go. I do not want to get caught on the cliff face, too tired to get back up, or get down. There goes about $30,000.

BACK IN THE boat I take another look up at the tusks, wishing I had a partner! Foil had volunteered. As I rest again I review that. I had gone to pick him up at my homestead. He wanted to do something, go someplace, have a river adventure. I suggested we go further upriver and check out a spot I know for finding tusks. Maybe I could use his help. He was all ears, and just in it for the adventure.

"Whatever we find is yours, Miles, I just want to be part of the adventure!"

So we find a tusk, and sure enough I could use his help climbing. He does his job. We get the tusk. But now he figures he did all the work and deserves at least half what the tusk is worth for payment. I give him a thousand dollars up front for his two to three hours of work. He is silent, obviously not happy. It is four years before I get the tusk sold. Meanwhile Foil blackmails, intimidates, extorts me, saying my activity is illegal! He will turn me in, saying I should pay more 'or else' and I would never have got the tusk without him and what not.

*Who needs that.* There were other similar incidents over the years. I'd think someone would be happy with $1,000 for a few hours work, gone perhaps three days. My experience is, when they hear the tusk might be worth thirty grand, they want half. Three hours work I need from them is just not worth half. My mind wanders on the money aspect of a find....

Partner issues are partly my own undoing. I grandly speak of big dollar numbers, grin, exclaim, "Beats working for a living!" Listeners may get the impression I bring in a tusk and tomorrow get handed cash. I'm trying to impress people. Reality is, I have to prepare the tusk. Clamp it, coat it, let it dry slow for a year. I then have to sand it, possibly display it. I then have to advertise to find a buyer. This requires knowing where the buyers are. I may have to ship it or deliver it, requiring a custom built box. I may get ripped off, or not get as much as I hoped, or the tusk does not dry well and must be cut up for scrap. I maybe have fifty hours of time invested. I may have thousands invested in advertising, travel shipping etc.

I still do not call it 'work' as I enjoy the time spent! I simply frown when someone expects their half, $15,000 right now. They are doing what in return? It's my boat, skills, gas, financial risk. I want a partner for just two to three hours, paying for their youth, as unskilled labor. Sadly, yes, Foil is partly, or sometimes right. I have to leave material behind I cannot get. So parting with half would be better than having zero all to myself. However, more than the money, is the fun and thrill of being Indiana Jones. If I feel ripped off and mad, what good came of having a partner? In such a case, then yes, I'd rather have nothing, then blackmailed out of half.

My mind gets back to solving the issue at hand. *Maybe there is a way up from below.* Maybe the tusks will fall while I am here. I try to review in my mind all the likely options, and also likely consequences and likelihood of such an event happening. The good news is, there is a landing spot. The tusks will probably not fall in the river if they come loose. *Maybe I can help them come loose!* In the past I have shot bullets at the frozen mud holding the tusk, and had them come down. Once I had to shoot one in half, and let the other half fall later on its own into the river. That half a tusk was still eighty pounds of ivory! I'm a felon and cannot have a gun. But the idea sticks, and I adapt it.

I write notes in my laptop computer diary

### Wednesday, June 29, 2016 next day

Day four ends alone on the remote upper Cosna River. I have not seen a boat or heard a plane. It's like another world, another time. I left the tusks in the cliff at the end of yesterday. One more night of warmth may thaw the frozen soil holding it. I may get lucky and find the main tusk in the mud at the river's edge! I go to bed early and wake up at 3:00 am, ready to go to work. There is twenty-four hours of light. I can work and sleep as I choose. I'd rather work when it is cooler and less likely other boaters or planes will be around. So goes the lifestyle of Indiana Jones.

The tusks look just the same. I decide to try a frontal attack. See how far up the cliff I can get. Partly, I am less scared when looking up not having to look down. The closest I can get in two hours is within ten feet. This is hanging on with no free hand. I decide to come down to the boat so I can use my fishing pole with heavy weight to toss up to the tusk. If I can wrap the line around the tusk, I can pull up parachute cord. This might be strong enough for me to climb up.

Within half an hour and a few dozen tosses, I have the cord around the smaller of the two tusks. The big tusk has both the base and tip stuck in the ice, with a hollow space, as a staple might look, not all the way in. If I can tie off to the smaller tusk I can work on digging out the bigger. Depending what I see, I might wrap a larger rope around the big one, and try to pull it loose with the boat. Or I may need to chip away at frozen mud. I carefully select what might be needed in a backpack when climbing up this rope. I try the ascender from my climbing gear. It seems to work all right on the double strand of parachute cord. I have my climbing harness with another ascender hooked to me. I can use one hand, and rest hanging by my harness when I need to.

I get within ten feet of the tusks again! This last part is the steepest. I had been hanging and digging foot holds with the ice pick as I worked up. I have my feet secure, but much of the weight is on the cord, helped by the ascenders that locks me to the cord. I'm too scared to go the ten feet and be fully, totally, dependent on the cord to hold me! No way to cut hand holds. I notice it is not so easy to loosen the ascender when I want. This requires creating slack. I did not carry the tool I need to descend. I was told this might be an issue with the ascender, being unable to reliably come back down using this tool in reverse. The cam works only one direction. I'm sure with experience, I could understand it better. The tool either totally grabs or is totally loose. To loosen, pressure has to be off it. I assumed I could simply turn the tool around! There is no friction resistance, as is best when coming down. I do not want to get the experience here and in this situation!

Darn! It is only 5:00 am and I am almost at the tusk! I struggle a little to get down. *I cannot use the ascenders to come back down!* I have read a little on the techniques. I wrap the line around a carabiner and use it as my resistance to slide down. This works. There is a special tool I did not put in the pack. I think it's called a pretzel. Just for descending control. I now know to have it with me! I may have once known and forgot. I had it in my hand when packing and could not recall how to hook it up, so did not wish to depend on it. There is a knot, a prussic I could figure out in an emergency, but no need here, I have the carabineer.

Now I have to come up with a different plan! I am scared to totally trust the cord. The cord is strong enough, but small in diameter, so can jam inside the wheels of the equipment not made for it. At the boat, I decide to pull larger rope around the small tusk using the parachute cord! With bigger rope I can feel more secure, and maybe try the foot ascender. My concern is, if this behaves like the hand ascender, I may have to take it off my foot to come down! *Am I limber enough to get my hands down to my feet and undo something complicated? When I have trouble being limber enough to put my shoes on?* At sixty-five years old and already had a heart attack, I am absolutely sure any advice from a sane person would be, "Why are you even alone in the wilds out on the river?"

I need to pull the entire small cord around the tusk with a big rope on one end of the cord. The cord jams on a twig near the tusk after tossing a dozen times. Maybe all it needs it a small tug? Otherwise, I can yank with the boat or use a come along once the large rope is in place. The cord comes off the tusk! I have to start tossing the fish line all over again. Some mud has fallen, so there is more of a hollow between the ends of the bigger tusk. If I can first toss the fishing line, then pull up a cord, then pull up large rope, I can go for the bigger of the tusks. I may not have to climb at all if this works! Get the big rope around it and yank like heck! Use the boat to pull with.

The weights I use on the fish line snag on an already broken line on the small tusk. I lose five weights! I have no more! I have padlocks on the boat, so bring three locks partly up the cliff to where I can see the tusk and use the fishing pole. The toss is 100 feet. I lose the locks as well. This is very frustrating because hours ago I had a rope hooked up there! Why can't I snag the big tusk? Why is the line so often snagging the same obstacles and not the big tusk I want.

I finally hook the large tusk! The lock is now hung up with slack in the line. I take a chance I can work it loose and move to tie the parachute cord to the fishing line. As soon as I go to tie the cord, the lock falls free, pulling the line with it, through my hands and gone, with no parachute cord tied to it. Darn! I'm now several hours into this project, with nothing to show.

I spend the next five hours trying to snag that tusk with the fishing line again. I try tossing from one spot, then another. Huge hunks of mud cliff are falling on both sides of me. I feel a little secure in this exact spot, protected by a large already fallen house size chunk of frozen mud. I have to ignore the falling mud and concentrate on what I am doing. I have a very narrow hole to hit. I keep snagging the smaller tusk, which now has gobs of tangleed fish line all around it that cannot be yanked out! The big tusk looks ready to fall with the help of a small tug, if I can get the strong rope around it.

I finally get the rope around it! I try pulling with the boat. I discover soon enough this is too dangerous to keep trying. The river is swift and full of whirlpools and the boat has enough trouble just staying stable facing into the current. Yanking on the rope causes the boat to change angle fast in the current. The boat almost capsizes. I was successful once before on another tusk doing this, but recall it was not easy to get a direct steady pull. Jerking worked better, but requires rope slack. This risks the rope drifting into the propeller and tangling. If so, the boat stops and cannot be fixed or control gained until I climb in the water and cut the rope free. Not going to happen in this swift river. No control of the boat is disaster here, with house size hunks of mud to bang into. I unhook from the boat.

Cliff swallows dart in and out of holes in the mud oblivious to my financial concerns. They have nests in more secure sections of the cliff. I disturb the environment; mosquitoes swarm up around me. Swallows dart in and around me, grabbing bugs inches from my face. A peregrine falcon watches for swallows to make a wrong move and be an available meal. The falcon has a nest up on top someplace. Feathers along the cliff face show where swallows have been grabbed mid-flight in the recent past. I see all this as I once again, rest. My energy is failing and I am not recovering well. I have to quit.

I tie the rope to a huge chunk of mud the size of two houses and wrap it all around to secure my come along winch. I find this anchor difficult. The rope tightens into the dirt without exerting enough force in pulling on the tusk. I crank twenty feet of cable and simply dig the anchor rope into its support. I'm sure the force is also on the tusk, but not the full ton of pulling power the winch has to offer. I have a second winch to tie in tandem! Each new idea offers new hope that this is it. *I got it! The tusk is mine now.*

The day is passing, now 4:00 pm. I have been at this thirteen hours with a half hour break. I ran out of fishing line and realized I have another spare fishing pole back at camp with a full spool of line. While fetching this spare pole, I stop to fix lunch at camp. More mud had fallen, so I think the ice front of the cliff is sloughing off fast, and will release the big tusk soon.

WHILE RESTING and contemplating my situation, one huge cliff section falls, far enough away I am not concerned. It sounds like dynamite or a propane explosion many blocks away. I watch and think, "Cool!" This chunk of cliff is huge! I am in the boat watching. I feel the boat move. I keep an eye out, but see no wave coming. Yet the boat moves and begins to rise! I am puzzled. I know it must be related to the cliff falling. I am not sure what is happening. Not a wave, more like an event. Possibly the cliff falling created a vacuum, or a temporary dam, or simply changed the flow of the river in some way. A full fifteen seconds passes and suddenly the river comes up four feet! Gently, the river sets the edge of my boat on a chunk of mud the size of a table. The river drops leaving one edge of the boat high while the river side drops. A terrific quick lean causes totes to fall out of the boat. The boat begins to take on water. *I'm going down. This is it.* I leap to the high side of the boat and hang on, as I watch a red gas can float by and head overboard, followed by plastic totes.

The boat stabilizes as the water stops dropping, and the river comes up again in another swell. I am able to dive for the totes and save them… partly anyway. I grab them back into the boat full of water before they can sink. There appears to be no current as the river rises. I suspect items are missing, but am more focused on saving the entire boat and my life! The swells come and go in a rhythm. Each time the boat threatens to go down, but gets less severe. My boat is tied short to the cliff, and would be better off free of the cliff to stabilize on its own without getting yanked short by the rope. There is no time at this point to start the engine and cut the rope loose. I'd have to run the length of the boat over all the spilled cargo and bilge water. Eventually the boat settles as the ebb and flow of the tsunami stops.

I have always half-jokingly called such experiences, a 'High Jesus Factor.' *As if I were being watched over.* There had been 'The Squaw' back when young, when I should have died my first year in the wilds. [2] She lived with me a whole winter, kept me company, who knows what else? I handled it in a way I felt appropriate for my beliefs at that time in my life. *She is a figment of my imagination; has to be.* A thought crosses my mind. *Flower looks familiar!* There are bone buttons on the doll she carries that look like buttons I made for Squaw. I do not recall that Squaw and I slept together. But my unconscious may not have wanted me to remember, as at the time I thought sleeping with a ghost would be too weird! It would make her alive and real! How could Flower end up with a doll with buttons on it I made? What if Flower is my grandchild. Or Papa is not her real father? Or? My unconscious only gives me a flash of possibility and wipes the thought off the clipboard, 'copy, paste, backup, delete.'

It only takes me two days to forget and lose track of time. When I walked out of the wilds years ago, I had been marking days off my calendar, yet was a whole month off when I got back to civilization! Where had a month gone? What could I have been doing for a month? I assumed at the time I went bonkers for a while and survived it. Does my unconscious think that now we are much older, I can glimpse the truth? Or, in my old age, am I reverting to 'bonkers'?

Eventually my heart gets back to normal. The near sinking experience is why so few dare to come and stare straight up at tusks. I have the two winches putting pull pressure on the tusk. I hate to simply undo all this work and leave. Clearly nothing I can do now is going to help. I'm tired and out of ideas.

# RETIRING WILD

*Camping spot where I meet the cave woman.*

# CHAPTER SIX

## MAMMOTH TUSK HUNTING, MEET PAPA AND FLOWER IN A TIME TRAVEL.

I believe in a day or two the sun's warmth on the ice will free the tusks. If I lived alone, I'd simply stay here and wait! *For twenty to thirty grand, why not!* But there is no communication with home. I am out of cell phone range, unless I invest in a satellite phone. No can afford! Ham repeaters have been neglected over the years due to the poor economy, so no more links. Iris will be worried. I'm now tired. Frustrated that all day long, a full seventeen hours, and I am no closer than when I started. This is a lot of hours sitting in the danger zone. This is why I took one look and said "forget it!"

There is tension on the come along rope; maybe overnight the tusk will come loose. I can wait until morning to leave. I need a good rest before starting a fifteen hour boat trip home. *Home is like time traveling!* My unconscious always has these ideas! I smile at the notion. It is true. *How would I know if I jumped time?* I so rarely see anything civilized! Others feel the change in time maybe. It is hard to bring anyone here and not have them freak out! They want to go home! Too much for them to handle! It takes special people. Civilization uses the term 'Weird people'! As in 'not all there', like one eyed Crazy Lawsen. Time travelers. Those who resist the warp can't come here. Non-believers who see ghosts, and react by screaming their heads off do not reap the rewards.

I am not sure I am up to another trip back here after just a day at home; these trips are wearing. I like a week's rest, especially with all the physical activity; wearing myself out.

When I lost the tote, a bag of rope floated away. The rope tangled in the prop. I do not know that. I start my engine to leave. At first, I assume my engine is hung up on river bottom mud. If so, one solution is to push myself into deeper water. I feel overly wary and cautious in this area of violent river turbulence. It is absolutely essential I have full control of the boat. Some chunks of mud the size of bulldozers are undercutting the river twenty feet or more. A boat could get sucked under the ledge. It would not be good to get in a whirlpool, then slammed into the side of an obstruction, or snag a root wad, spinning around suddenly. Any of these situations could sink the boat. Especially my long, low sided boat.

The depth gauge indicates there is five feet of water under the boat at the engine. I shut the engine off, and trim it to see if there is an obstruction in the propeller. The bag of missing rope got tangled and wound up, bad enough to stop the engine. I try to fix it by leaning over the back, as is sometimes possible. When trimmed, the propeller is four feet behind the boat. It is too dangerous to lean over and hang out over this turbulent water. I try using my boat pole, then my ivory spear. The prong comes off the spear and snags in the rope, pulling the prong off the spear's shaft. I should move to shallow water and stand in the water to cut the rope free. Luckily I have the spare engine on the transom, hooked up and ready to go in just such an emergency! In a few minutes, the five horse is fired up! I have to fool with it a little because it has not run in a year.

There is enough power with five horses in harness to control the boat in this bad water; at least to get out and away from the danger zone. There is a lot of water still in the bilges, but I prefer getting away from the cliff before cleanup. I give a good shove to get well clear of the first chunk of mud, then steer out to the middle and head for the other side. The boat is listing to one side. I see a good place on the other side with sand to walk on and shallow calm water. There is a method to getting the back of the boat in to shore. I can't just back in because of the current, and the back can only get in two feet of water. I also have to be able to quickly jump out and onto shore to secure the boat. The stern is not made for jumping over.

The bow can reach shore. I have had experience, so know how to easily do this. I hold the bow rope and another rope on the stern while I stand on shore. I shove the nose out into deep water and let the current take the nose around… but not too far, as I can control this with the rope as I pull the stern into the shallows. The stern bumps bottom and stops. The water is deeper than need be because what stops the boat is the propeller of the five horse engine. I can let go of the stern rope, and anchor the bow rope ashore to hold the correct angle to work.

I take my pants off and get out in water about knee deep. I was unable to tilt the small engine up easily and out of the way from the big engine. A lot of gear in the back of the boat is in the way, but I finally tilt the spare engine up and lock it. Now I can pull the stern closer in. I will save the larger pieces of rope I unwind from the propeller, but all of it gets tossed into the boat so no scraps get sucked into the propeller when I get going again.

With an empty coffee can, I bail water, then put wet gear on shore for sorting and partial drying. I had hoped to get to bed early and rest, get an early start in the morning. I think I need to head home. The permafrost around the tusks needs a few days to thaw. It is hard to know for sure how long... maybe one day, maybe a week. I cannot stay a week. I'd be listed as missing in action. I'd like to rest a day at home, then come right back! Likely not going to happen. I have Saturday Market to do. I have the greenhouse to tend to. I have fossils already acquired to tend to. The boat needs gas and replaced supplies. There is a long list of things to get caught up on. I have the 'Clark Kent' by day, 'Superman' by night complex. I can hardly arrive home and announce I am off again tomorrow! Or the next day! I might be paranoid thinking someone might try to follow, get me in trouble, rob me as I return, etc. Only if I follow a schedule. No one knows for sure when I leave, when I get back, where I was.

Where I am is an ok place to dry my goods... on a treeless sandbar bend in the river. Not river silt but actual small pebbles that are clean and attracts the sun's light. A nice breeze is warm.

It seems every day of boating cost another day of preparation and repair. I'm frustrated. Another trip is another $500 cost. If I return and there is nothing to come back to because the tusk was found by someone else, or fell in the river, I am out $500 more dollars for nothing.

I spend a night, feel rested, and most of my gear is dry enough to load back onto the boat. I leave the rope and winches hooked to the tusk. If anyone comes along, they might take a try at getting the tusk. *Maybe they will assume someone is coming back and leave it alone.* But this is a lot to walk away from. It might be assumed, abandoned. I'd rather not draw a lot of attention in Nenana either. Where is Miles going? He's been gone a week, now off again? What's up?" Some may guess, "Something lucrative on the river no doubt, we all know what that means with Miles!" Wink wink. I do not want to give others ideas of where to go to get rich. Foil for example, already says he wants to do what I do, and get rich fossil hunting. He wants to go where I took him, on his own. Why share the money?" He'd follow me and says so. "Nothing illegal about that Miles!" Big greedy eyed grin. Said by the guy who swore he was my friend, respected me, and wanted to help; take care of me in my old age.

Thoughts are interrupted with the sight of a duck caught in the open. My trusty bow gets dinner. It's a small teal, so only flavors a noodle dish. While it is cooking on the small propane stove, I think of all I might have lost if the boat sank.

Probably not my life, I could have jumped ashore. Without my camp. No meds, food, tent, or sleeping bag. It would be rough. I have a good twenty three grand tied up in the boat and engine alone. I have my laptop computer, $1,000 worth of climbing gear, GPS, camera, all kinds of 'gear' I'd just hate to lose. On top of this, rain and wind comes. I'm dizzy with my diabetic sugar count messed up by missing meals and working hard.

**From my laptop diary Sunday, July 03, 2016**
Tusks no fally over nighty. I leave them behind and head home. It is now 5:00 am. Rain, wind, drift wood and foam in the river. I had hoped the drift would slow up, but with more rain there will be only more rising water and drift wood. This means I need to run the boat standing up; face into the wind and rain, not behind the protective windshield. I have to see the drift logs and read the water. I cannot see my gauges well standing up, so have to sit long enough to read them now and then. The water pump is getting weak; I noticed this on the way down, but assume it will hold up for the trip.

*Do I really need a $150 water pump every single trip out?* I have a new one with me and could put it in if I had to. I daydream as I eat up the miles. The water pump issue. In the old days we ran outboards for years without needing a new water pump! I assume pumps were larger, but the main issue is chrome pumps being available then, but not now.

The local guy at Reeds told me, "Because the chrome industry is toxic. No one wants to go through all the permits and safely issues in order to be legal."

Consequently, no more chrome work anywhere in the country. No chrome, no last long time pump. I think I could come up with something to coat the tin pump with what would hold up better than the existing housing. Few outboard owners are running such silty rivers. Silt is rock dust; extremely abrasive. Hard on water pumps, but who cares? It's a small market. Big companies are not concerned with Alaska. To them, a foreign country populated with bears, not customers. What is a few thousand customers to them?

Nothing. Fool with the water pump and void my warranty. That is what it comes down to. I bet I could cut up a plastic jug and make a simple liner in the pump that I replace each season. I'm also sure some of these new epoxy paints or glues are very wear resistant. Or try enamel, melted glass on the inside. Should I try any of this? Or, just pay the $150 every trip like everyone else has to do. The good news is, hardly any boats on the river. Who can afford to be out here?

So, yes, it cost me $500 a trip! *I am not doing that on $600 a month Social Security.* So fossil hunting is just part of my ace in the hole, a card up my sleeve, a grin on my face, to not be between a rock and a hard place crying the blues of the poor with bowed head at the back of the bus of life. 'Only if'! I solve my own 'only if'! I have a different story than crying in my beer at the bar! A story of,

"Yes, I go out on the river as often as I like!" 'Ooo! Ahhh! How do you manage that Miles?' Thus, are my positive thoughts as I face the rain and wind and drift. *Life is good! I am blessed!* I watch fuel per hour and total gallons burned on the gauges. This is a common pastime as I eat up the hours, playing with my rpm and calculating gas. I decide to try to make the trip in one day instead of the usual two. Doing the math, I decide to set fuel consumption at 3.6 gallons an hour, making RPM 3,900. I'm doing land speed of twenty-one miles an hour, fighting a seven mile an hour current. I should be home about 7:00 pm. I forget my favorite fishing stops.

The rain stops. The water is glass smooth. I am easily reading the water and having a fine trip. I can sit down now. Actually I have traveled far enough to be in another climate zone. I look behind me. Way off in the distance, I see bad weather is still there, a snake of black cloud at the horizon. I'm a hundred miles from where I started this morning.

Hot soup, fixed this morning for lunch, is in the thermos within reach so I do not have to stop to eat. As I burn gas and get rid of weight, I re-trim the boat and watch my speed go up by a quarter of a mile an hour. I check time as I pass checkpoints. One item I forgot to bring is my wallet! I have no money on board! Not a good thing if I have to get gas, or fly out to get emergency parts from a village. I add to my list to put spare money in the first aid kit! I might stop in Manley and spend a night, eat at the lodge as an option. But no money. I could stop anyhow if tired. Manley Hot Springs is a major checkpoint because it represents getting to the first road. This is part of the road system to anywhere I might need to get. I spent a winter here ages ago on my houseboat in the hot springs water. I know people here.

I get to Manley on schedule; I am not tired and can be home tonight. It's best not to stop anyway as there are people who know why I am on the river. They might figure out what is going on and backtrack to get the tusks, follow me, and, or figure out where the hot spot is. In the past, this Hot Springs village was a sanctuary, a safe haven, a good place to hole up and enjoy 'my tribe.' I'm sad to recognize the change.

MEANWHILE, back at the ranch… "Lawsen, where is Hornet? She took off, owes me! She gave me some GPS positions guaranteed to produce mammoth tusks and it looks more like a place someone would go fishing!"

Crazy Lawsen smiles, "So you believed Hornet, Foil? Good luck on that!" That's all he has to offer and does not want to know any details. Another one of Hornets normal, 'only if' deals. "Good luck getting paid if Hornet owes you!" Crazy finds humor in this statement. *The scammer gets scammed!* Crazy believes in karma, fate, what goes around and stuff like this. Foil figured, being wise to city life and knowing all the ways to fleece the sheep, he'd have easy pickings in this quaint naive village of Nenana. Miles is still a sucker, so this Kantishna land deal might go ok.

"Miles needs me more then I need him! He's getting old!"

This is what Foil also says to Mad Jay who agrees, Miles is pretty incompetent.

"Foil, I heard Miles spent $150 on some unheard of pitch propeller and it didn't work! Ha! What an idiot! And can't even tell what day of the week it is, or remember how old he is! I'm surprised he knows how to put his pants on right!"

"Yeah, Miles went out with me at break up into the pack ice to teach me about boating. We almost died! Then another time he fell asleep at the wheel, or at least nodded off."

"Foil, people that stupid and careless should be weeded from the gene pool, right? If he has an accident it would be better for the rest of us."

"Have you got anything specific in mind?"

---

MEANWHILE, out along the river someplace…

I dropped fifteen gallons of 'get home gas' at Moonshine's camp on the way down. I had hoped he was home! He must have left not long ago because his lawn out here in the wilds is freshly mowed. The garden is planted and growing. I left a note with the gas, letting the family know I stopped by. I hoped to show his little girl my fossils. She collects rocks and fossils. She remembers who I am, and I often surprise her with gifts for her collection. *They must be up at fish camp on the Yukon.* King salmon season is just getting started.

I load my gas that has not been disturbed and keep on going. But at my next checkpoint, Tolovana Roadhouse, I feel tired. I'm nodding off to sleep as I drive. *This is my usual nap time at home.* Due to my health, and not wanting to fall asleep at the wheel, I decide I can take a power nap for an hour. I think my blood sugar is off. The door is unlocked. There is a bed where I can simply lay down and snooze for an hour.

I enjoy the large log building full of antiques from the old days when this was a roadhouse for mail carriers and travelers who arrived by dog team. There is a large enamel stove in a central room where guests gathered to keep warm and dry out dog harnesses, and caribou sleep robes on racks over the stove.

The Tolovana River is out back. This is a large river, not much current, with clear water, full of fish. In winter, fish nets and traps were set under the ice to catch pike and whitefish for feeding sled dogs. Fresh dogs were kept for the mail carrier. Much like with the Pony Express out West. Wore out dogs could be dropped and picked up in a month. Fresh dogs would get put in the team. There would be fresh straw for dogs to sleep on and a hot meal for the dogs as well as the humans. Trappers and miners stopped here as well. Manley Hot Springs and Minto people visited back and forth with Tolovana as a resting spot between. This is where my friend Josh walked to sell a single lynx pelt to a fur buyer when he was twelve years old. The buyer told me the story of paying Josh $10. A lot of memories and stories here.

Everyone headed up the 300 mile long Kantishna River could stop here just twenty miles away. In fact, the purpose is similar today, a roadhouse; a resting place for travelers like myself. If someone lived here, they could run a business. Sell me gas, a hot meal, a place to sleep, maybe a hot bath. Travelers now, pay as I do, by cutting some firewood, cleaning the place up, leaving something useful behind like bug dope, food, gas, canned food, anything of value to other travelers. The door is open to anyone who stops. The owners mow the lawn and keep it nice. The new owners are dog mushers and river travelers who believe in the old ways.

I'm grateful, as this can be a lifesaver, better than me falling asleep at the wheel and crashing. I could stop and set up camp, but often, all I need is an hour of rest. There are times in the past I'd stop and go inside out of the bugs and simply sit and have my hot soup, get my ears away from the monotony of the sound of water on the hull, drone of an engine, and forever view of water. I need a break looking at something else! I get up, walk, stretch, take a pee break. Maybe put on or take off a jacket, add gloves or wool hat if I feel chilly. [1] I have never in my life had trouble not waking up when I wish. I cannot recall even one time I overslept. I tell my unconscious to wake me in an hour. My unconscious sits up and keeps an eye on me, watches for bears, bad weather, etc., then wakes me up. It's that simple. *One of the rewards of being friends with your unconscious!* I am out like a hibernating bear in ten seconds. I wake up from a deep sleep in exactly an hour, refreshed. I know I am good for the rest of the trip home. I have about three hours to go. The water pump is failing. It is a rubber impeller, about three inches in diameter, spinning on the drive shaft, no gears, as a turbine.

No engine alarm yet. I keep an eye on the gauges. I find if I stop and run the engine in reverse, and then forward, this realigns the rubber impeller in the housing to a new position, like a new grip that gives me pressure back again for half an hour or more. I can trust my engine because I have gauges that tell me what it is doing. I can watch the wearable parts wear out on the gauges. In idle I only have one pound of water pressure. Running it is about seven to eight pounds. A new pump is up to twenty pounds. I drop rpm's to help save the water pump which changes my expected arrival time by an hour.

I arrive home at 8:00 pm as expected. I load the boat on the trailer and drive home. I have seen no one. I'm up at 5:00 am Friday, and begin changing the boat water pump. As usual, the outboard lower unit comes apart easily and fast. Getting it all back together is the tricky part. I call Reeds to ask if I might have done it wrong. Maybe I have it right. I launch the boat in the river to test the water pump and shifting. It shifts forward with no problems, but in reverse the engine acts like it is not going into gear. I will have to take it apart and see why the shifter has not lined up, and hope I did not break it. A drive shaft and separate shift shaft have to line up to matching splines in the exact configuration as when it came apart. This happens inside where I cannot see, so has to be done by feel.

After getting the boat back home on the trailer, I begin to gather the tools to take the lower unit off again. I stop alongside the park fence across the street where there is plenty of room on a field of mowed grass. I have no propeller! *Dang!* I know what happened. I set the propeller on the shaft and did not put the nut back on, or put it on loose. In a hurry, tired. The engine is probably going into gear, but the shaft is turning with no propeller so I feel no change. A $500 propeller. It just scares me a tad, that I am getting older and forgetful. This could be more serious. Bad enough to lose a propeller. I have spares, but it is still a loss.

Friday is spent getting ready for the next trip. Iris washes my dirty clothes and dishes. She did a good job of looking after the plants while I was gone. I do a little of the major weeding. Move a few plants, transplant emergency needs. I have a few plants still in four inch pots that will die if I do not move them to bigger pots.

I press on with the boat work. I unload the fossils, hose them off and leave them in the grass to dry a little. My yard is surrounded by trees and bushes planted as a hedge fence; designed so it is hard to see into the property. The growth gets thicker every year. Eventually I want the hedge to be all edible plants; vines like raspberry. I tie branches together and they grow this way, an impenetrable barrier for 200 feet around a corner and another 200 feet. Much nicer looking then a traditional metal fence. A lot cheaper and not so intimidating, yet effective. No one can see the fossils drying in the lawn, not even a dog can find it easy to get in and take a bone. A metal gate in the design of the sun allows us access to the yard.

I flip the fossils over, wash that side, then coat them with a mix of Elmer's glue and water. This seals the bones and ivory so they dry slower, and are less likely to crack while drying. The tree hedge stops direct hot sun from shining in the yard. This treatment is vital because small cracks affect price a lot. The price drops in half with the least crack. While fossil bones dry enough to handle, I move on to other projects.

The tent comes out to dry, trash picked out of the boat, rope gets replaced, another come along winch tossed in the boat, better rain gear located, used propane bottles are filled and replaced. I can buy gas when I am ready to go. I consider leaving soon after farmers market Saturday. I can hang out near the tusks at the cliff until they fall, even if it takes a week. Iris will know not to worry.

Saturday farmers market has a glitch. It is the same day as a music festival that will be using the civic center where we set up. We get told we will not be in the way, and it may increase customers. But we open late. I am tired, so hard for me to set up. I have some plants, things different from the others. I bring cat nip and herbs growing in pots. No one but me brings cucumbers as I expected.

"Miles, it is not worth selling cucumbers when we can make pickles and have the add on value." My thinking as well. However it is an exclusive market. I have my books and art work. There are not as many customers as we hoped, but I make about $200. It is 3:00 pm before I am home. I am beat, not ready to go out on the river. I had considered leaving right after the market. Instead I can leave early in the morning after a good rest.

---

Sunday I get up at 2:30 am, leave by 3:00 am. There is rain but no wind. I'm dressed in rain pants. Right off, I have to walk knee deep in water to get the boat launched. So begins a long day already wet. It also looks like someone has been rooting around in my boat! *Kids looking for a fishing pole or some weekend warrior forgot to bring a paddle and wants to borrow mine?* Everything valuable is locked up. At 3:00 am I see no one around. I stand so I can see over the windshield into the rain. I get to Tolovana with no incidents, and change into dry clothes. Again, grateful to have this dry roadhouse stopping spot.

As I am sorting gear in the boat, I hear a boat engine far in the distance. I hear the rpm change. There is no reason anyone would slow down exactly now, as Tolovana comes into sight. I hang around a little and watch as a small boat goes by. Foil thinks I will not recognize his boat. It appears to be listing to one side by unseen weight in the boat. Possibly Hornet. It makes no sense Foil would be traveling in this rain. He hates bad weather and has little experience.

However, I notice he runs his boat better than when he started, so is learning. He has no power trim, so must balance the boat with the load. His speed is only ten miles an hour, barely up on step. It is possible he is following me. If so, he thinks I am going up the Kantishna, as this is the word I spread around. There are places for him to hide to wait for me to go by after he gets to the Kantishna River in twenty miles. He will wait behind one of the islands. He can stay out of my sight as he follows after I pass. However, I'm not going up the Kantishna.

I add my note to the ongoing log book kept at the roadhouse. I can read who has stopped, when, where they were headed, and see the adventures various visitors have had. Most names are familiar. River people as myself. Dim has stopped, 'Checking on my cabin, lots of drift.' Tom and Lana from the Kantishna River, 'Hauling sled dogs, going to fish up the Kantishna this summer, hi to everyone and stop by if you can!' The Collins Twins from Lake Minchumina, *they are a long ways from home!* Various friends in Manley. Most stop as I do, transfer gas, leave a note that all is well. I notice Foil has not been leaving notes when he stops by. Wants to be secretive.

No other stop. I run on adrenalin, and a little faster speed just in case I am being followed. I can get amazing fuel economy if I am willing to settle for about the speed Foil goes. At fifteen miles an hour I only burn two gallons an hour, maybe less. Doing thirty miles an hour way more than doubles my fuel consumption. Maybe five gallons an hour. I usually settle for three and a half gallons an hour and twenty-six to twenty-seven miles an hour. I never had the engine full open, but briefly saw it is possible to burn twelve gallons an hour doing about sixty miles an hour. I once again play with the fuel numbers in my head for something to do as I put in a thirteen hour day.

I would be upset to get to the permafrost cliff, and find the tusks went in the river or someone else found them. Likewise upsetting if anyone guessed where I go, and heads there ahead of me using my knowledge or taking advantage of my work. I do not mind so much if there is another mammoth hunter and they are doing well with their own knowledge in their own area. The truth though, is there are not enough good areas for the number of people who would like to be in business. People who think they will make a quick easy buck go out, get disappointed sometimes raid someone else's finds.

Like gold mining or trapping! We dream of big nuggets waiting to be picked up with nimble fingers in shallow streams, or that pocket of 100 marten in one creek drainage! It happens, just as I once found a complete perfect mammoth tusk at the water's edge, and all I had to do was roll it into the boat! I once had a great trapping season and filled my dog sled with so much fur I had to make a trip to the cabin to unload and come back. I once found a half ounce gold nugget six inches in front of my eyes when I put my face in the creek to get a drink. All one time experiences, not to be counted on.

I reminisce on such times as I eat up the miles and smile to myself. It would be hard to actually make a living depending on just the best times! I often joke to the public, "It beats working for a living!" Yes, I suppose 'rub it in' and it bothers some people. Of course I am working, but get people to think in their own mind, *what is work anyhow?* Many, of course, have visions of what my life is like boating, camping; life as one long picnic. *Only sometimes it is like this!* Like the times we choose to recall? *Ha.*

I zoom by Manley without a glance. The village is way up the slough, but there is a road with parking lot to the river. I look, but see no cars or boats tied up. As with any occupation, there will usually be good days' worth talking about! As with all of life! There will be a price to pay as well! Nothing is free. This parking lot and road is what Foil likely came to when he was trying to find the Kantishna river and missed it by fifty miles. As he found out, gas is $7.00 a gallon. Instead of paying this, he calls his good buddy Miles to come rescue him and bring gas he does not have to pay for. *Or Foil got here, saw no people, did not understand, and left.*

SOMEWHERE ELSE ON ANOTHER RIVER:
"I thought you said we'd see Miles go by if we wait here?"
"Yes, he has to come by, there is no other way."
"Would he be smart enough to give false directions?"
"I think not, but maybe he knew it was us going by the roadhouse and decided to hide and wait to go up!"
"I do not think he could recognize us, it was too far away! You were down low. I had a hat on! How would he know one boat from another at that distance!"

Hornet is dubious, as she has been around on the river, and knows we all recognize each other from a long distance. She and Lawsen used to spend more time on the river back in the day. She is now not sure what to do if the trip is wasted. "I guess we can take some fish home, Foil!" Hornet is easy to distract or more like moody; unpredictable. This is when Foil realizes it is his boat, his gas and what did Hornet invest? Besides, "Ra, ra. Let's get rich quick?"

FAR AWAY...
I'm at the mouth of the Cosna River! This has a very twisty steep cut bank with dark water. I now travel with lots of foam and drift logs. I see no boats to speak of, just two tourists in a canoe. I used a higher engine rpm with a lower pitch propeller which burns a little more gas, but is easier on the engine; much like choosing a lower gear but higher rpm in a car pulling a load.

I am anxious to see what transpired while gone from the tusk site! I arrive and see my rope. A section of tusk is lying in mud about to go in the river. The rope I had tied is hanging loose. *Did someone shoot the tusk down?* No bullet damage. Why not pick up this big chunk and leave what is still up in the bluff? I look thorough my monocular to see what the pieces in the cliff look like. It looks like the smaller tusk is more visible with the possibility of getting it down! Two pieces of the big tusk are ready to fall. I may as well go make camp. It is more dangerous to be close to the bluff in the turbulent water with a heavy load in the boat.

I can dump some of the heavy load of gas to lighten the boat and make it safer. I decide to pick a different camping place, because the one I often use is in direct sun, gets too hot to sleep, and is unprotected from heavy wind. I also do not want to be as visible from a plane or near where boats go by. I may be camped here a few days and want my activities to go unseen. My agreement with all land owners is that I have permission to find and keep fossils if I keep it low profile and exact location secret. No land owner wants those who hear of good fossil finds to come around the property, or have people camping, cutting trees, digging, stealing. No one wants the risk of the university declaring the property or mine an archeological site and confiscating the land; which they can do, and have done.

I'm not very worried about Foil for now. His engine is burning five gallons an hour doing ten miles an hour, and worse coming back upstream. He can haul fifty gallons of gas with a passenger and marginally get on step as I observed. I do not have to be a rocket scientist to figure out he will run out of gas trying to make it this far. My concern is, he may learn, get a new boat and engine. Or decide to bushwhack me. I want to choose a camp easy to defend. Me with a bow and arrow. Foil with high powered scoped rifle.

Looking for a new camp, I decide the mud hunk that the tusk had been tied to slid down toward the river. Instead of pulling the tusk free, the force broke the tusk. Only a ten pound piece came down. Now just scrap ivory. I'd have been better off doing nothing at all. No climb, no tie up. This is why I so often simply walk away and forget a tusk in the cliff. All this 'hoop-la' for ten pounds of ivory. Even a special trip to come get it. I may not get any more ivory. Before I accept this, I will make camp, calm down, think this over. This is a common way to make tough decisions. Wait.

I find a secure camp spot and hide the ten pound chunk of ivory. It could easily have gone in the river, glad to have retrieved it. I take a nap until 6:00 pm in the boat. Lots of fish here, but catch none. *Must be whitefish, not going for my big pike lure.* I smell it first. A field of mint! A good, flat place to set up the tent. I breathe it in as I trample through it setting up the tent. I unload the boat to get rid of some weight before I fool around in dangerous currents. The boat is sluggish with gas weight that has to be transferred with a siphon hose between containers so I can move the portable containers and get fifty gallons off the boat. I leave the running tanks full. I have to stop what I am doing and use the boat to fetch a fishing lure that got stuck on an underwater snag. I was going to cruise looking for other fossils that may have been exposed since my last visit, but change my mind as mist turns to heavy rain. I'm tired, and can use more rest.

If I am lucky the pieces of big tusk and smaller tusk will thaw from the cliff and fall on their own. Rain will help free the process. I can then just pick them up off the ledge at the bottom of the cliff. The longer I wait, the less ice I will have to deal with getting them loose. If they fall on their own they could roll into the river and be lost, but not likely. I'm undecided what to do. I set up the larger tent with computer and chair to work on book notes. I also have a book to read. There are two factors most important in a camp for me - comfort and entertainment. If either of these factors does not exist, I tend not to be patient, and leave. This can get me in trouble. The opening of the tent always faces the water as adventure will come from this direction. Even bears would most likely walk the water's edge and approach from this direction, but also dinner is this direction, like beaver or ducks.

Even if I do not get the tusk pieces, it would be well to wait a day here before heading home. Wait for drift to settle for a safer trip. I can use the time seeing about the rest of the two tusks and/or fossil hunt for more. I can be adaptable with the plan.

Time should also ensure Foil heads back home without being a problem to me. It is possible it has been Foil messing with my boat, even took my propeller, drained my oil earlier.

Rain. Things out drying, wet again. Report was for rain to slow up. *Good to have rain though!* Few other boaters and overcast means no satellite view. I want to hunt after normal working hours, like now, starting at 7:00 pm. Maybe go to bed about 10:00 pm and up as early as 4:00 am. No one else would be about. Rain changes the plan.

---

I HEAR voices outside the tent. I frown—puzzled—and tilt my head. Yes, I am sure I hear voices, but not language I know—it is the language of Flower and Papa. I heard no boat, it is pouring rain. I unzip the tent flap and in scurry Papa and Flower, saying, "Burr it is cold out there!" In the language of all cold living creatures. They look around, not afraid, just curious. The thin material of the tent, the zipper door, propane heat source, seem to fascinate them the most. I wish to offer a place to sit, but I have only one folding camp chair. No problem, they plop themselves down on the carpet of mint and sit cross-legged. I have plenty of food to share. There are pike from the mint creek and ducks of all kinds.

I have food on ice in a cooler that we found on eBay and Iris ordered. 'New and improved', it is advertised to keep food cold for five days. I'm unhappy with the twelve volt cooler that runs on the engine - just too wussy. Does not have the volume and is not cold enough. Or it simply wore out. Papa lifts and closes the lid. He seems curious where such pure ice comes from. Permafrost ice is filled with silt and debris. I have no way to explain beyond, "Ahead... where I come from." I use a gesture and these words enough they know what it means.

As we eat, and the two dry out, the rain stops. Through sign language and the few words I comprehend, mixed with the little English they pick up, I understand they are with someone else—the Chief. They have killed a mammoth within walking distance of here. This area is good for mammoth hunting. I would guess this, since this is where I find tusks! I hand Papa the chain saw. He has no idea what it is. I hand Flower a full thermos for lunch, and knife sharpening tools. I bring rope and a winch. I am old, so we walk slow. A twig snaps when I step on it. We hear a sharp voice, filled with authority. The equivalent of, 'Halt! Who goes there!'

Papa answers, "It is Papa and Flower, with Soonshine." Out steps a short elder with round face, beaming smile, bowing a greeting. He reminds me of a munchkin from the Wizard of Oz. He is covered in blood. The bloody knife I gave Papa is in his hand.

Before him is a dead mammoth. The guts are being pulled out before butchering. Experience tells me that if the guts are away from the meat when the meat is cut up, there is much less risk of contaminating the meat with gut juice. Even cave men do not enjoy crapola mixed with meat. From a practical standpoint, contaminated meat has bacteria, so meat can spoil in as short a time as in one day. Clean meat could keep several weeks.

The gut pile stretches out thirty feet and is still coming out. Very impressive. Intestines are the diameter of Chief. I see how the winch I brought will be handy here. Rope is tied to the gut. I pull with the winch from a tree nearby. Chief cuts the membrane holding the guts inside as I crank. The stomach is the biggest and most critical mass to extract. This is about the size and weight of a fifty-five gallon drum, at 300 pounds. I understand why we do not want to cut the stomach open and drain the contents to save weight. We would be walking in it as we butchered the meat. Not fun. Flower and Papa bring roller logs to roll the stomach out and away as I winch.

We sit on the mammoth's hairy legs during a break leaning back against the warm skin of the mammoth...there is room for four of us to sit on one leg. The inner heat mass has been removed from the mammoth's body allowing the meat to cool down. Getting the bacteria and heat away from the meat is vital for preservation of meat. Flower had cut a trench downhill for the blood to drain, otherwise, we would be ankle deep in it. There must be a hundred gallons of blood in a mammoth. Chief explains if it were closer to winter they would save the blood as food in sections of gut. I know how fast blood spoils! As with me, one of the favorite parts is the heart! We all go "Yum!" and rub our bellies. The heart is still inside. Getting this organ out is important in getting more heat out of the cavity, but a short break is required before this monumental task.

Sticks are used to prop the pried open ribs to form a cave someone can crawl in to cut the heart out. Flower has her ever present doll with her. Once again, I wonder if I have seen it before in my past, why Flower looks familiar, and why we seem to understand each other and have a special bond. She likes to be next to me, watching close as I work. She hands me the doll to hold as she decides to be the one to dive into the opened cavity to fetch the heart. Her entire body disappears into this cave of flesh. Papa reached in and clasps her feet. We hear cave girl cussing sounds as she cuts the lung membrane, separates the heart sack, puts a hole in the heart and ties my rope through it.

"Thurk!" I hear from inside, and papa tugs on the rope. This lets Flower know what still has to be cut. As papa pulls she cuts the heart free. "Tarse!"

Papa stops pulling on the rope, grabs Flower's bloody bare feet and pulls to help her get out.

"She-waa!" Chief and Papa point and laugh. I repeat the word and laugh as well. We are commenting on how she looks all covered in blood, but there is no time to wash up. Flower helps wrap the thirty gallon 150 pound heart in a large piece of leather. She can easily lift this to set it aside as 'taken care of'. Now we can begin the task of getting the hide off. This is a two person job, so Chief and I begin pulling out fat to save. A shake of the head tells me they do not want the lungs, kidneys or liver. I know why. Lungs are mostly air and not worth much, Kidney taste like pee unless you spend a lot of time soaking and rinsing, and liver begins to spoil within an hour. The interest is only if we are starving.

Already scavengers are gathering. I get the impression the six wolves I see nearby follow the tribe, pay attention to their activities and reap the rewards of the regular gut piles and garbage the tribe ends up walking away from. Wolves are within sight and no one is paying attention or cares. I say nothing because on the trapline I have seen this; had this relation with a bear one summer.

I ask Flower how big her tribe is and if this is all the members. As she is about to explain, another figure comes out of the brush, greeted by the Chief first, then Papa. He appears to be the spiritual leader, so I simply call him Shaman. He has taken credit for the killing of the mammoth using his magic. Shaman looks at me, but is not pleased. I can guess why.

Much of what I bring appears to be magic much more powerful than anything Shaman can conjure up. The Gods I deal with are more powerful than the Gods Shaman speaks to. I am a threat. There is no interest on my part in power plays, or disrupting the dynamics of the tribe. It helps for me to be showing respect to Shaman. I take him aside and show him how the Bic lighter works and give him one that is half empty. I explain this is a gift of power. A secret between us, so he can impress the tribe. If I pass on special powers, he sees how I can be useful, a friend, and not his enemy. I teach him sacred words to say before spinning the spark wheel. Sacred words ending with… "This Bic lighter is fifty cents! Behold my ego, as I impress!"

Flower is watching with puzzlement. I give the 'shh, quiet and just observe' sign. She wanders off to leave Shaman and I alone. I imply, off and on I can pass secrets along to help the powers of Shaman. Later, Flower gives me the puzzled sign and points to Shaman.

I reply, "In about two moons." I hold up two fingers, while pointing to the emerging moon. "Shaman will run out of Bic magic. The Fire God will not be so happy. Shaman will get greedy and abuse his power. I have no control over this. Power comes with responsibility. Shaman will wish me harm. For this reason I do not share all my knowledge."

Flower wishes to be an apprentice; do good for her people thorough knowledge of herbal medicine and any other way. There are healing crystals, predictions of the future. This could be her partial attraction to me. I could teach her a lot about medicine of the future. How to get rid of worms in the body, make disinfectant, new ways to tan leather, and such. Unlike Shaman she is not as interested in power. She wishes to know more of the why and how that will keep her people safe.

This could also be part of issues with Shaman who has taken Flower as a helper. Now she is less interested. I think, as well, Shaman sees Flower as a potential mate with me as a rival. I'm not sure what my role in the tribe is, or could be. I value respect, acceptance, love, protection. Not necessarily power. Power requires body guards. If I am interested in control, it is only control of my own life and fate. I stand up for choices. My conscious attraction to Flower is as a bright young eager mind, wishing to learn what I have to teach. Possibly an interest as a parent or grandparent, as if we are connected by blood somehow.

"Does the idea of moving forward or back in time bother you, Flower?"

"Soonshine, if I hope to be a Shaman, if I truly wish to help my people, I must be ready to make sacrifices and not be afraid."

I'm thinking I could use some help getting the tusk and looking for fossils. There is the possibility of finding the tusks of this mammoth her tribe just killed. Would the land look different? This could be a tentative introduction of Flower to the future, to see how it goes.

I take a compass bearing off the rock formation and another bearing from a far off mountain top to triangulate. Magnetic North will have changed. *Magnetic north is closer to true in the past.* My GPS will not work, of course, because there are no satellites. True north at this high latitude is twenty-six degrees off magnetic. I beckon Flower to follow me into the ice fog. I do not know what will happen. Possibly neither of us can find our way to my boat in my time. God may allow only me. All we can do is give this our best shot.

Clutching her doll, Flower bravely follows as the rest of the family watches in silence. Mud is sliding and dangerous. We work our way down a narrow ice cut to the river in silence. I am filled with spiritual relief when I spot my boat and camp as I left it. Flower seems ready for anything, and knows this will be a different world filled with strange and exotic things. I am impressed she is neither afraid or filled with endless questions. She steps on the boat but I can tell her people do not know about them. The boat wobbles as she steps near the edge. By sign language, I show her how to walk down the middle to balance the boat better.

She knows to clean up before the blood on her dries and becomes harder to remove. I wash up as well; though more fat, hair, and mud come off than blood as I cover myself with water. We both know to use rough sedge grass as a scrubber.

Flower finds the peanut butter sandwiches strange but fun food. I'm trying to get a compass fix on our location. The river has shifted and silt depth has increased by fifty feet. Flower does not know what I am looking for because her tribe does not value mammoth tusks.

When I find a tusk, Flower does not share my excitement. She stops mid-sentence when I point out marks on the tusk she made. Her eyes go wide. Here is an ancient blue tusk, no longer bloody and white. *She could look like this tusk!* Neither of us knows the consequences of what we are doing in time travel. There has often been speculation of dire consequences, like altering the future, coming back to something different! Often, a huge mistake! I think all of us already time travel. But we do it all together and stay in sync, so do not notice. Time ebbs and flows like a tide. Only an outside observer would know.

The earth rotates, passes through space with the solar system, passing through time zones together. Einstein tells us our perception is of staying still, staying in one time, in the way all passengers on a train traveling together do not know unless they look out the window.

A porpoise finds its food buried in the sand by picking up 'brainwaves' the animal puts out. How far can these waves go if they do exist? As far as a radio wave? My friend Bean is a ham radio enthusiast and explains the low wattage ham club. Operating on no more than five watts, and bouncing off the moon, or reaching a space station.

Of more concern than altering the future, is the knowledge I could bring disease to these people. I could as well bring lost bacteria forward. After Flower helps me get the tusk in the boat I take her back to her time and people and return alone to my boat and camp. There is, of course, the strong and more likely possibility I'm mentally breaking down.

---

THE BIG TENT IS OK. I can sit in a chair, and work inside. I still cook and eat on the boat. Fewer supplies have to be taken off if I live more on the boat. I now notice there is a bear bait station where I used to camp near here - not posted as required by law. Few bear baits I run into in the wilds are marked. Not that I care, but nice to know there is one near my camp! This is grizzly country. I want to know if someone is deliberately attracting bears to this spot, possibly pissing them off, or having a wounded bear around. Even a strong possibility bears are being taught to hate humans, angry, and wanting revenge. I have lots of food, lots of gas; I can stay a week if I like.

I think about things at home that need to get done as I rest in the tent listening to the patter of the rain on the fabric. There are art orders to fill. I recall a special custom knife blade and guard. It sold for $200 at farmers market, so that time was well spent. The garden needs help. Iris fills in when I am gone, but it is really my garden. True gardeners say plants know what is going on. 🖤 ☺ :)

There are people plants like, and people they do not like. That sounds nuttso, until you have observed over several decades. Some call it a green thumb.

THE RAIN STOPS while I rest so afterward, I head downstream to cruise the cliffs. I'm too tired to go deal with the frozen tusk pieces. I am not up to climbing and it is not good to hang out in the dangerous area. I find two small bones. Two canoeists I saw days ago upstream paddle by. "Did you find a tusk?" They were told to look for a tusk on the Cosna, as someone saw a tusk with a rope on it. I'm glad I hid my ivory so it is not visible. If I had arrived even two hours later, they may well have seen my rope, found the tusks, and taken it all. "Can we see it? Is it in your boat?"

I do not let on there is anything interesting going on. "I just found some small bones." I show them a few dime and quarter size scrap bones. They are all excited. After they leave, I find two big bones right out in the open, that these two guys missed. I watched them look high. They drifted too far away from the cliff. It is not so easy to find fossils, unless you know how. It took me four years to find my first tusk. It cannot be expected someone can plan to come here and find a tusk, or depend on making enough money to even pay for gas. I can, but I have been at this game forty years. I daydream about all the finds over the years. I am cooking a stew made of mammoth meat, potatoes, peas, dry vegetables that I put up, spices, rhubarb, mint, and macaroni - good healthy ingredients. I have learned to make meals from what I have on hand when camping. It does not have to be good, just edible. Being outdoors and hungry helps. I have trouble tasting the difference between mammoth and moose meat. Maybe more fat.

**From my laptop diary Monday, July 04, 2016**

Up at 6:00 am to a cloudy calm day. Looks like weather could clear. The smell of mint is all around me. I watch swallows taking mosquitoes off a beaver in the water; the beaver does not mind, even seems glad. It is not common, but sometimes I witness cross species interacting in a positive way. Looking around in the soft morning light, there are baby ducks with parents. Later a cow moose with twin calves jump in the water.

Early morning finds me on the river, hot food in a thermos. The boat handles much better and is more responsive with much of the gas weight gone. I can travel heavy if I go fast without the need to have the boat react quickly, or deal with sharp sudden turns. Fog still shrouds the terrain when I arrive at the location of the tusk pieces. Since I cannot have a gun, I shoot arrows at mud around the tusk and manage to expose more after arrows chip at the ice. I am sure the tusk is coming down soon, but am tired after all the shooting and climbing around for a better position to shoot from. I decide to leave the tusks, and see if the warmth of the day thaws enough ice around the big tusk to set it free.

I cover much of the same ground as a week ago, focussing on newly fallen sections and find two nice bones. One under water can be seen as the prop wash pulls water down an inch along the shore. The other is sitting on a sandy ledge that had been underwater yesterday. The good ground and extent of the legal property I am searching comes to an end. I am satisfied I found the easy to get fossils and mostly want to assure myself I have not missed some important object out in the open while I struggle for another! I can now focus on the tusks I came for.

I come back to the tusks, and yes! The major portion has fallen intact. Possibly the pieces found earlier could be put back together through restoration to sell as a specimen tusk. Then I think not. I am guessing this section at sixty pounds which still has a lot of raw ivory value. If I make knife scales from the outer blue bark I may get more money, I think, as I haul the tusk to camp.

This mint camp is a new location from the last trip; a few bends up a side creek. In this way an inquiring airplane cannot land and ask questions as Fish and Wildlife once did in the Yukon River to ask if I was commercial fishing. Wanted to check my boat registration, see my ID. Ask questions. "Why are you here?"

"None of your business," is not a good reply. If I mention fossils I could be asked whose land I am looking on. The officer may not have ever heard of subsistence. I replied, "I live on the river, visit people, enjoy being out." That worked... that time. If anyone like Foil tried to follow, maybe bushwhack me, there is a hairpin turn and I can hear any boat across the pin with enough advanced warning to be ready. I return to where I got this tusk section, to deal with other parts of it, or the other one.

The second tusk now seems exposed enough to consider going after. There is the possibility of getting this one intact. It looks smaller, so not a match to the bigger one that was next to it. I work hard climbing, digging footholds in the mud. I need a break, so leave again for camp and lunch. The day gets breezy and cloudy, looking like more rain. Wind is worse for me than rain, in my unsafe boat.

Time to slow down and enjoy the roses! I'm at mint camp. I hope to pick and dry enough of the mint to make tea for a year! Maybe dig up plants to grow. On my break, I write book notes and doodle on the laptop computer.

After my break I go back to work getting the smaller tusk down. On the way I see where a bear had tried to climb the steep cliff unsuccessfully. I assume the bear crossed the river. This is a bad section to be swimming. The tracks indicate the bear was in a panic; claw marks in the frozen mud show his attempts to get out of the river, but there is no indication that he was successful. The tracks were made since I was last here two hours ago.

The small tusk is not up in the cliff when I arrive. I have to hunt around looking below until I see it landed higher up than where the larger one fell. The whole beach front has changed in two hours and there is no longer any way to tie up the boat. Leaving the engine running in forward gear to hold it in place, I slowly make my way up the sliding loose dirt with the boat rope in my hand to drop along the way. If the boat begins to move, I can make a mad dash for the end of the rope. I can tell from a distance that this is a small tusk, guessing around twenty pounds. Not in the greatest shape, but I am retrieving all there is here. I think it can be restored to be sold as a specimen.

One good news flash is, I get my locks and fishing weights back that are tangled around this tusk! When first seen, this smaller tusk was sticking out only a few inches! The fishing line kept hanging up in a crack in the tusk! It had been so frustrating.

Years ago, I had found a huge leg bone frozen in the ice and decided not to waste time chipping it out and maybe damaging it with the ice pick. I left, and came back a few hours later. The ice had thawed and the $2,000 leg bone had fallen in the river. All I saw was the outline of where it had been. Total loss. Decisions are not easy to make.

I could get Flower and her tribe to help me. I might do well, get rich even, bringing items forward in time to sell. I'd be using them. But not if I pay with modern technology. I choose not to exploit this option. If I came forward with too many unaccountable artifacts looking fresh, there would be a lot of questions I do not want to answer. Questions that could jeopardize the tribe. *Well, myself too.* A stronger reason I could sum up is simply spiritual.

God has allowed or caused this time travel to happen. I am not to exploit and make personal gain. I do not know what the purpose is. But spiritual is rarely about greed. Or shouldn't be… if we care about our creator and the life given us. I might fool with a little extra profit on the side though. *After all, it takes some amount of money to survive and make things happen. I need to be able to afford gas and supplies just to get here. But, we want to stay qualified for subsistence if possible.* I daydream as I dig the small tusk out of the mud. The blue clay is very sticky, with a suction to it as I pull. I have learned that if I pull and let go to rest, the object sinks deeper. A slow steady pull works best.

When the fossil fossil is loose, I have to drag it in the water to get all the mud off, just to see what color it is and what shape. Not perfect, but a lot of blue and green.

A rainstorm hits as I return to camp, so I stay in the tent working on notes; take a nap, read, and wait for the river to clear of drift. I could leave, travel slow, but I can use the rest. *Mission accomplished, so not on a time limit.* I told Iris it might take two weeks! I want the option to hang out here due to weather, or waiting for a tusk to come loose without her worrying, or calling in a rescue. I watch an eagle at her nest near my mint camp, but do not see young this year. It looks like last year's fire may have damaged the tree and nest. She does not seem protective of the area, as she just calmly observes the odd behavior of humans.

I take my nap at 11:00 a.m. as I've been working since 4:00. I'm lucky once again, that I seem to be able to easily go to sleep at any time I want to. In ten seconds I'm out. I wake up refreshed at about two p.m. There is more wind than I expected, with clouds that could get darker or could go away. I decide I may as well load the boat, taking the time to carefully balance the load. If I can get the ivory under the front deck, its weight up front and on the left side should balance gas weight. The boat appears to be heavy on the right so I move cargo around under the deck to make room and a tusk fits. Having the tusks low and out of sight is a big consideration. If the sun comes out, or the day gets warm, I want the tusks below the waterline where it is cooler, as well as out of sight of prying eyes. I am not being evasive because I feel I am doing something illegal.

I do not trust the authorities. Other hunters have passed on bad experiences. A government official confiscates the fossils and gives no receipt or ticket; makes no charges. We can only assume these fossils end up in private hands, not in an evidence locker, mostly because it is not illegal. We assume the officers sell it and keep the money, or it is on display in their home. Best not even let him see it.

Next decision is where to store one empty gas container. I find a place behind the driver's seat for a full container and siphon one tank into the other so I do not have to lift a full tank. Two empties are tossed way back near the engine. Everything else can go where it fits as the weight is not enough to effect how the boat handles. The ivory weighs about the same as the gas burned to get here, so I know travel time will be fine. I am not overloaded which has been a lifetime issue. It is hard to realize those days are over, with twice the horses I had then. Whatever the load is, the engine picks it up and leaps it forward like a gazelle.

"How sweet it is!" As Jackie Gleason used to say.

Now I just have to wait for the wind to die down, and hope drift in the river runs past as the water changes depth... hopefully dropping. I usually have a better handle on fossil value and who my market will be once the drying process has progressed past the most critical stages which is the first two weeks to first month. It can take a full year to dry well. The Tucson show is six months away, so I am already thinking ahead, deciding how to handle the ivory and fossils. I can dry faster, but need to keep an eye on the process better.

The wind has dried my clothes and tent cover so I may as well pack this in case it rains. Time to think of another meal. I have not decided if I want to heat leftovers. It is hard to warm up something with the propane stove. I either burn it, leave it as is, or put in so much water it is thin soup. I smile, thinking of my meal. While I like the meal situation well enough, I'm used to it, it brings back a lot of good memories on the trail over the years.

There is some method to my madness. My mixes of strange ingredients are not quite haphazard or random. I am not following a recipe. After forty years I have a feel for what goes together and what does not. I have a mental list of things that go with meat that I might run across or bring. Another list is based on testing which spices work, even if unconventional, based on what is in the spice kit on board. I know what goes with fish as well, which ingredient has to go in first and get cooking, and what items to add halfway through, and at the end.

With meat or fish I put in the water first and start meat cooking. One reason is, meat can often be made more tender with longer cooking. But also, if it has gone a little bad, is not totally fresh, or is wild and could have worms, I want all that dead and killed! So well cooked works! Items such as rice or dried vegetables go in at the same time. Noodles are added about halfway through the process. Dried herbs go in halfway through. Fresh herbs, onions, some vegetables - especially canned - go in at the very end and hardly have to cook at all. Sometimes the reason is more cosmetic, or varies the textures in the final meal.

I often use canned creamed soups as a thickener when the rest is almost finished. Boiling and soups is by far the cleanest, quickest, easiest, method of eating on the run in my lifestyle. It is difficult to control cooking heat; boiling ensures no burning or sticking. Cream of celery, broccoli, or mushroom are my favorite thickener cream soups that go with almost anything I might have.

I deal with unknowns. If I hunt, I do not know what I will get - duck, rabbit, fish, beaver, etc. They are all different. Beaver is greasy and can be gamey, so has its own blends of additives to make a meal; curry powder is good in it. Duck cooks faster, and has a milder flavor that does not need strong cover up spices. I acquire the meat first and then decide what can be picked from the wild that compliments it. Marsh marigold grows near water where beaver and ducks live, but not fish. Water where this plant grows is too warm for good eating fish. Cattail roots might be better with fish. Mushrooms go with many things, but different kinds that grow in specific locations. Harsher mushrooms with strong, even bitter flavor compliments or masks some gamey meat flavors like summer bear.

I can bring food in a cooler, but it only stays frozen about two days... after that, anything can happen. Leftovers need to be used. Food is valuable enough out here I am reluctant to toss it out for no other reason than I am not in the mood, or it does not fit my recipe! So odd things get mixed now and then. I learn what is horrible and never mixes! *Never ever put catsup on cereal or pancakes.* Potatoes, onions and carrots keep well, taste good with a lot of things, and are at least fresh compared to noodles and rice. Squash would keep well, but I am not excited by ways to fix it.

While sitting in the boat eating, thinking of all the many ways to prepare foods, I admire this camp site and area. Such a pretty spot. I forgot about the mint of the past spot. Fish splash all around me. I'm tucked in the trees out of the wind. This spot would be horrible for mosquitos if there was no breeze! *I do not have to own this to enjoy it!* I was raised where, if you like something, you should own it.

"White man thinking," my Native friend, Josh, would say. There are other cultures and ways to enjoy without ownership. It might require sharing though. I learned of the issues when owning the homestead and using it as a getaway place. I spent so much time outfitting it and hauling supplies, it is hard to just relax and enjoy. Most of the few days' time are spent fixing it up after being away a year. Various levels of disrepair, weather and animal damage to deal with. Did I get to go fishing? Read? Relax? It was always, "There! Now I am ready for next time!" Next time never arrived. The joy factor is affected. This is the mood Foil took advantage of, "Miles, I can help you do the hard things! If I am here I can keep up on all that, you just come to enjoy!"

So here I am living in my boat; everything arrives and leaves with me. I bring little ashore and do not need to. I leave nothing behind and have no need to so it is just as it was when I arrive again. Nature has taken care of itself. What is it I need to have a good time? A shelter from the weather, food, something to read. A comfortable place to sit out of the bugs; a place to sleep. I bring all that with me as a travel package. I can then enjoy anyplace, different places at different times, for different reasons. My home being the great outdoors.

I am reminded of a translated ancient Indian poem I was moved by as a child...

> *House made of rain.*
> *House made of evening light.*
> *House made of dawn.*

I understood the poetic beauty of the poem, but never understood it as a way to live. Now I get it. Well of course I am white, raised white; have that upbringing, and enjoy my 'house' not made of rain! I like to store stuff under lock and key and out of the weather. How else could I do my art? How else could I have a computer? The alternative is to live like Rooster. I like my electricity, hot water and 'stuff'. It's just that I can also enjoy a getaway time and place, without having to own it. This is a big part of my 'getting fossils' story. The entire process - memories created while fishing, sitting with swallows darting around my head eating mosquitoes, the slap of a beaver tail, scream of a peregrine falcon at the cliffs. Fossils as a reminder the scene is timeless. Forty thousand years ago a mammoth had the same joys. Someone may buy the tusk and dream of the story behind it. But I have the reality of the story behind the tusk.

I reminisce over times past as I read a romance western, enjoying the evening. There would have been other mammoth hunters sitting around a fire telling jokes, exchanging hunting fossil techniques and stories a decade ago. There would be barter and trade of items. There are few of us left now who hunt. I have seen no one else's footprints in several years. I am sure a few people cruise by for a cursory 'looksee' while headed someplace else, like the guys in the canoe. I see boats slow down looking for the easy finds. Huge tusks glaringly, obviously, sticking out of a cliff you can see half a mile away. Such finds happen, but are not common. Some tusks are sitting, waiting on the beach to be rolled into the boat. But most are hard to spot. So hard you can step on them and not know. So hard I have to touch each possibility with the thunker rod to test it. Hard enough, the average person would give up.

It is ok to be alone hunting? There are other changes now. Secrets. What good is a great adventure if you can't talk about it and share the story with others? I smile. Ages ago when young, I was alone and glad to be so - totally alone. No relatives, no friends I acknowledged. The world could blow up, civilization wipe itself out and it would be all the same to me. Even fine. No one to bother me! I loved being with God, free to live my life as I saw fit. Good memories and times. After twenty… maybe twenty-five years, being alone as a way of life… this does not work well.

I have met far too many hermits who spent their life alone who are odd—beyond odd—insane by civilized standards. I'd also say - not happy. Incapable of coming to civilization, even for supplies once a year. Most… but not all… become odd enough to be dangerous. They view civilization as the enemy, not as human, or worthy of respect.

Jeremiah Johnson, played by Robert Redford, is a fine example. I got a lot of letters after that movie came out. Oooo! I want to be with such a man!" The 'alone in the wilds guy' we hold up as a hero to emulate. I repeat it again, his nickname was 'Liver eating Johnson'. He loved raw Indian liver. Now if you live like an animal, liver is pretty yummy! However, civilized people frown on being hunted down for their liver. That's not polite in hoity-toity circles. It's not a good way to end up on the 'invited to the party' list. I reached a point after a pile of years, I had to make my choice - stay out and be an animal, treated like one by civilization, or come back to society, understanding we are social beings, and civilization is not my enemy, nor on the menu. Some remote hermits I know got carted off in body bags by swat teams. Not all, no.

**Past Flash**

Like Oliver, spent most of his adult life alone in the wilds... certainly several decades. Karen lived with him a short time before we met. Nice enough guy. Quiet, filled with wilderness knowledge. He is writing. I suggest a 'how to book!' He preferred to write poetry few wanted to read. He had little to no money. I had dinner with him once. After the meal in the remote log cabin he built himself, he told me it was sled dog meat. I'm not sure what society would do knowing some guy living out remote kills and eats sled dogs.

Would he get hauled in, or is it no one else's business? He may even raise dogs just to eat. He gets in no trapping wars because he does not trap. No one wants his land... it is swamp. no one wants anything he has. He had nothing civilization values. I assume he sits in his cabin, gets out to cut wood, haul water. He may have a garden, assume he hunts. Never comes in for supplies. Got visited perhaps once a year. Lived and died out on a remote lake with no name.

That is simply not my destiny. I want more events as memories than Oliver settled for. When I meet St. Peter on the other side, I want the scroll of my deeds to fall off the end of the table and keep unrolling across the floor. "You did all this?"

I want to tell St. Peter, "Indeed I did." No regrets. No 'Wish I had, dreamed about, could have but didn't, was not allowed to, scared.' Hold my head up and say I at least tried. St. Peter can decide what my best is worth.

**Past flash ends.**

Sometimes I enjoy the attention and hopefully respect of sharing my adventures. I told myself at first, it was the adventure that was the priority! I'm thinking about this, eating up the hours on the river to the drone of the engine. Telling of the adventures was secondary and not even necessary. Meaning, I was not going to change the adventure, alter it in some way so it would be better to tell if I did it a certain way. Nor would I seek anyone out specifically to brag. I tell of it, and if you like the story, great, but if not, oh well! I had an adventure, I know it, and you do not need to. A decade or two goes by. I depend on my books and continuing stories. I depend on selling art by telling the adventures. Making a living has a lot to do with my image and stories. So now a change.

The stories in fact 'matter'.

"No story, no make-e da living." Money pays the bills. People pay to hear the story. Now the story as perceived by others is how I can eat. It is not easy to decide to not keep a secret, or tell a fib. Alter my ethics in order to, on the one hand eat, on the other hand not get lynched.

This changes a lot. The public and I are now engaged in a relationship. Like all relationships, we have to get along. We depend on each other. I provide a dream and vicarious adventures. Will these same people I entertain, seek to stop me? Are there those who do not care for the reality of their own lives? Do such people support me;

live life through me? I have always received mixed messages that have been confusing. On the one hand I provide a vital service with stories and adventures the public pays for and says it wants - not just wants, but demands it be 'real'. Puts in glaring headlines. While on the other hand, these stories and activities can be illegal or immoral, not polite. So are there two different groups? One wants the stories and pays me, while another group does not want me to have adventure and seeks to stop me?

Or is society two faced, double standard? The same people who pay to hear my stories, the same ones want me in prison? I think about this as I travel in the rain. I have always tried to focus on the positive. Forget the negative. Move forward, overcome.

I was raised 'privileged'... with a silver spoon in my mouth so to speak. Somewhat like the son of a plantation owner. Or privileged German blonde blue eyed Arian of the SS. However, I left it voluntarily. My generation was known as the rebellious antiestablishment movement of the 60s. I was a teen; no one told me the price and consequences. I could have waved my hand hoity-toity, "Waiter?" Looked around, "I dropped my napkin!" Ten waiters would have leapt forward to retrieve it for me. That is not the life I want. I saw the rewards of the rebellious, but not the price. I should have seen. Kent State protesters got beat up and locked up for example. However, the protesters who were the heroes; the cutting edge of change brewing.

The change I made was an adventure... and life dream. I was taught in school we are all equal. I stood and proudly pledged allegiance to the flag every morning. Ending "With liberty and justice for all!"

It has only been later in life I add up all my experiences and cannot ignore what these experiences reveal. I left my station in life and acquired the rewards of a new lifestyle with different sorts of people. I talked about this before, even often. The subject keeps coming up. How it affects me gets mentioned. While I reap the rewards of Indiana Jones, I pay the price of Robin Hood. I'm wanted by the sheriff of Nottingham for snaffling the Kings deer. While it is nice having friends and tribe that include Little John and Lady Marion, we are hunted down like dogs by the King's henchmen who ride horseback through the forest seeking out our lair. Trying to extract taxes, fees, permits we cannot come up with, or worse, treated as an enemy of the King.

Crocodile Dundee is a great story all right. In reality if such a type shows up in New York with an exposed hunting knife that has an eight inch blade, such a person would not be welcome, nor would his behavior. It's just a lovely romantic movie that sells and makes money. I consider how it would be for Tarzan in civilization, a felon within an hour.

So here I am living a different life now. No fossil hunters to mingle with. I chose a tent spot that cannot be seen from the air. Cloudy days keep satellites from working well to see all. My boat is checked for bugs and cameras. I keep my story off the internet, working only on a laptop off line with external thumb drive I keep hidden. I hide my GPS and camera. I keep a false diary and make up place names and details when speaking to others.

My rights are less than a civilized person's pet. How can I not be affected? How can this not affect my happiness level? How can this not affect my relationship with society? How can I tell cute funny or adventurous positive stories when I know how you feel? Likewise, it is why others like me keep their mouth shut. There are not many stories like mine that ever get told... for good reason. Others I know who are like me will not own a computer and keep a low profile. You will never meet them. They may well insult you if you seek them out.

*I MAY BE GETTING DEPRESSED from diabetic low sugar level.* I become aware of my unusual mood, so know I better eat something; get my mental strength back. I warm last night's dinner for tonight with a few added things. The plan is to get to bed by 8:30 p.m. If all goes well, I expect to wake up and be rested before 3:00 a.m. I have to wait until after three to leave for enough light. No one should be out on the river. I have not seen anyone but those outsiders in two canoes. No one was out in Cosna when I went by. I may stop in Manley if I am tired and the time works out. Get a hot meal, visit friends, even set up the tent at the lodge campground. However, I do not know if I want anyone to know if I was on the river, what day. My loss visiting friends, but the community's loss as I normally spend $100 or so. Not a lot, but in a community of twenty people it is a day's wages for a twentieth of the population. Helps keep the lodge open. If I have concerns stopping, other travelers must have concerns as well. If travelers do not feel safe and welcome they do not stop. It seems to me, like-minded people should stand together and make the village a welcome and safe place to stop. *As it used to be.*

## Photos

*I made a cooker out of a tin can.*

*A place to store all these show displays is difficult in a subsistence lifestyle.*

*These are turnips cut and blanched, ready to be dried or frozen.*

*Three tiers of starter plants in the house in March, before transfer to greenhouse.*

Such a moose provides 500 pounds of meat. I bring a tarp to keep meat out of the mud.

RETIRING WILD

*Ducks in the garden, early arrivals looking for food.*

*Woodpecker tries to get into hummingbird feeder just outside the window at our Tucson home.*

*Iris cleans cabbage. We put up about 100 jars of pickles, relish, sauerkraut.*

*View of boat while digging a mammoth tusk out of the cliff.*

*Easter celebration at the library. Introduction to the wonders of the world!*

*Tolovana roadhouse on Tanana River, where I stop a lot.*

*Tiger swallowtail butterfly on lilac in the yard.*

*Running out of room storing fossils and interesting wood used in crafts.*

MILES MARTIN

*River level drops and I can't get the boat up the side slough to the homestead.*

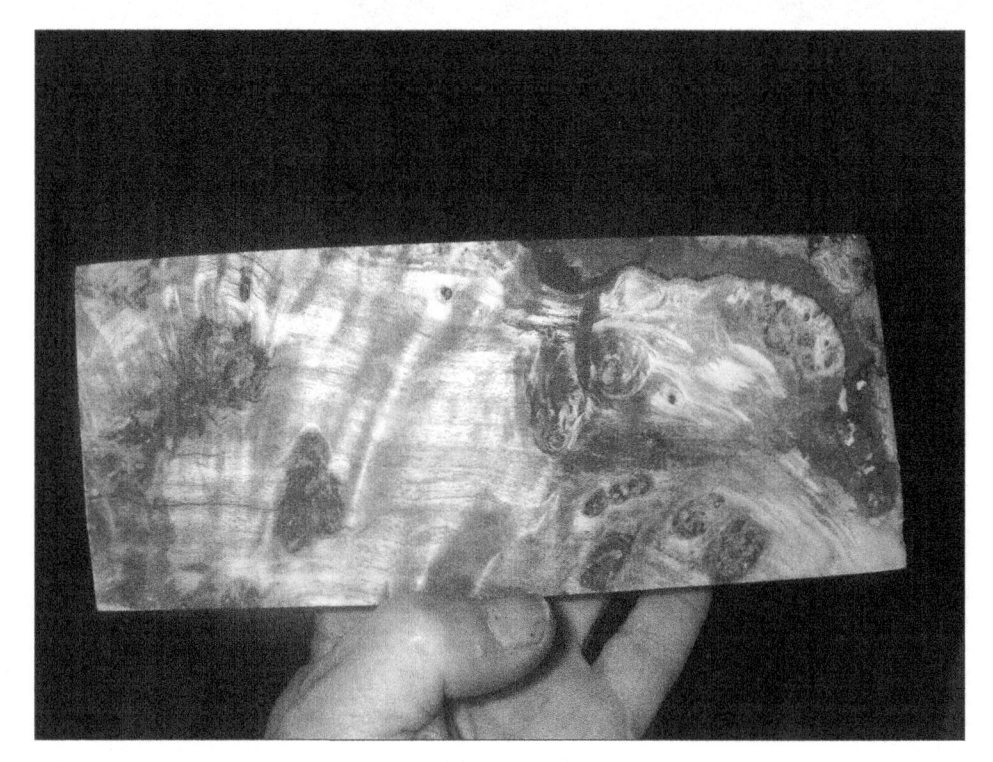

*Piece of my stabilized wood.*

*Our home in Tucson.*

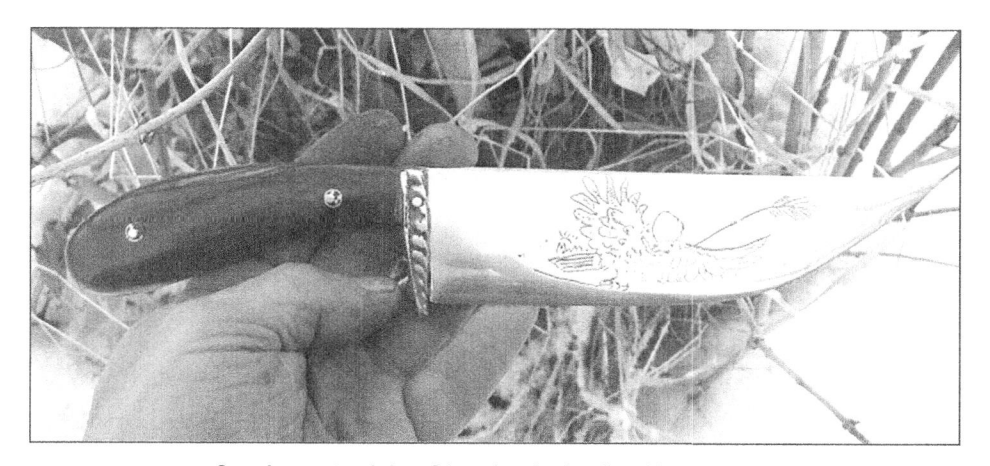

*One of my custom knives. D2 steel, amber handle acid etched blade.*

*My mammoth tusk for sale at Tucson fossil show.*

*My banner at the Tucson fossil show.*

*Mammoth ivory pistol grips I make & sell for $400.*

*Mom and I. She still makes apple pies from scratch.*

*It's not much, but serves as shelter on Kantishna land now in dispute.*

RETIRING WILD

*Ground squirrel newly arrived species in Nenana. Part of global warming.*

*Iditarod dog race team camped on the ice, with Nenana tripod and bridge over Tanana River.*

*Barging across Nenana river. Bridge issues.*

RETIRING WILD

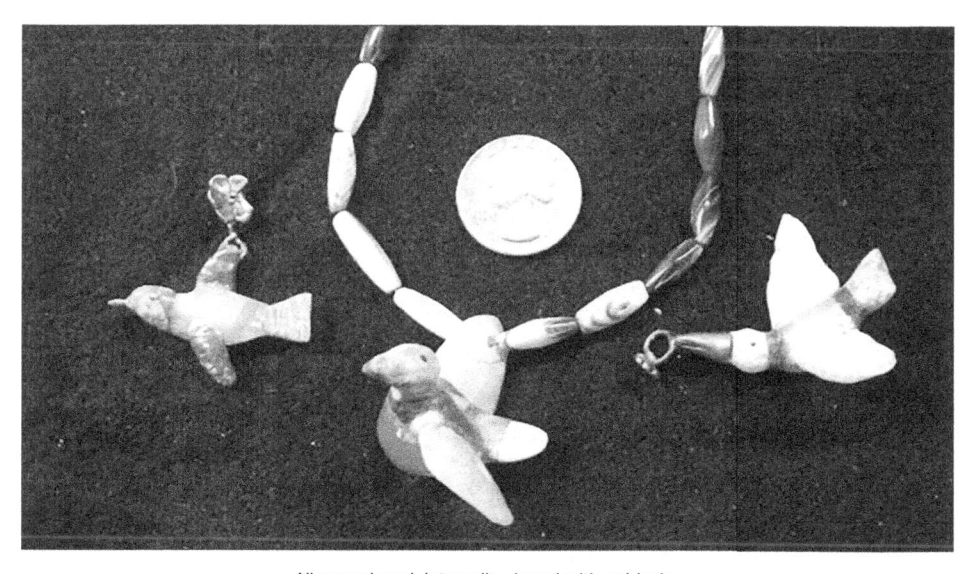

*All carved opal, internally pinned with gold wire.*

*A batch of forged blades after etching.*

*Longtime friends Helm and his wife on one of our annual picnics.*

*Kitty and her love bear.*

*My sister and I growing up. We moved a lot.*

*Mom was still good looking in her 60s.*

*Rocks of all colors good for slicing. A tax write off when boating.*

*My custom art goes to Denali park.*

# CHAPTER SEVEN

## MAMMOTH TUSKS AFTERMATH

Flower is filled with questions. "Papa, who is Soonshine? He looked familiar!" She knows Sunshine can't be familiar because he is from the future - a long way into the future! Papa knows of such things as time travel, and she is trying to learn what he knows so she can help the tribe. There are twenty people in the tribe of mammoth hunters. Papa is not sure himself who Sunshine is.

"Even though we were able to call him and bring him in, he was very receptive. This is for a reason. We have never succeeded before, but I have heard of such things in the past. It takes connected people who have a strong reason." Also a high Jesus factor! Even before the days of Jesus, having an in with the Gods and being saved was understood. Even though this is all before the time of Jesus, Flower giggles. She understands a little bit. How the Gods have to favor you. Gods do not protect just anyone! Nor just any purpose.

Flower was very touched Soonshine fixed the sacred doll. She is not sure if the doll came from her mother or grandmother. She got it so long ago.

"Soonshine saved your life, Papa!" They are both grateful, and talk about the interesting things he brought with him. *Strange things get done by medicine people, that is for sure.* Flower is not so sure what part is magic, slight of hand, use of smoke screens, noise, and drugs, and how much is real! She thinks even Papa does not know, and just calls it all 'magic'. However, there are things she knows of her past, her own instincts, and what she has been told as a child that are connected to Soonshine. She is not saying anything about this. "Someday you will understand!" She remembers being told.

Without magic, the world is a harsh place to face. Without the belief of an 'in' with the powerful forces all around us it would be easy to give up, be afraid, maybe not survive so well. Magic is what makes the people press on, so is a necessity of life. Shaman still has the Bic lighter. He says "Bic!" every time he flicks it and gets a flame!

She frowns, "Shaman, stop playing with the magic! It may lose its power if you abuse it!" She is pretty smart for such a young one all right. Shaman scowls and takes off on his own to study the power of the Bic, and how to steal it, get even more power.

Papa knows Flower's mother was an ordinary woman, not especially bright, but good natured; liked to help others and had a connection to nature, with the ability to live in dreams. A bit odd though. There was some experience she had that she did not talk about much. No one knows for sure who the father of Flower is. The mother has passed away now. A saber cat got her. She was rescued, but died of her wounds.

"There should have been a camp dog around!" Maybe her magic simply ran out. Or she had fulfilled her purpose. It is not for anyone to know. Her body was put on a high scaffolding wrapped in leather with flowers all around.

There is much that troubles Papa concerning various past events.

**From my computer diary at home Thursday, July 07, 2016**

I make it home in a week. Wind, rain, drift. Waiting did not help. Left at 3:00 a.m. home at camp at 5:00 p.m. with an hour nap at Tolovana. Stop in Manley, lodge closed, no water. Snag a log. GPS steers me wrong in a thick fog. Ding propeller bad, tear off trim tab. See no one on the river.

Spend $900 on parts, repairs etc. at Reeds.

Ah yes, the trip home. My diary helps me remember the order of events, and jogs my memory so the details come forth. I get a vision sometimes when viewing my diary

"Ships log…" Captain Kirk; Starship Enterprise. Alone. In the background, sonar pings seeking out other civilizations. So far, twenty-five years and nothing. But any time now…

When I leave, I have hopes of making it home in one long day, but that is going to be difficult. I am traveling faster then I should in a pouring rain with limited visibility and drift logs in the main current. I can stay away from drift if I get out of the channel, but then risk running aground. This increases the distance traveled. Usually when traveling upstream it is better to stay out of the faster moving water while at the same time not choosing a long distance to go just to avoid the current.

I hit a dense patch of fog so different and odd I slow up to take a picture and ponder how to handle this wall of fog so thick it is like cotton candy. This reminds me most of scenes from Star Trek, out the window of the Starship Enterprise at a dense galaxy ahead. Or an ammonia cloud on Planet Z. No. I do not go back in time in my mind. The purr of 115 horses run by a computer at warp speed is more space age futuristic than past. I can barely see the front of the boat. No drift here out of the main channel. I will trust the GPS. *"Beam me up, Scotty!"* I am following an easy long curve of the river into the future. I slow to about fifteen miles an hour and can sort of make out the shoreline off and on. The GPS map is working all right to let me know where I am. Another boat in the water, or a moose swimming across, would be the biggest danger.

At fifteen mph, I am likely to survive impact, so am confidently purring along the shoreline when I see land right in front of me! I back off the throttle and slide up a side creek not showing up on my GPS map. The water is shallow and there are rocks below the boat as I coast to a stop luckily without hitting anything. Now I have to back out and figure out where I am. As I back the propeller clips a big rock but I only hear a slight 'ting'. The thick fog makes it difficult to get a good look at the propeller when I tilt the engine up. $500 stainless propellers usually hold up well and are hard to damage. Two of the three blades look fine so I take off without giving the incident much thought.

Out of the dense fog bank I can travel faster. There is a slight vibration to the engine that smooths out at high RPM. Sometimes dragging a stick for a short distance will do this and with the amount of drift, I assume this the cause of the vibration. Speed has not changed in relation to what is expected from this throttle setting, so I move right along with my next check point up ahead. I'm pleased with the boat performance. The engine just purrs, the hull a faint hiss of star dust on the surface of the time space continuum. It is annoying to have to stand up and steer over the top of the windshield due to rain and keep a good eye out for drift logs as I come out from behind an island into the main channel.

Suddenly the boat bangs, slows and almost stops. I know right away I hit something big! The engine lower unit has snagged a huge log. I assume this log is heavy, and floating just below the surface where I could not see it. This is a strain on both the engine and boat. I have to power trim the engine up out of the water to clear the log. A visual inspection does not show any crack in the lower unit. The spinning propeller never hit the log. The bolts holding the engine to the transom are a little loose, but the encounter with the tree did not tear the transom off, nor bend or damage it that I can see.

I feel lucky I had braces welded on the back, knowing I have an over amped engine with power beyond what the factory transom would safely hold. Surviving a thirty mile an hour crash is asking a lot. Having a ton of weight come to a halt from that speed down to almost zero in two seconds is a lot of force! *"Dang!"* I mutter. This is not a good thing to have happen! I am on my way again, and it's just another incident of many in the life I live.

If I had been going five miles an hour faster, say thirty five, I think the transom would have been damaged enough to sink the boat, or a following wave would have capsized me. So while some aspects of my trip are not safe, like the boat design, other aspects I am more in control of; I create a safety net.

There is a stop in Manley, six miles out of the way up the hot slough. Tropical ferns grow along the edge; vegetation is lush in this reddish mineral water. A lot of memories here, lot of years in and out of Manley. Karen is from here, with property still along the slough. I stop to see Frin, his wife Eggs, and their son, Way. If I needed fuel I could get it here. *At seven dollars a gallon!* I look forward to a burger at the roadhouse. Maybe visit with locals I know.

"Miles, the water is out at the lodge, no drinking water for a few days."

"Of all the luck ,Mr. Way! Dang! Oh well. Hey, tell your dad I could use some more Nowitna agates if he collects them!" These agates sell well when sliced and polished. They are not commercially available, so those who want them depend on trappers and fisherman like Frin.

ON THE WAY TO FAIRBANKS, I stop to visit my German friend Helm. He still keeps his summer home in the U.S. I show him pictures on my laptop computer of my adventures. He has some experience as a tree topper, maybe mountain climbing. He tells me what I should have done, what I did wrong, and what I need.

"Miles, a special ice axe might cost you a few hundred dollars, but can save your life! Special light pulleys, You might have to spend a few thousand dollars!"

What he says makes sense, but to my mind, simply does not work. I laugh, "When I went to Beaver Sports to buy what I already have, they did not even want to sell it to me! They told me I did not know enough. They did not want to be liable for my death!" I explain to Helm I am not even sure I know how to put the harness on correctly. How do I explain the best equipment would be a waste of time on me. My competitors and friends in the old days, like Crazy Lawsen, went off the top with nothing but a length of five dollars' worth of poly rope! No one had a harness. No one had ice axes or pulleys. I'm way far ahead of them.

I am reminded of those who ask me about my profession of trapping, hunting, or working with rocks. I advise them, "No use getting the best equipment to learn with. Get something used, something basic and just fool with it until you understand. Your learner car should not be a Rolls. Buy a $50 simple rock tumbler to learn about rocks. Eventually you may want the thousands of dollars of cool stuff available as I now have. Maybe a decade from now after you fully understand the $50 tumbler. The time will arrive you understand the limitations of the tumbler you have, and why you need something better.

I do not feel I need multi thousands of dollars' worth of climbing equipment. Just knowing how to use a rope would be a great help! Help? I could not figure out how the rope goes in my pretzel. I ended up wrapping the line around the carabiner, hanging on, and using that friction resistance to let me down. It worked. For now I only need to go down or up twenty feet. The rest is climbable with toe holds.

"Helm, I used parachute cord. I was told that is impossible and absolutely should never be attempted. I did and it worked. If I fell it would only be twenty feet into soft mud. My weight held, the grabber took me up. My biggest concern was not knowing how to get back down. The grabber would not let go." Not entirely true. I did have problems! But I tried enough I was convinced eventually it could be done… with practice.

A winch I understand. The cable goes in, the cable goes out. The remote has only two buttons to remember - up and down.

"Helm, what about a simple come along hooked to my harness?" A hand crank come along.

Helm says, " No. Use a rope winch." I only know of one that uses big rope. I never heard of, or saw one for sale, small enough for parachute cord or thin climbing rope. Helm thinks most climbers would use a pulley system instead. "Use a double pulley you carry with you, very light!"

Yes, maybe, I understand pulleys from pulling my boat and moose, and know the line easily tangles. I often have to undo the load to un-twist the ropes. How do I do that when I am hanging off a cliff and am caught in tangled jammed rope that will neither go up or down? I can install special thin strong aircraft cable in a crank come along that gives me three times the distance to pull as normal cable. I might be able to crank a full 100 feet. As I get older, I run into a lot of things that need equipment to help move heavy loads. I can design a portable winch that pulls my boat out of the water, or the canoe, a moose, or a log. Maybe even me up a cliff.

Helm is not as creative, and thinks it is best to follow expert advice. Helm did not think I should be experimenting with bullet reloading data on my own."Miles, the experts have worked it all out and know more than you do. You cannot think you are smarter!"

On the bullet issue I said, "Not smarter, Helm. But the expert reloading data is based on certain priorities, like safety." I wanted a bear stopper load for my 357 pistol. It did not have to be safe, accurate, or have the load work in a variety of guns. It needs only to work in my gun in an emergency, and deliver maximum stopping power.

"I came up with something you cannot buy off the shelf and there is no data to support it. It produces the energy of a forty-four magnum in a 357." Yes, I could just buy a forty-four magnum pistol! For $700. That I might find handy twice a year. When 99 percent of my pistol needs are met with the 357. Anyhow, it is fun, rewarding, and what I like to do. Fool with stuff. Hardly anything I do is 'normal'. I had loads worked out for all my firearms; some for special needs, not worth reminding Helm of.

My methods of climbing do not have to be the safest standards. The first priority is not 'safe'. The first priority is $30,000. It is understood there will be risks, even high risks. There is not likely to be data for this. I listen to expert ideas and see if they can be adapted. I may hear about a piece of equipment I did not know existed. Helm had told me about foot ascenders that I did buy. He is exasperated with me because I ask for advice and an opinion, and do not take it! Yet I do take some of it, and consider the rest. *But which of us is coming home with tusks off a cliff?* Helm may feel I am being like Foil. In my view, the difference between Foil and I is, I am accomplishing what I wish, on my terms, bothering no one, expecting nothing from anyone. Therefore no one else's business. My mistakes do not cost anyone else. Foil is expecting favors that cost others, not paying his way, nor getting the results he wants.

MONDAY, July 18, 2016

"We should go to the park and visit Glitter, Honey. Collect the money he said he'd have, and check on our goods, deliver some new material." One of the issues concerning consignment art and raw material sales is that I have to be responsible as the owner. I deal with repairs, prices, theft, loss, and making sure I get paid. Even though I should see more money when items sell, there are costs I do not have when I sell outright. Like travel costs going to collect money and check up on my goods! We enjoy the drive to the park though, and have other friends to stop and see. Last season was the first time doing business with Glitter.

He's a tall redhead dressed like a cowboy, only he always wears his polar bear fur vest. He's cocky and stands out in a crowd. Glitter has offered legal advice saying he has had legal training. I have no reason to disbelieve him, but do not take it to heart because he says so. He has big dreams, with big talk. I'm not as bothered by this as most people. I do not mind people who know how to dream big, and do something about it. Glitter could be the stereotype of people who mess up my life. Someone who does not have many friends.

This deal last season only went marginally well. Mostly talk of future potential. "I'm still setting up here, Miles, wait until the two shops are done up how I want, and I have it figured out better!" Glitter had to take off a month early to get his wife into the U.S. She is from Bali or somewhere in the Philippines. His long time side kick was in charge. There was no last paycheck for me as promised at the end because Glitter is gone.

"We'll send it in the mail about Christmas, Miles."

No check. It's now spring.

"Miles, my partner sold me out! Kept all the money! Now I owe everyone, and it's a mess! Be patient, you are the first on my list to pay back!" Later, "Well, the records are a mess. I'm not sure now what is owed!"

Iris and I have our copy. The records seem clear and uncomplicated. We are only looking at one month. Glitter has the same copy, it's $1,500. Not a lot of money, relative to the consignment goods he has not sold yet.

'Well Miles, he misspriced a few things, wrote things down twice, I need to review it." Glitter pays some amount each time he sees us; no receipt, not telling what this amount is covering. Sloppy bookwork to say the least. But nice to get cash. Just hard to know where we are financially. *This must be why people like paperwork? I understand, however it is all about intent. Lack of paperwork to deliberately scam people is not what I believe in.* I assume at some point, Glitter and I will get it all straight. I notice he keeps meticulous records for himself, and is not lacking for knowing what is coming in and out. I assume as I observe, he is simply busy, and when he is not, he will clear this up.

We have lunch with Glitter, and maybe have worked it out. He at least agrees with my concerns. Acknowledges they are legitimate. Glitter, with great ceremony, pays the lunch bill.

"Business between us will be slower this year, Glitter, until we settle up for last season!" I give him a few things, sell a few things, but not like it could be. I'm doing about as well at the Nenana Culture Center with a lot less inventory tied up over there. I had been more optimistic about Glitter Gulch when my work first showed up there.

---

MORE WORK TO do on the boat. Running in the drift during my last trip was hard on the lower unit. When I pull the oil plug for a routine check, water runs out with the oil. There is still oil, but it is black. It should be clear and grime free. I assume the bent propeller caused vibration that wobbled the shaft and gears. This affected the oil seal around the shaft and let water in. Maybe there is excessive wear on bearings and gears. It's like putting a thousand miles of wear on the lower unit. I will drain all the oil and maybe flush it, then replace the oil with new. I can ask Reed if the seal is easy to replace, and do that as routine maintenance. I'm guessing the leak will stop once the propeller is balanced. My concern is more that river silt got in the oil. Silt is powdered rock, extremely abrasive. Another concern is if the prop shaft got bent a little. If so, I need a $1,000 fix job. I change the spark plugs while fixing up the engine is on my mind. I made some money using the boat. I'll put some of the profits back into it.

The good news is, if I had not done a routine check and left the lower unit with water and silt, I would for sure have had an engine problem. Possibly stuck on the river between villages paying thousands of dollars for my mistake. That used to be called adventures that I do not have anymore because I take care of stuff.

After a week in the shop, the fossil bones are drying enough to look at. I put the word out among a few friends. They all want the best at low wholesale prices. I'd like some immediate cash to pay for my trip, so I offer the mammoth tooth and jaw for $800. I tell Glitter, "I could take it to Tucson and get as much as $2,000! But that is six months away. I have shipping costs, have to dry it out and stabilize it. It may not sell as high as I hope. So I sell it now at a good price. I can use the money to fix my boat." I'd feel differently if I was not the source, and I had to buy this material. I have three eager buyers, but Glitter comes by first and says he will pay cash. He later hands me a check, and considers that cash. A lot of people see no difference. Cash discount is not a check discount. He knows it and I notice.

I forget I have a seven foot spear with an eight inch hand knapped point - genuine sinew wrapped. It would pass for an antique, except it does not look old enough.

"Obviously some modern day reenactment type person made this, Miles, but still, an amazing job!" I get $700 for it.

"Where did you get it, Miles, it would help it sell to have a story."

"Oh, you know me by now! Not good with names, places, dates, paperwork. I'm not really sure. I get a lot of things that come and go." *You'd think I'd remember.* "Oh yeah. Someone in Tanana found it... an old Athabascan. I traded it for something or other. I forget. Just say, Old time Tanana spear, off the mighty Yukon River!'"

"Whatever!" I had another buyer ready to give me $800 in cash who is mad I sold it before he could pay for $100 less. I'm simply not going to show my good stuff to Glitter anymore. His loss more than mine.

---

THE BIG TUSK is in broken pieces. I hold them together and think I can fit it all back together as a display tusk. With this in mind, I begin the process of basic structural fitting and getting it to hold. Cosmetics comes later. As a single piece it measures out to nine feet. I have not weighed it yet, but it will weigh less when dry, so I guesstimate it at 150 pounds. If sold by weight, it is about fifteen grand wholesale at the show. Maybe not because no matter how well it gets restored, it is a damaged item. I get an idea that I explain to potential wholesale buyers.

"I might be able to make a display, not just on a simple stand. You know, like how trophy animals have a setting around them?" I envision a fake cliff, a creek at the bottom, rocks, mud, ferns, and hidden lights. Duplicate how I found it, and how it is in the photo of finding it. Few who sell tusks were the one who found it. In fact, no one I can think of can say they found and dug up the tusk they retail at shows like Tucson. I can use this to my advantage. Write up the story, include this with the pictures. Frame it all, and have it on a stand as part of the display. In this way, I do not have to do a total restoration and make the tusk all pristine shiny and perfect. I could be different, and offer the tusk 'as it was found' which in its way, is more exciting than a tusk taken out of context as a stand-alone shiny object with no provenance.

I'm unsure if I can handle the logistics of getting the tusk to the show. This requires I build a special crate, then get it to a private carrier which would be harder with a display. In the past, the buyer came and picked it up and that was my deal, let them take care of packing, shipping, retailing. I had two partial tusks sent down to the fossil show ages ago which could go on the airline as excess baggage. At that time, regular baggage on the plane was four suitcases each with a 70 pound weight limit. Now reduced to one free carry on, period. Check in limited to fifty pounds.

One local who restores tusks, told me he sold a restored one in a stand for $35,000. Geez. Twice what I get when offering it 'as is'. *Maybe I can get higher value if I create the display?* Create a centerpiece for the show to get interest and traffic into my selling room? I can play with this idea. In this way as well, I could sell both tusks in one display! The smaller female tusk is about thirty pounds, not as big a deal, but toss some bones in and make it look like the find area.

Sometimes my ideas are over thought out. I may not have the ability to pull it off. I have never made a display before. I think it would cost thousands to have one done by a pro. I recall Tusk telling me he paid $3,000 got a custom stand once. Out of my league. I do not fit that role of high end seller.

Helm tells me in his German accent, "Boot this is one thing vee likes about you... dot we tink is admirable, Miles! You take a da risk and do tings! Here you can lose thousands of dollars, but you are not afraid!" Yet he is often says to me, "Let an expert do this, Miles, spend the money!" This is what he would do. If I fail, "You should have listened! You need to learn to listen, Miles!" If I succeed, "Good for you!"

Someone I know says, "This is not going to work, I know it. I'll try, but you know me, bad luck follows me. It is what it is!" Sure enough, a high percent of the time this person gets what they speak. It's an interesting process to watch. The reply is, "I tried!"

I'm not sure what my response is expected to be. I hardly ever try things - I'm not going to. I'm going to do; though the outcome may not be as expected! I repeat how my art turns out, like casting. "As much as a third is not sellable and gets re-melted!" Said with a happy grin.

With a common shocked, "You admit that? That's awful!"

"Why is that awful? Awful to who? Not me!" *It's all fun. I learn from mistakes, it's exciting. I still make profits I am happy with, because when my experiments work, it's off the chart! Worth way more than the loss of the toss outs!* Much of my thinking may be the result of not working well with others, not having much experience.

Helm understands, but many do not. I noticed tourists at the last farmers market. At my table they look at my art and books and walk away - I blew it. No sale. Not interested in the least in talking to me. Not interested in the source. Indiana Jones holds no interest for them. Wearing alligator shoes, but not interested in the one who wrestled the alligator. No wanna meet Crocodile Dundee. They walk over and spend their money with Hornet. She is all dressed up. She buys the material and even art from me. Customers willing to pay twice as much, from someone who does not have dirt on them. *Interesting.* I smile at the ways of the world.

I would care if I was making less money than the other vendors. In truth I usually make more. I'm interrupted.

"Prescriptions arrived we were waiting for, Miles." Iris is back from the post office thinking it is the medicine refill we were waiting on. She opens the package from the VA.

"Oh great, it's the old dosage and not correct; the doctor just doubled the dosage and the pharmacy wasn't told!" I can take two doses instead of one. We want to call and change it to the correct amount to save an extra $100.

Iris calls. Nothing but recordings; no possibility of talking to anyone. I hear her scream in the phone.

I've toyed with ways to give it back tenfold. Do they ever need to reach me? Indeed! Maybe a government agent, lawyer or doctor made a down payment on a piece of my jewelry for a Christmas present for their wife? Press one! Press two! Here is how to reach the cat, here is how to leave a message I will never respond to, have fun going in circles and getting no place!" ☠ "Get in trouble with the wife, have your inferior gift from China we never agreed to by next Easter!" Hear the recording again, "Customers are important to us, please hold! Thank you for waiting!" "See how far you get trying to sue me. Discover who cares. No one!"

But I try not to do that. Try hard. Because I believe pay backs are not an effective way to change people in a positive way. It only perpetuates the anger and damage already going on. One reason I think this? "Iris, I think of my own life and what changed me the most. It is not memories of the stick. It is memories of those who forgave me when I did not deserve it. Remember that contract I wrote out for the bank?"

"The one you never sent, Miles?"

"True, but it was fun to write, putting it down helps me understand more clearly what the issue is." The bank sent a notice that our contract and agreement has been changed to suit the bank's needs better. The reason stated, was because the bank needs more profit.

I learned in school, and believed, a contract, by definition is a signed agreement between two parties. They both have to agree, or it is not called a contract. Otherwise, it is an ultimatum. Now I understand one of the parties can change the rules and still call it a contract. I wrote to the bank — "Now that you have shown me what a contract is, here is our new contract which will go into effect if I do not hear from you." *Good luck getting hold of me, har, har.*

I changed the contract because I'd like more money. I'm struggling. I have bills to pay. I'd like more profit. Just as the bank says. "Due to unforeseen regrettable circumstances…" I add to that after thinking. "I hope this does not cause any inconvenience, thank you for being understanding!" 😀 :)

Similar letters could be sent to the electric company, the water people - everyone who simply makes up new contracts and sends them out informing me there is a policy change that I never signed or agreed to. Saying 'No' is not an option because they know they have a monopoly on a necessary service needed for modern survival. What are you going to do? Change electric companies? Yeah right. Even change banks? They are all in Ka-hoots.

My unconscious and I are not as great a friends as we used to be in the wilds. Civilization unravels my unconscious because it is the old part of the brain that does not do the reasoning, nor has subtle emotions. It only knows survival stuff. It knows how to roar and charge, run, or provide raw data to the rest of me. Advertisers know how to reach the unconscious mind with repetition and subliminal messages. To the extent civilized man tries hard not to listen to the unconscious.

My blood pressure slowly creeps up over time. *Well Dah!* My unconscious keeps turning the screws. But there really is no amiable solution. Guessing I am going to die a decade or two before I need to in the name of keeping peace. I'll put my hand out and politely, quietly, keep taking increased drug dosages. The doctors smile and nod at their increased profits. In the news almost daily are stories of how we mistreat our elders. How we fleece them and leave them to die. Nothing they can do about it.

Iris reminds me my three days in the hospital cost the VA over 35 grand. I wonder what people do who have no coverage, were not in the military or can't afford insurance that cost half their wages. *Try to count my blessings.* I have more than one doctor friend. The high prices are more complicated than the doctors being greedy and corrupt! A shocker was a dentist friend who told me that by law, all his equipment has to be replaced every five years whether it needs it or not. All the computers; all the high end equipment worth thousands of dollars. It cost him a couple hundred grand every five years. Insurance is also outrageous. There are a variety of levels of permits. He too, misses the days of the traveling doctor who made home visits. You fed him, he thanked you, gave him what you could. No dealing with machines, unable to reach a human concerning your medication. My doctor friends tell me this vision is why people like him want to be doctors in the first place! To be needed, save people, be respected.

My dentist friend said, "I looked over all the possible career choices, and decided on being a dentist as the best way to support my love of free time and travel." The biggest bang for the investment. He had the brains to do it; be anything he wanted. In his free time, he rides his motorcycle with a backpack down the least traveled wilderness roads. He sees a kindred spirit in me which might be why we are friends. He and his wife are kind and caring. He volunteers, runs a scout wilderness camp for kids. He takes poor, or handicapped children out mountain climbing, skiing and such. He may only have to work a couple days a week.

I change the subject, but it is still about spending money.

"Remember those plastic totes I got for the boat for $25 each?" Iris had questioned spending so much money on simple plastic storage containers. They are only two ft square x six inches tall. But they are waterproof, and sturdier then five dollar totes with the usual snap on lids. I have had trouble with items I store up under the front deck of the boat getting wet when I keep thinking it is not going to be possible to get wet. I have ruined a lot of supplies getting them soaked.

We have a lot of rain. I pulled the boat out of the river and have it out front on a trailer with the truck. However, when hooked to the truck, the front is low enough that rain builds up in the bow and everything that is not waterproof gets wet. In the locked front deck, the two new $25 totes have dry goods. Everything in the old tote is sitting in water.

"So Iris, I am simply investing profits back into my business. Along with other necessities, I got these storage totes. Should be good forever if I do not lose them."

Seymour, my survey boss taught me a good lesson. I hated to spend money I did not have to. I would show up for survey jobs with my camping gear and goods in a used cheap three dollar military canvas duffle bag. A lot of times my gear would get wet; sometimes ruined. Or, I'd be sleeping in a wet sleeping bag.

For years Seymour commented, "You know, Miles, if you invested in a good waterproof bag, you'd save money on the loss of water damage."

I'd only grumble about a cold day in Hell before…

Seymour bought me a top of the line LL Bean waterproof bag as a gift.

"Cost $80, Miles." I grumbled at such a cost. No way was I spending that kind of money just to carry and store stuff! Thirty years later I still have and use that bag. Many times it was nice to have everything dry when I knew my old methods would have had me in a wet camp. Once the bag fell in the river. The bag floated! Everything in it was still dry.

A CONVERSATION ACROSS TOWN:
"Hornet, that boat trip cost more than expected! You should chip in more for the costs."
"Not what we agreed to Foil! I thought you were big on sticking to agreements? We agreed I supply the food, you supply the gas."
"Yeah! But you supplied $20 for food, while it cost me $500 for gas!"
Hornet shrugs with a smile, "That's your problem."
Foil adds, "Well it wouldn't have cost so much except…" Hornet only smiles because she has lived with Crazy and been on the river enough to know it always costs, and the unexpected almost always comes up. But that's Foil's problem if he can't figure it out. The wonderful world of not adaptable binding contracts.
They still did not accomplish what they wanted. Foil suspects, but cannot prove, Hornet knew his costs would be a lot more than hers. He feels set up! Foil still has to buy a new propeller, and his tent and sleeping bag need repairs. *"But once I take care of that, it will not happen again, the costs will get down to normal! Once I deal with Miles and get him out of the way!"* He will have to come up with a better plan. Getting Miles back in jail would help.

BACK AT THE RANCH:
"What are you thinking, Miles?"
I smile, "Just long ago lessons, from the days when I was king of the forest!" Seymour and I used to call ourselves specialists; an elite team. Doing jobs few would or could do. "I got dropped off in the wilds by helicopter, picked up in a week at a different location. No communication. Bear stories, all kinds of adventures." I pause, "I was the first man on the ground—to scout it out— before building the state power intertie. Blazed the trees where the line would go. Seymour and I would come back and survey that blazed trail. That is the job I hurt my back on, the last serious survey job I had." Long ago days. Iris and I are chatting while we read the paper. This is our together time…set aside every day.

**News Miner July 19, 2016**
"Agency proposes to limit some hunting practices in dispute with state."

I read further...

"This isn't a warm and fuzzy question about bears and wolves. It's a question about the state of Alaska's previously granted authority to manage wildlife within its borders. And about a Federal Agency's determination to erode that authority."

The article reviews the agreement between the state and Federal Government and the history of the eroding state powers, being taken by the Federal Government. I explain to Iris, "This disagreement has a big affect on residents, especially subsistence—off grid people who depend on the land that is in dispute." Subsistence is defined differently between the state and Federal Government. There is state land, and federal land, with no lines in the woods. "No fence, no marker, no good way to know for sure which side of a constantly changing arbitrary line that we are responsible for knowing!" I think there are GPS machines and maps to buy that give an exact location... if you have such an expensive machine... and keep it updated. Alaska maps are not the most up to date and GPS maps are over $100. Cell phones are not the rage yet and no service here.

There are situations where a moose gets shot on one side of the river on state land... and the animal jumps in the river and ends up on the other side... another jurisdiction with different laws. Suddenly the wounded moose is illegal to harvest. Or, how would you prove it, if caught, "Really officer! I shot it on the other side!" Yeah right! The burden of proof is on us. I know scared people who left the wounded moose on the other side and walked away. Went and killed another moose and let the first one die in the woods. This is actually common. It is hard to see this going on, and call it good management—saving the moose population. There are other better examples of problems, but I focus on this one at the moment.

"Miles, did you get your subsistence fishing permit yet?" The application is still sitting on the table. I resent permits, but know I better fill it out.

"Last year I filled this out. I was not sure if I was going to fish or not, but figured if I was, I better have a permit!" I bought a net and later find out the law changed again. I can't legally use this new $300 net on King salmon. I decide I might fish at the old homestead on the Kantishna and bring legal silver salmon home. No, I need another permit for that area. I can't fish in two districts.

Part of a subsistence life is being adaptable, mobile, having traditional areas for various activities, like remote fish camp. Districts and lines on a map are not going to accommodate traplines, fish camps, berry patches, moose hunting areas, etc.

I fish a little, but give up. It's cheaper to buy fish or trade for it, than catch it myself. Fishing is no longer a fun experience. I got asked which days I fished, how much of what kind of fish did I catch? How many people did I feed? Were any fish fed to dogs? Failure to answer in a timely fashion can result in criminal charges.

Why did I fish on only two days? How is it I caught a Shee fish? What did I do with it? I do not feel these questions are the governments business, when I think the government wants to stop subsistence, and may use our answers as leverage to do so.

"How would you like to fill out paperwork to go shop for carrots? How would you like to be asked who you fed them to, asked why you got only a pound, please explain yourself! By the way we notice you also got celery! Why?" Again, if there was trust I would not care. But the government is not my friend. The government does not divulge all its activities to me or the public, and is not held liable if information it withheld, or purposely inaccurate. The government does not trust the public with information either. I look over the rules connected with my fishing permit...

### Subsistence Permit and laws
"Dear Subsistence Fisherman.
State of Alaska regulations require fisherman obtain a permit prior to fishing." Reading further, "When summer chum salmon become abundant, subsistence and commercial opportunities will initially be provided with selective gear... that requires the immediate and careful release of all Chinook salmon alive."

I see on the reverse side...

"For additional information: Subsistence fishing schedule call..."

I have to use five inch mesh net if I wish to net non-salmon species on off days. I need another $500 seven inch mesh net. Having a permit is ok unless the permit is late, denied, filled out wrong, or lost. As has happened. I'm not great at following rules and instructions. For a time period I need to watch the net and notice every time a fish is caught, then go out and release any king salmon alive. I am told by biologists, gill netting fish kills them, period. They rarely survive damaged gills. If so, turning them loose does not save them. I'm required to call to hear the days schedule, which can change. This requires I have long distance on my phone. I do not. If I fish on the Kantishna river, there is only cell reception if I climb fifty feet up a tree. I have not succeeded in this yet.

I know many subsistence people here in the village do not have any phone at all. Some had their power cut off, not affording the electric bill. By definition, subsistence means poor. We are encouraged to stay that way. People generally like to live up to what is expected of them.

I get out the separate hunting regulations book, 135 pages long. I look for changes and general rules to remember.

"Beginning July 2016 all bow hunters in any hunt must have successfully completed a department approved bow hunter certification course."

Bow hunting is how I am expected to hunt for my moose. I am in Nenana, courses are offered in Fairbanks... for a fee. I do not drive. When Iris and I do drive, it is $30 in gas per trip. I wonder as well, *suppose I do not pass the course?* I do not know what the course entails. My bow or arrows may not be approved. Quite likely, as I am someone who can be counted on to design and build my own arrows, and tweak the bow in some way. This is all stressful. The rules are designed for sport hunters who trophy hunt, have money, and live in Fairbanks, are in the majority.

I read further...

"This publication is an interpretive summary of Alaska Hunting Regulations and contain rules which affect most hunters, which have been simplified for your convenience. It is not a legal document."

We are supposed to contact someone else, consult with others, (a lawyer?) if we wish to know the law.

"This is a summary of changes adopted, not a comprehensive list of all the detailed changes. It is your responsibility to read the hunting regulations carefully for complete information."

There is no information on where these regulations might be found. They are not on the internet in a Google search.

As Moonshine told me long ago, "It is impossible to be true subsistence and not be an outlaw." All I want is maybe ten fish. It's easier to simply snaffle them, and keep quiet about it. Or more, it is less risky than trying to be legal, getting on the radar, and getting it wrong by mistake. Also, once I am on a list, if laws change, here is a list of people to go pay a visit to who depend on the land who can be blackmailed, extorted. As one subsistence person put it, "It's like registering as a Jew before the extermination." While it is on my mind, I make a copy of subsistence laws for the Mayor.

The Mayor is interested in some of the subsistence laws he was not aware of. The part where some amount of money is legal as part of a definition of barter and trade, even some amount of cash. For example selling the fish or meat received, to get cash to pay city bills.

"After all, some of our poor citizens with no money get 1,500 pounds of moose meat, or salmon they have no freezer for. They owe back taxes and..."

The mayor grins. "Leave it to Miles!" I look puzzled. "Headlines in the paper" 'Nenana Residents Pay Bills with Fish.'" We can see it now.

The issue would be this, quoting the law, "Of a small, noncommercial nature." Yes, I agree, not for profit, just to break even." It is brought up that if people paid bills with wild game, the wild game could get wiped out in short order! I have given this considerable thought, so say,

"Maybe not. There are laws already in place. There is a limit on game per household. It might only stop the serious wanton waste now going on." Reality is, those poor who do not have freezers and get a moose, or 2,000 legal salmon, very often waste it. Much gets drug around town by loose dogs in the street, eaten by ravens, left in the backyard to rot. How much can they eat before it goes bad? Few do the work to make jerky, or put up in cans. Many cannot afford canning jars, or are not functional enough to understand or follow the directions.

If such people had it together enough to own a pressure cooker, have storage space, had the knowledge, and were willing to do the work, they would most likely have a job and not be poor. Where getting a moose is often just luck and not work, taking care of it is work. Let others who know how, and are willing to do the taking care of, participate in the harvest.

We see it, we know this is the truth.

"Instead of letting half the game go to waste, pass it along to those who own freezers, who want and need this high quality food source!" The bottom line being, the city is going broke, failing, falling apart, struggling for its very existence, while there is illegal wanton waste going on all around us. "How much fish gets wasted? For sure, thousands upon thousands of pounds. Less moose waste, but moose is worth twice as much."

My concept goes even further. The city does add on value, and smokes fish, makes coveted fish strips and valuable canned smoked fish. "The city takes salmon off your water bill at $2.00 a pound—the commercial value. Crazy Lawsen pays his back taxes by smoking fish for the city as an employee of the city, or subcontractor. He's already set up for it and has customers. Jarred kipper salmon sells for $20 a pint. Not for profit. But to pay off the $200,000 the city owes the school." Native children would love traditionally handled salmon; the school could acquire it through the city as school lunch. I argue, "No more fish or moose are missing from the environment. We are simply dealing with what is already taken, in an efficient way! This converts to serious trade.

"Let's say 10,000 fish at five pounds each average times $50 a pound smoked. It's more than I have fingers and toes to count on."

"No wonder the government wants to shut you up, Miles!"

Yes, bad news for the government if we could take care of ourselves. We'd be getting uppity. Forget our place, take jobs away from government people. The last thing the government needs is, "No thanks, we do not need you." *Take your grant and welfare and shove it.*

"I recall a case in court, and a ruling for the sale of subsistence fish. There was $500 cash per household that is acceptable and written into law." Some of this is of interest when selling or buying craft items using animal parts like feathers for the culture center gift shop. There are even more exemptions when it comes to culture centers, scientific, and or displays in a museum." We have a section in the museum marked 'not for sale' where we educate the public on our culture. It is legal to purchase and display a lot of items that cannot be sold. "We could purchase or trade with the poor - up to $500 per household. Items like mounted ducks, white ivory, bear claws and hides." The display will bring in more tourists who utilize the gift shop in the other room." He reads over what I handed him. I have a brainstorm, "If we do subsistence life and trades, people like myself could trade art for city bills. You take my art wholesale, sell it at retail, and make a profit at the cultural center. Why hand out precious cash you do not have to local artists?" That statement is ignored.

"Miles, a big issue hinges on 'tradition'. It would be important to keep records and prove a tradition." This is do-able.

"Notice there is nowhere the tradition has to be Native. There are white man traditions connected to the land, hunting, and use of resources for subsistence." Supposedly these subsistence laws cannot discriminate by race. The mayor agrees this is how it reads.

"Why, then, does the government act differently. Miles?"

Yes, I wonder as well.

"If a case went to court… I agree with you, Miles, the government does not want a subsistence case to come to court, they want subsistence to go away."

In prison I met those who believed they were railroaded, so fought. Instead of the community service they might have got in a plea agreement, they got felony serious years of jail time. Hearing some of the stories, I had trouble imagining guilty people willing to risk their life savings, and years in prison to fight. When all they had to do was some community service. The message is, do not try to fight the government.

One of the newly hired city workers wants to live subsistence and admires me for having succeeded. "Miles, is there a way to live this lifestyle successfully… without ending up a felon, without problems, getting left alone? I mean… I just want a peaceful life, and do not mind being poor. It seems to me you have been high profile and hostile."

I have to think on this. I do in fact know some subsistence people who are left alone. I have talked about it before.

"The biggest issue you face is how to make a living. Where is your money going to come from, even the little bit you need for survival basics."

"Miles, what if I need no money at all and plan to make it all myself, live simple, no money."

I answer based on my own experience. "What would you do if you get a jury duty notice? You need transportation, a phone to respond, a way to get mail. How will you deal with your fish permit, a class required to use your bow? I feel there are some basics that have to be bought... no way around it. Stove pipe, matches, canning jars, pressure cooker, salt to preserve fish or other foods, mosquito poison."

"Well, Miles, mosquito poison and stuff like that you can make from the land!"

True. However you simply run out of time to do all that needs to be done.

"Miles, what about our ancestors?"

"Something about our ancestors you need to know. They did not live alone as hermits. There was no equivalent of the Mountain Man. It takes at least five to survive. In this way there is a division of labor and talents that are shared. Even today in remote villages, there are no residents who go off and live without village help. Everyone has to come in for supplies. Supplies will not be free. Something is expected in return, if not money then trades and obligations that amount to the same thing. Skills are shared, someone hunts while someone else tends the fire and cooks, while another tends to the sled dogs."

"THIS IS art you are bringing to the gift shop?" The mayor looks into the box I brought. I nod yes, and pull out a knife I made.

"Dang, this is nice, Miles! Mammoth handle?" I tell him it is scrap ivory I stabilize, that would otherwise be junk. In fact I bought it as junk from Tusk in the $15 a pound garbage pile. Cheap ivory is usually $50 a pound. Decent, not even great is $100 a pound. Primo might be $300.

Johnson smiles and repeats my often said line, "The garbage man!"

I nod proudly, "Yes. Well, look, material that would normally get thrown away! I have resurrected it into something not only useable, but beautiful!"

"And worth something!"

I laugh, "Yes, of course. Paying the bills with recycled material." I ran some numbers when I stabilized this last batch of scrap mammoth ivory.

"Lot of profit, Miles! How much time in it?"

"It is hard to count hours as I run the vacuum and let it set a while. The vacuum pump is on a timer now, so it turns off and on every four hours. Not much of my time involved. I have to sand and stabilize cracks with Star Bond glue and then hand sand and polish each piece." *It is not a good idea to let people know exactly how long the art takes to make. Part of the magic dream and story is how long it took, not how short.* I haven't figured out how to tumble, and sand ivory, as you can't get it wet. My present tumbler method requires a liquid. I tried oil, but the tumble grit gets hard to clean. I'm not trying hard to tumble ivory yet. Just fooling now and then. It does not take that much time to hand sand. I think it takes years of practice to get fast at it, to know when each step is done.

The mayor and I talk often of ways to make money, wages, retirement, what things are worth and such related topics.

The mayor has some inside information, "Bear was buying used heavy equipment for the village from the feds. Surplus they had mothballed and needed to get rid of; sold cheap."

We think the bidders had to be working for a government entity to be eligible. "My information is probably at least 3rd hand, but as I hear it…"

Bear bids on more than what the community needs, gets a dozen bulldozers and graders, and the village gets the half it needs. Bear invested his own money on half a dozen more, then sells them privately on the side to miners who need heavy equipment. No one but the government complained. The village had no problem. The village did not fire him, paid his salary while in prison, and he has his job when he gets out.

It's equipment the government no longer had a use for, got the money they wanted for it and did not get ripped off. Who did Bear steal anything from? But yes, I'm sure if Bear asked first, the government would have said, "No, of course you can't buy a few extra for some side deals! What right do you have to make a profit?"

Ha! Yes I understand people in privileged positions with an in cannot use insider information or privilege to personally profit. It hurts someone someplace when insider trading is done: buying land, stocks, using privileged information, stuff like that. But it happens all the time. *Look at Martha Stewart, the famous cook* I'm suspecting more than a lot, as in most of the time. That may not make it right. It's one reason I stopped working for the fire fighters and BLM.[1] I'm guessing our mayor makes more than his forty-five grand with his connections and information. It's only a guess. In the same way I guess, so does the President of the United States. Who arrests him? So my little shell game is, "Now you see it now you do not," making garbage disappear and return as a new coveted material.

"Miles, do you tell everyone it is dyed stabilized rejuvenated enhanced material?" We both know I do not… well not usually.

"If I am asked, or I feel it matters to the customer." Otherwise I just call it mammoth ivory. I do not say natural. Is that misleading? I shrug my shoulders.

Ninety five percent of all turquoise is sold the same as my art. Is the gallery you buy turquoise jewelry from lying to you when simply calling it turquoise?" There is no information, often no knowledge the turquoise is reconstituted turquoise dust and glue. We still call it turquoise, and it is. But worse, is other cheap soft stone dyed and hardened to look like turquoise. Few consider this a big deal. Natural high quality is not even mined anymore, and does not exist, there is no source, except as old jewelry, and is worth ten times the price most want, or can pay. There are few experts on the subject.

I explain to the mayor, "The mammoth story is similar." The really high end colors—solid material that knocks your socks off—is gem grade. This is not even one pound per thousand pounds I find. For the most part, when people see it, they do not even want to look at anything else. I have no trouble selling gem grade. The problem is getting diamond prices. The bigger problem is, how to unload the other thousand pounds that is not naturally as flashy. I'm sorry reality is as it is! I'd love it if lots of people felt as I do, that it's ivory and exciting, even if it has a crack, even if it's tan and not blue. It's awesome because of what it is. I wish more people understood that is what you can afford. It's a lot of work to treat the ivory to enhance it. I'd love to get ok prices for B, C, D grade. Other sellers—most—toss those grades out in the garbage. It's hard to sell, and who wants to bother? So here I am being the garbage man.

Much like with food when I was young. Our family tossed out three times the food we ate.

"Oh it's day old left overs." I hated to see waste, so I would ask to eat that so it would not get wasted.

I'd hear, "Oh give it to, Miles, he'll eat it!"

Obviously I can't take care of everyone else's garbage without becoming a blimp! It was difficult enough to take care of my own garbage. I was just viewed as a nut. I do not ask for gem grade prices. That in my mind is why I am not ripping anyone off.

"Miles, some people may not know the difference, but they want the difference to be there and an expert to tell them, grade it for them and they will pay!"

I agree, and that's why you go to a reputable expert. You would not go to the muddy miner who found the uncut diamond to buy a AAA diamond expecting him to grade it for you. I'm the expert at finding. I can grade better than the citizen on the street. I'm not the guy in the lab with a microscope.

Anyhow, the mayor is impressed. "Opal rivets, you notice?"

"How could I not, the color leaps out, really nice! How did you do that?" I get asked that a lot.

"I set up a hollow tube, then fill it with black Star Bond glue. I put big pieces of lab opal chip in - left overs from my large carvings." He nods. No one else has created a look like this. Very unique to me. It looks like black opal was cut and inlaid. I have 'Alaska' acid etched into the steel along with a log cabin and some trees.

"For only $350 you, too, can own it!" He laughs, "Looks like a $1,000 knife, Miles."

"But at a price more people can afford."

I've asked how competitors can sell knives similar to mine for $50! It's not just made overseas. - its deceptive. The opal is broken shells, or paint, or plastic. Engraving is a clear label stuck over the steel that looks like engraving but peels off the first time you use it. Glitter, with no substance. Worth what you pay. Such a knife may be fine for someone showing it off and cutting string.

I'm working on selling my books, looking around on the internet for groups of customers to advertise to. I find 'outdoor and survival skills' as a topic at the end of the trail in my searches. There are blogs, followers, tweeters, and twits. I have investigated this outlet, and wonder if this is a good market? After all, the key word on my writing is survival. The key word of the group is 'survivalist'. A subheading I end up at is a web site for patriots. This is advertised as information and books on skills to survive. I glance around, but am not excited to be connected with this group. The overall feeling is anger, even hate. Get even. Fight the government, revolution, be ready to kill people.

"Why you need a knife" has as its first reason - "To defend yourself, learn how to kill people." I frown. I heard the phrase when young, 'A man with a knife is master of a thousand tasks.' I'm not sure even one of those tasks had to do with killing anyone. I have carried a knife since I was thirteen. If I do not have it with me for an hour, I reach for it, go to use it, and it is not there, and missed. I only once used it as a defense tool… well twice. Out of 100,000 uses I put a knife to - cut string, rip open boxes, cans, use as screwdriver, carve pointy sticks to poke for things, make garden stakes, tent poles, trim electrical wire, cut most of my food….

I see a headline and link to an article to get information about what this group of people stands for and wants. Headlines about a family in Missouri that the government is stopping from growing a garden on their own property! The article wants me to be outraged, alarmed, ready to do something about it. "The government wants us to stop being self-reliant, does not want us to feed ourselves!" Oh my! Alarming! *Hmmm, well let me read for myself…*

The impression I get is this family is doing something no one else in the area is doing that is bothering the neighborhood. Talk of being forced to clean up the yard, with the family only partly complying, saying they have rights. So sounding like not just a pretty flower filled garden, but a mud hole, manure piles, flies, tools, junk, when everyone else has a nice green lawn. Maybe an attitude of, "This is my land, screw you, I will do what I want!" So an ordinance was passed to ban gardens within thirty feet of the road. Which only applies to this one family. I see both sides. My view would be, if you believe in gardens this strongly, choose to live in a neighborhood where gardens are the rage. Be someplace where you get along with your neighbors! Or far enough in the suburbs that what you do does not bother your neighbor. An astute person would pick the right neighborhood in the first place. A lot of neighborhoods, like Nenana, have a community garden, or one could be started, as a compromise.

I understand in some cases it is the neighborhood that changes over many decades. In such a case I think 'old timers' should be grandfathered in. They arrived ages ago in good faith. At the time all was well. Your great grandpa was a farmer. The area was farms. Over 100 years of time, the famers sold out, the community changed, and now, 'your place' is the only farm, and now just a truck farm. Let it be!" I hope most communities have 'grandfather rights.' I do not feel much like getting involved fighting the government, so this family can have the right to piss off their neighbors. I believe in 'judged by your peers'. Survival to me is partly getting along with your neighbors. When I wanted to be a Mountain Man, I did not choose a suburb and start snaffling the deer and turkey in my neighbor's yard. I moved to the wilderness where this is a common practice, and there are like-minded people.

---

AT THE FARMERS MARKET. My neighbor Atomic has a brother visiting from out of state who reads and bought my, 'Going Wild'.

"Different environment, Miles, but I think we share the same thinking and skills."

I'm not sure then what it is he does for a living. Very vague. Works alone. Makes good money in the innovative adventure industry. If he is much like his brother, my neighbor, he is smart, way out there off the chart. Atomic, I think I mentioned before, has one of only five atomic clocks in the world, set up in his garage. His time is as accurate as Greenwich. Time is measured by the vibration of an atom.

"Miles, I love your book. It explains a lot of things in clear language, that takes the reader right where you are."

Atomic is supposedly very rich. He commented he could buy all of Nenana if he wished. He bought a fire truck. It appears to me, just to drive it around because it is cool. He keeps it parked in his driveway. The brother has a life of danger and adventure of some kind that he isn't going to talk about. I'm one of the best writers he has ever read. So, how do I reach more people like this? What group is he in? How would I advertise to his tribe? How would I convince Einstein he'd love my book? No, there would be no need of convincing. More like, how do I get my book into Einstein's hands?

# RETIRING WILD

Barge on Tanana. No more roads to the outside world, 1,000 miles to the ocean.

# CHAPTER EIGHT

## A BIG ART DEAL, VISIT FRIENDS, TRY TO COLLECT MONEY OWED

Yin and Yang meet in the circle of life, and are closer to each other. Opposites look alike and actually touch to close the circle. Maybe as the cop and the criminal are closer to each other than either one is to the rest of the world. Maybe I can smile that one reader screams I am nothing but another cheap poacher! To others I am a shape shifter. I can be a lot of things. *What would you like me to be? Seek and ye shall find.* It's not so much I mind being called a poacher as in the first western ever written, 'The Virginian' - "Smile when you call me that." Considered the beginning of the one liners. The days before the Clint Eastwood one liners. 😀 :)

I notice Atomic has invested in security cameras as I have, that watch the perimeter of his property. He's not very trusting of the human race. The subject comes back to survival, ending times, the fate of the human race. Certainly 'poaching' defined as 'snaffling the Kings deer' *or surplus bulldozers,* could be a useful skill in ending times. I'm not sure believing in the King, his rules, and being a snitch would save your bacon. We shall all see. In other countries that had difficult economic and political times, we can study what worked. Likewise past history.

"Those who do not understand history are destined to repeat it."

The patriots I am reading about on the web site, are saying we should examine Venezuela right now. I assume something is going on there. Is the government collapsing? More than Greece? Is there a country doing well right now? In a study, I read what was done in Germany during the war. Something about how the Gypsies survive. I know, studies are dubious, *Who did this study? What was the purpose?* Still, I hear something and decide if it fits my own data base or not.

Money did not help as much as we'd like to think. Precious stones and art worked better as things to hang on to and to try to escape with. Gold was not even the best. Money and gold are generic. If someone steals it, it is hard to identify it as yours later. If you have a long way to travel on a hard road, gems and art are less likely to get stolen. Pawn shops and low life road bandits want gold, money, and car radios. Those who got to their destination with gems to sell, did well. German money was not worth much. Gems could be disguised as cheap glass costume jewelry worn by the poor openly. Art and gems are hard for thieves to put a value on. Hard for the average illiterate thief to convert gems and fine art to food, drugs, guns, things they want. I have seen this in my own world. Someone stole my hypodermic glue needles and left behind gem stones. Being adaptable and blending in helped gypsies survive, from all I have read.

Those who trusted their leaders and followed the laws were not as likely to be among the survivors. In the past, those who simply left, did well. Some took off into the wilderness and waited it out. If you liked the life of growing gardens, and hunting for food, you could avoid the insanity. Some Jews lived in the wilds of Poland, I think, and lived rustic somewhat happy lives. Many hated the very idea of this, and went off the deep end. As during our own countries economic collapse of the 30s. Farmers did ok, not those who stored their wealth in the bank, or trusted the stock market. Those who waited for their leader to guide them through the hard times had a long wait. Those who had useful skills desired in any culture, could offer those skills. Jewish watchmakers who made it to New York from Germany for example. There were cooks, those who sewed, shoemakers, skills that deal with the basics are in demand in hard times. Skills you can move with work well. Skills you can begin with using simple inexpensive tools helps. I look at those who went west to find furs, or seek gold, who instead, set up practice in remote towns.

I recall the history of the gold rush days in Alaska. Few made it rich finding gold. More did well plying a trade in the outfitting of budding communities like Skagway. Blacksmiths did well in the gold rush towns, as well as those who knew how to keep chickens alive in a harsh environment. Eggs sold for a dollar each in a day when a dollar was a week's rent.

I personally think knife makers would do very well anywhere. A meal cannot even be prepared without a knife. It is the knife that alters material goods to shape them into useful things we need. A knife can carve other tools. Eating utensils, spears, tent poles, trap deadfall triggers, clothing, useful in any setting, be it the wilds, farm, or the city. A knife opens stuck doors, trims pants and shirts that are too long, opens cans of food, cuts wires, hot wires cars, and as I say, makes you master of a thousand tasks. Knives can be traded to anyone, for food, shelter, services.

In a rabbit cycle low point, we cannot predict what rabbit will make it, and which will die. So goes my rabbit cycle theory of civilization. There are far too many factors not in the rabbit's control. It is just as often simply luck. None can be smug. I thought I had all my bases covered a decade ago. Cabins, a lifetime supply of food, tools, ammo, fish nets, everything I might possibly need in ending times. All burned up in a forest fire. Twelve cabins and all the linking trails. Twenty years of work. I forgot about forest fires. Health issues can grab any of us at any time.

"Josh?"

"Mr. Mawtin!"

"Hi Josh, figured it was you." Josh has been in the habit of calling every day at about the same time. He tells me he went upriver to paint the logs of his remote cabin and is tired now, ready for bed at 8:00 pm. "Did you have your gram crackers and milk, Mr. Josh?" Josh laughs. "I remember your phone number now, Josh, 5530, both our ages." He laughs again. "Guess which one is older!" I can tell he needs cheering up. If I do not go out of my way to force a joke on Josh, the conversation normally turns negative.

"I'm tired too. I took a small sixteen foot lake boat someone gave me over to this lake I know about. I hauled it with the big boat." I explain how it was not the easy project I thought it would be. I had to jump out of the boat and drag the two boats through the weeds. "Josh, the water was chest deep! The weeds were too thick to run the engine, or pole through."

Josh tells me of his own struggle. "I got to the spot on the river where the cabin is without trouble, but the overland uphill trip was hard because of that last forest fire." Josh told me on his last trip he spent two days cutting burned fallen trees out of the way. A big task for someone seventy-two who had heart surgery.

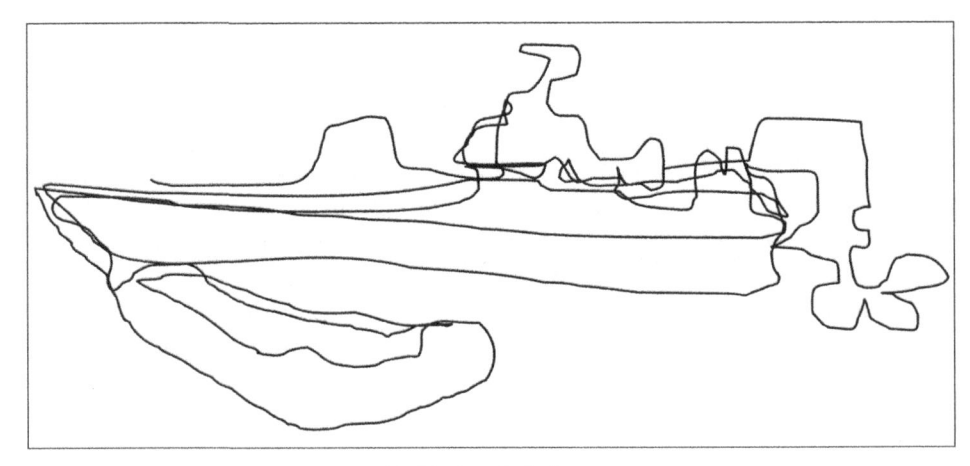

*Upside-down in the boat.*

"Well Josh, I had trouble getting back into the boat in chest deep water! I bet it took me half an hour! I tried climbing up the lower unit of the engine; that did not work!" I had a stool in the boat, took that and sank it for something to stand on, but the stool sank in the bottom mud and was little help. Josh chuckles picturing that. "I finally got in head first and could not get up. So there I was head first in the bilges of my boat waving my feet in the air!"

"You always entertain me, Mr. Mawtin!" Yes, that is my job, to entertain Mr. Josh. Now my dinner is cold and I do not understand what has gone on in the program we watch in the evening. Iris does not know either.

"I couldn't hear over you two talking!"

Yes. Well. *So what is more important, TV or humans?* Josh has few others to talk to. He has burned his bridges. Not even allowed in the senior center, according to him.

Valerie says, "Nonsense! He just can't come in and insult everyone as he does!" Ah yes, the incident where Josh told Bonnie she could not give the blessing to the food because she is in a cult, a Bahia. Josh tells me he has to tell the truth! He cannot be part of anything Satanic! Well, then he can live an isolated life.

I try to visit. Josh still will not get a door bell that works.

"Miles, just come to the window and bang on it like I told you before! Use a stick!" There is no handy stick provided nearby. The window is over my head, and under it, is his wife's flower-bed. I'm not going to trample her flowers or risk breaking the window with a stick banging on it. I assume they do not really want visitors. I mentioned a wireless remote door bell I saw for ten dollars. No wires to mess with. Fasten it to the door frame, take the ringer part and set it anyplace in the house you want. Last time I tried to open the entry door to access the main door to knock, the door knob came off in my hand. How hard or expensive is a door knob to fix? So I do not come around to visit, and hear no one else does. Well Jade does. He tells me he carries a cell phone, and calls Josh on the phone to tell him he is on the doorstep, please open the door. Josh wants the latest news, so I tell him about the cultural center.

LATER I RUN into the mayor and bring up the subject of the Cultural Center since it is on my mind.

"Miles! We need to give people a chance! There are plenty of city employees with felony records!" He's looking for employees.

Oh. I did not know that. Or never gave it much thought. Mrs. Assembly has gotten in contact asking me when I can be available for a meeting. There are issues with the cultural center gift shop. The city council decided a committee should investigate, come up with a policy to follow. I volunteered to be on this committee because I am an artist who understands shop policy and was the buyer and on the committee that started the cultural center when it was built. My friend Josh remembers, and knows the issues his own son had with his art.

We are talking about Dim's wife being hired as the manager, and how that is going. One issue is, the mayor and city clerk do not fully trust her, so check up on her and do not let her exercise the authority of her job. It's said she's been sober and appears to me to be serious about turning her life around.

"As long as she is not scamming us all" - her past reputation was being a scam artist. "This is the issue, Josh, the city thinks she is back to her scamming ways!"

There are two smart moves for the community running the gift shop. We specialize in local handmade crafts. I'm the one who insisted on this policy when we got started. We also offer the lowest prices in the state. The mayor started this. Most important, the tourists want our product, and the Nenana experience. The bus drivers noticed, and like us now. When I visit the center, I see the lines. Fifteen to twenty people lined up at the register waving money, eager to hand it to the clerk. As fast as the transaction can take place, non- stop for two straight hours. While the nine busses unload and offload. The stops are longer now. The bus drivers now say, "Buy it here, if you like it. There will be no better price anywhere we stop."

---

I RUN into my friend Crafty in Fairbanks and ask how business is. I have not seen him in over a year! He used to supply the Nenana Culture Center with his standard local crafts, as seen in most Alaska gift shops. Alley, a first long time cultural center manager used to buy from Crafty.

"Crafty, I dropped a few of my books off at your store, did you get them?" He got them, but does not push my art like he used to. Our friendship is cooler than it once was. Crafty has a long list of former friends who are no longer welcome in his shop or his life. I'm not in that group, but we may not be spiritually on the same page. This has to do with his Scientology beliefs. I accept he is very rigid when it comes to these beliefs. I forgive this, but am careful around him.

I respect Crafty for what he has accomplished. He still looks like Grizzly Adams, and still has the picture of himself and Grizzly Adams taken together. However, Crafty does not support the outdoor lifestyle. He moves his million dollars' worth of local crafts a year supplying the entire interior of Alaska with his tourist items. He is a mover and a shaker. That is amazing.

Crafty and I have lunch together. "So Miles, how is the Ex, your son, and family?"

Mitch was born in Crafty's store with the assistance of a midwife that cost $800 that I traded art for. The Ex and I lived in a room upstairs during this time we worked for him, to help pay to send her and my son 'home' to California where she wanted to be. Mr. Glitter is having another child and tells me minimum thirty grand to have a child today. That's if everything goes well. *Is it true? Dang.*

"Oh sorry, Crafty. Not a good subject I suppose." I pause. "Have not heard from the Ex in a lot of years. I hear from Mitch every couple of months. Good weather blue sky. Hard to know how he is. I only assume fine. That's all he seems to want to hear from me so that's all I give, 'everything is fine.' No serious details." I do not mention anyone else. I have not heard from my sister in a lot of years. She may not be alive anymore. My Mom is ok, I see her in Tucson when I go." Crafty is no closer to his family. Never married, no kids, no family he is close to. An alcoholic sister who tries to mooch off him.

Crafty just nods, but has a look that tells me, "Well what you tell me confirms this is what happens when you choose the life you have! Too bad you are not a scientologist like me."

Yes, I understand. That has its merits all right. I can see what the religion has done for Crafty. It's been his life raft. When he needs answers, he can look it up in the only book that matters. Crafty still spends eighty percent of all he earns on Scientology. He goes to their facility in Florida, 'Flag' he calls it. He has taken every class offered so he can help teach, takes lessons, and he is now 'clear.' Gets told he is way up at level five at the top. Not at all like the number two people. People who score a two are not worth helping or talking to.

That works for him, so I say, "Good for him." I'm glad for Crafty, so I smile back. He has a tribe, a group of people who forgive his sins and help him feel good about himself. That must be nice to have. I do not envy him though, because I think he has been brainwashed. He is one of their puppets. He'd disown me if I said that though.

Crafty pays for my lunch, seems genuinely glad to see me, as I enjoy seeing him and getting caught up on his world. I know all his helpers who come and go, and came back again, over the past forty years.

I learned from Crafty about all the Alaska artists, as well as about antique beads, various collectables we both deal in. Crafty is a gifted seller who could sell snow to Eskimos. I prefer not to talk people into buying something they can't afford, do not really want, and do not need. Money is not my God. I do not think I am any happier then Crafty, nor have more friends, or even respect. We simply have a different approach to life.

Iris says he still dresses like a slob. I only shrug my shoulders because so do I... or so I did... and still would... if Iris did not keep me in line. His shop still looks like a bomb went off inside a vacuum cleaner bag. I still marvel at customers sitting on the floor pawing through endless trays of beads looking for bargains. The nonstop flow at the cash register all day long of his regular customers. most of who know me. I could probably tell you the price of every single item in his shop. I'm comfortable here. I still have a key to the back door. I can come and let myself in and have a place to stay if I ever get stuck in town. That says a lot, that all these years, he trusts me with the key to a million dollars' worth of inventory.

So, while we have our differences, we are bonded by tribe, and by time. If I was down and out, Crafty would give me a job right now, or lend me money, no questions asked. He'd give me a place to stay and enough food to live on, keep all the rest I earned. Which is more than the average person would do for a fellow human. Just tell someone you are broke, need work, a place to stay till you recover, and see how far you get.

"Miles, you need some mammoth teeth?"

Maybe. I have a customer asking." *Never tell Crafty you need something, it will cost twice as much.* I glance at what he offers as if I have only half a bored interest. This is not how I deal with all sellers! Mostly Crafty. "Hmm, north sea stuff huh?" I mention that because, obviously the quality will be down, thus price way down, compared to local material. Crafty is sorry I noticed, and sighs. He admits no one wants it. He's trying to dump it. Has to get rid of it cheap to make room. He did not have to tell me. I know all that in a glance. I have a secret. I specialize in junk like this. I will never be known as having number one material. But I can put these rotten falling apart huge teeth in a vacuum with resin and stabilize them. I have dyes, and other methods. I can take a $20 tooth and make it worth $400. No one else is bothering. *Do I keep repeating that?* I may have to slice it and sell slices to knife makers, or to jewelers. I have a market in Tucson Crafty does not have.

I think about my son, Mitch, after getting asked how he is. He seems to be doing all right and happy. I'm told he is well liked. Gets along with others. He is at the University of Hawaii. Mostly as a student. Now in some kind of professorship program I do not understand. In a program where he ends up with a PhD. If I understood him right, he gets paid something to teach now. I only think so, as he has no other job he speaks of and tells me things he does that cost money. He's twenty-seven years old.

I review his last email to me.

**Hey Dad,**

Hope everything is going well with you. I have read in the news that there are some crazy things going on with the Iditarod. People intentionally attacking the dogs and dog mushers with snowmobiles.

Things are going pretty well with me. This Friday I will be presenting a paper I wrote on whether or not the first person pronoun actually refers to anything at a philosophy of language conference that is being held at my university. I am actually co-editing the anthology of last year's conference with another of the PhD students in my department.

After the conference I have spring break, which I have been looking forward to. I always end up getting a little burned out halfway through the semester.

There has not really been much else going on with me lately. More of the usual stuff in regards to course work. Slowly chugging away at getting all of my course work done and thinking about what I will write my dissertation on.

One interesting thing that we have been dealing with this semester in my department is that one of our Chinese philosophy professors is retiring. We pulled a lot of strings and have managed to get permission to hire a replacement. The pool of candidates was reduced down to two and they both had a week where they came to meet with the faculty and students, give colloquium, and guest lecture classes. It was really interesting to see the process I will have to go through when I am applying for jobs.

Well I hope you are having a nice daylight saving day. Talk to you later. **Love, Mitch**

This is a nice communication to get, informative, chatty. Mitch sounds happy. When I respond later and try a follow up, ask about someone specific, or how a certain paper went, or project, I do not get a follow up, but some new subject. That is ok. I mean I am hearing from him and we get along well enough. I see his newest letter,

**Hey Dad,**

I hope you are having a nice Father's Day. I sent you a gift through Amazon and it says that it was delivered to your mail box. It is the first season of a TV show about a sheriff in Wyoming. I was surprised just how good the show ended up being. It is based on a book series that I have also been reading.

How are things going with you? Everything is going alright with me. I am finally on summer break. Immediately after school got out I was the main person helping to run a weeklong conference of over 200 academics. It was a lot of work and 12 hour days. Luckily there were no huge problems. The biggest problem was that we did not have wi-fi access in the building where all the panels were being held. So all the presenters that had emailed their presentations to themselves or had assumed that they would have internet access were coming to me bitching about it. But I managed to get every problem solved. And then got really drunk at the closing party on the last day at this fancy members only club.

I do not really have much going on for the summer. I will be visiting my mom in San Diego for a week in July. It will be my first time leaving Hawaii since I moved here. After that I am coming back to Hawaii. There is another summer institute I will be taking part in at the end of the summer. That should be exciting. The only other thing I will be doing is working on editing all the papers for the book I am co-editing. So not a very busy summer, but hopefully busy enough. I have enough money saved to make it through the summer without needing to get a job. Which is a huge relief.

Do you have any exciting plans for the summer? Well, I hope everything is going well with you. talk to you later.

**Love, Mitch**

I reply...

**Hello Mitch!**
Thanks for remembering me for Father's day! Looks like a lot of nice movies to see. The package arrived pretty early but I waited to write. Had an ok Father's day. Went out in the boat with friends. The Yukon 800 boat race is on. Today I go out again maybe for a picnic, and we watch the racers go by from a good location. Hopefully it does not rain! Looks pretty cloudy out. When we made plans it was record hot and clear out.

I found an empty shack on the river a nice travel distance out... Yes garden doing ok. I feel behind on getting it all how I want. Gardening takes a lot of time. Just had one of our farmers markets that we have every weekend. It only went ok. It's held in the same building you might recall, the civic center where we set up to do shows like tripod days.

My book seven is with the computer guy getting set up for Amazon. So working on book eight! Sales getting a tad better. I sell a lot on my own at shows. I am not sure if I said, we are going to be setting up at the fossil show again in Tucson. I'm off probation finally and can sell fossils like mammoth ivory but not animal parts. So my wood, rocks, fossils and books. Not sure I told you Iris used her retirement money and got a place for us in Tucson. My Mom is getting older and cranky and does not like our stuff in her house. So it is good to have a place of our own and we can visit her without interrupting her routine or depending on her for a place to stay or store stuff. We have friends down here now so it is like another home to us. Nice change from the dark and cold in Nenana middle of winter.

So wondering how life is for you. Sounds like the big picture looks great. Off on a career and making progress to where it will not be a struggle and even enjoyable. Maybe summers off, ending up teaching and getting some travel in. I keep thinking, much like the life my father had? I was worried about your student loans! But believe if you stay in the system the loans never have to be repaid. I guess you make up for it or pay it off by teaching and working for the university. You mentioned maybe travel this summer, was it Japan? I know it is interesting to get overseas and experience how others live and have a break! So no class over the summer and start the routine up again when school year starts? You must be making some regular friends you have known a while now? I assume! Always good to have a few friends in life.

I think back over the years and there are a few friends still around I have known forty years. Will is still around, but has MS now so can barely walk. You may remember him? We are all seniors now! ha! We all talk of our latest ailments and trips to the doctor and medicines we take! Yuck! Never get old! Ha! My survey boss Seymour is doing ok. Retired, has his plane, gets out to a remote wilderness postcard perfect cabin to recreational trap in winter. He is in the trappers magazine sometimes. I get a call or email telling me the latest. I see him sometimes when I boat past Manley and spend a night there.

Well thanks again for remembering me and have a good summer! **Love Dad.**

I tell myself once again, Mitch is a lot like his grandfather, following the same path. I did much the same. I have a lot in common with my grandfather, more than my father! My grandpa was in the working class. He moved furniture, and was a boxing manager. He is the one who wanted to come to Canada and help me build a remote cabin. Grandpa did not express problems with me wanting to be a trapper and subsistence person. I feel badly my son and I do not have more in common. But maybe it is not my fault? I mean… it could be in the blood? *So simply be glad and enjoy the relation for what it is.*

My grandpa's ideas may have changed in his life. I get the impression from what my father said, that grandpa had different ideas when young. He told my father, "To succeed and get anywhere in life you need an education!" I think that was a more pat phrase, and more true, for the era than it would be today. It seems odd the first thing my father wanted to be as a child, he told me later in life, was a gunsmith! I never knew him to have an interest in guns. I had the impression he stayed in school more to please his father than himself. His father paid a great sacrifice. I have early childhood memories of all of us together. Grandpa at Thanksgiving watching football, knowing all the players, and excited about the game. Dad not really following, not knowing the players. I understood that is not the life of a college professor. Grandpa had no idea what his son was all about. Dad used big words grandpa did not understand. He simply, proudly said, "My son the professor!" But the price paid for that is to not know his son or be close.

I wonder then, if later in life grandpa had regrets about telling his son getting an education is the only thing that matters. Lots of servicemen were getting out of the military after the Korean War and the last world war was not so far in the past. The economy was good. The GI bill allowed the soldiers to get an education, as my father did. Industry in the US was booming, growing, and we needed educated people to fill new manager, engineer, and scientific jobs. Machines were taking much of the labor work away.

Factory work in the 50s was assembly line slave labor. Smart knowledgeable people were needed to design and build the machines. I could picture feeling your child needs to escape a life of labor, and be someone further up the totem pole, higher up on the hog, near the eyeballs, or ears of that hog. At the same time I could imagine losing touch with your son and having some regrets. Maybe seeing a son ending up not so happy as expected. I recall grandpa as not very rich, but mostly in a good mood, cracking jokes, being happy, loving life in a way I never saw in my father.

I wonder something else now. *An interest in being a gunsmith?* Picturing slums of New York, his father a boxing manager, guessing connections with seedy people, organized crime, with related problems. Was my father on a path of joining a gang, guns, as in violence. His father saw the future of that path? Made his son get an education, keep his mind off the street? I see some of this life as being classified 'low class' while educated people have life different.

I have said to my own son, "It is not all it is cracked up to be, and not my choice, but if it is a direction you feel is your path, go for it!" Not encouraging him, but not holding him back either. The price would be us not being as close, which I find regrettable, but "Oh well." Life can be like that. I do not agree we could not have close. I was bonding and close in the beginning, wanted, expected to spend a lot of time with him, help raise him. I spent time with the two girls with Karen, so I know I am capable.

More important is what Mitch wants, and him being happy. Mitch mentioned having saved enough money to retire for the summer, but he has never mentioned ever having a steady job. I have no idea where his money comes from. Likewise, for all I know he is still a virgin. Never a mention of a girlfriend or a date.

When Mitch was a baby we were much closer. He followed me around, imitated me, walked, and talked like me, loved sled dogs and loved the outdoors. He laughed and screamed when he got put in the dog sled and we went mushing together. He sat with me selling stuff and was a good salesman! He grabbed loose balloons and brought them to my booth and sold them to other kids his age. He went around collecting free stuff from other venders. He knew how to look doe eyed and longing at something until venders laughed and handed him their product. He loved the water. We threw rocks, made bows and arrows and such. It is hard to imagine all that going away, him being embarrassed about it now. It is hard to believe all that change was his own, without influence. It's easy to believe the Ex turned him against me. I recall specific incidents. I wondered if one day Mitch would find out who he was and somewhere inside find an outdoor person, who likes working with his hands, being independent, and this is part of his blood he cannot ignore.

But no. At twenty-seven he is a man, probably on his life path and that is not likely to change. I worry he might discover, as my father did, he is not so happy on this path with no way out. I see relatives influence on him, pushing him as grandpa did dad. "The only answer is to get an education! Study! Work hard!" Mitch was put in a costly private school my father paid for. A school with no art, no sports or physical activities. Things that might have helped him be closer to me. No one said, "Hey his father is an artist maybe he should have some artistic influence?" I felt more like, "No, let's make sure they have nothing in common!"

Later, Dad's wife paid for Mitch to go to college. She chose what courses he would take. I heard Mitch, "I'm not sure why I am here, or what I am doing, just going along with the plan." That did not seem like a good reason to be in college. I saw other things going on.[1] I do not know if I should have stepped in and done anything. I felt I was not in a position where I could. My rights as a father were removed by the court.

I was a parent with minimal visitation rights. A baby sitter. My father played more of a role as the father as were the wishes of the ex. Everyone was pleased but me. Who am I? I went along with the program, as Mitch did. I wanted everyone to be happy. I have my own life to live. I find happiness elsewhere. I did not raise my son, so I do not feel like a parent. If he is happy, successful, or not, it has to do with others, not me. My ideas and ways to raise a child were not heard. I will never know if I could have been a good parent. I do not know if Mitch would have been as happy if I had helped raise him. I can only speculate. Our relationship is much like that between my father and his father. Which was not horrible. I assume grandpa had his own life. He seemed happy. But I never knew him well. He lived far away most of my life. We rarely visited, except early on, just holidays. Dad rarely mentioned him. Grandpa was an embarrassment. Much as I am. I smile at how that works. Grandpa passed away and it was two years before anyone bothered to tell me. No hard feelings. But yes, a great loss. I wish I had a son to be close to. *Mitch is still my son, I just do not know him.*

So what is this next generation doing? I do not know! But I can at least smile that I understand. *History repeats itself.* I recall when the Beatles came out and the world was coming to an end, that long hair, that crazy loud music. I suppose the generation before it was Elvis and the Buddy Holly group. Or the roaring 20s with flappers driving Flivers, emerging from prohibition. All those shocked church people. So! What has this generation come up with that sets itself apart? Whatever it is that I do not understand, it's all part of the big plan. Pierced unmentionable body parts, men becoming women, women now men. *Go forth! Explore the possibilities!* I understand! Does my son? Does the next generation? Maybe when they become seniors? Then will my son reflect back and smile, after I am gone? So no, nothing to get excited about. I have no wish to bother and embarrass my son. I smile and wave at Hawaii, from Alaska.

Sure I long for a young person to take out and pass on what I know, who loves boating, snow machines, fishing. Tons of youngsters do! Just not mine, or anyone I could mentor or be close to. They are someone else's kids. It'd be awesome to look for fossils together, do shows together. Perhaps Mitch has similar wishes, a father or father figure who is academic, that he could count gold stars with. I feel badly for that. Way back when he was a baby I wished his mother would find a guy she would admire and respect who had the same views as her and mine, who would see Mitch as a son. A husband and parent who would fit in! My son could then know what a functional family was, and have a role model to aim for.

I do not understand why she did not look for such a person when she was looking for a husband and father for her child, and turned to me. But she found my father, and suspect this was as good as it gets. She was not likely in her circle of friends to have access to dean of a college! Perhaps my father took Mitch on as his own, and was the father figure. Mitch, the son he never had. I do not know, because they did not communicate to me about such matters. It's a family filled with secrets. I do not like to talk about it. We should not control others. It is hard enough to be in control of what I do! All we can ever really count on and trust is ourselves.

I mentioned to my son that I was fixing up a remote cabin. I have not talked about it around town. I am uncertain of the status of the property, or if anyone might object. It is my get away place. Some men might understand it as 'my cave'.

I am in my canoe I keep at this cave. The water level is almost flood stage high. I can paddle many miles out into the swamps normally not available this time of year. I take advantage of this to explore, relax, take pictures. If a perfect opportunity presented itself, I might come home with dinner of some sort. Blueberries are ripe. I might be able to pick them from the canoe. There may be mushrooms, perhaps morels. Life of the hunter gatherer.

---

ONLY THE SOFT sound of the paddle splash. The gentle rock of the craft. The view is sedge grass, cattails, blue sky, Toghettele Hill. I can feel my blood pressure drop. I relax in a way that does not happen anyplace in civilization. *Well, maybe sometimes when I am working on my art.* I smile. 😀 :)

I day dream as I take a deep breath, enjoying the smells of the wild. I drift into one of my topics of interest. Trying to compare how the world used to be when man lived in canoes as I pretend now, compared to modern man. There is something about this scene that sends me back, spins me to a time 10,000 years ago. Almost as if this is in my blood, or a memory from a past life. Da ja vue. Been here, done this. We all have, only some remember. There is something comforting in the thought. I wonder once again where we, as a species went wrong.

RETIRING WILD

*While being a serious crime to misuse or sell a moose parts, this hide rots in the city street in front of the fire hall for a month, wasted.*

# CHAPTER NINE

## PATRIOTS, MY RABBIT CYCLE THEORY OF CIVILIZATION, MY SON AND THE EX

In prison I read books, looking for hints in my country of how it was founded, what the principals were, and if there was something inherently wrong in the plan itself. A design flaw. Today I go back thousands of years. If man was full of hate, or greed, and desired harm, Thor could wipe out Igor's family with rocks and sticks. Devastation would be a few dead people, and a couple of broken rocks and tree branches. There were no buttons to press that might wipe out the entire planet. Mankind had not created dangerous chemicals. Our effect was the same as the effect of any other animals at war.

At what point was there no turning back? At what point were we clearly different from any other species on the planet? One of my thoughts is that mankind is not from earth at all! We arrived as aliens sent to destroy the earth...or simply conquer it, did not and do not care because it is not our planet. That would explain a lot. I have given thought to the bible. Supposed to have been written about 2,000 years ago. The earliest recorded indepth history. If true, and it has been dutifully, correctly translated over all that time, there are clues to see.

A creator from up above. Mankind as His sheep. Perhaps then we are a crop, planted, to be harvested sometime when we are ripe. The creator is coming back for us. It is expected we will be glad, and give ourselves willingly to be taken someplace else, heaven, for dinner. If so, I wonder if we escaped, or there was a crop failure, or somehow the experiment went bonkers. As yeast taking over the sourdough pot and creating a polluted moldy goo that causes us to go dormant. Until more sugar arrives. Or as an unweed, not what the farmer expected or wanted.

The bible states that our creator was not pleased. Tried to wipe us out once already, and promising to do so again if we do not mend our evil ways. Much as the farmer deals with weeds using pesticide. A rival power has taken hold... Lucifer, who once knew the Creator, but had a falling out. A fight over the crop, we, the sheep. Imported seeds, like vetch, that strangles out everything else and takes over. Perhaps there have been harvests? Many harvests! The dark ages. A displeased God, because we dare not come to be served up as dinner willingly.

"Bringing in the sheep" We happily sing...

"Quack!" I have startled a mallard duck. The soft "SSSSSS" of the canoe parting the sedge as the canoe moves forward. I am only ten feet from the bird.

"Friend or foe! Who goes there!" The duck asks. I can tell there is a question. All speaking living things make sounds that go up in octave when there is a question, followed by tentative silence, expecting an answer. I say.

"Qvark" This is what the duck in the cartoon Hager the Horrible says.

"Hey! You aren't a duck!" I slowly smile. *Indeed*. But what am I, and where did my species come from?

As far back in time as we keep going, my species was recognizable as 'human' with human features, skills, and attributes. We can go back 50,000 years and that family found, dressed in furs could pass for 'today'. I'd find the woman attractive. The tools understandable. A bowl to eat out of, a knife to cut with, a bow and arrow. I do not think she would be a stranger to me. I have difficulty picturing a gradual transition from animal to human. I am not seeing anything that appears to be halfway between me and a gorilla. A hairy scary human. Fifty thousand years ago the female had a dress, shoes, beads in her hair, no hair on her arms or face. She did not transport herself bent over on all fours. She did not need four shoes.

Along another line of thought. For many years I gave mankind the benefit of the doubt. I was taught in school, "We are different and better than animals." I argued animals are just as destructive. They too have all the habits of 'us'. It is just that they do not have a thumb, so cannot act out on their destructive desires. Their hearts are no more pure or innocent.

The mallard does not fly, but paddles off into the thick tall grass to watch from a distance.

The older I get, the more I believe animals have all the feeling man does. I saw my cat walk by a flower, stop, back up, stick her nose into the center and take a deep breath with her eyes closed. Hold her breath, sigh, pause, and enjoy the smell.

"Life is good!" She walked away with a new bounce to her step and tail in the air. She did not see me, did not know I was watching. I cannot prove what happened. I saw, and I believe. I used to get told, "Oh you are just reading things into the situation that you want to see! Be careful of giving human attributes to animals. This is only an animal! Science proves through studies and experiments, how much superior mankind is."

This feeling that animals have things in common with us would be a contradiction to the notion mankind came from someplace else and has nothing to do with anything else on earth and has no connection. So it is just a thought that passes through every now and then, that explains why mankind treats its environment so poorly. As if we hate it. Perhaps we just do not understand our environment? If so, why not? Why would we understand our environment less than wild animals do? Are we more stupid? There are no answers to such questions. It is difficult to envision any other inhabitant of the planet treating the planet as mankind does, even if they could.

Other animals had rein over the earth at different times in history. The earth recovered from their destruction. Will earth survive mankind? I shake my head to clear it. Christianity taught me that it is not good to judge others. All I have control over is me, and what I do. Even then, only maybe!

This is a nice area. I am glad to have found an empty shack. I hadn't thought of it, but now notice there are quite a few such empty shacks and cabins along the river. As if there is less need, or less time to get out to stay in such places. Perhaps getting out is not encouraged. Most people have jobs that keep them busy. Time off is spent catching up with laundry, sleep, mowing the lawn and such. Mankind has more bills than ever, more stuff we feel we have to have. Internet, cell phones, that can eat up half a poor persons income do not allow for any 'free time'. Or that free time is used differently now. Not used to commune with nature. I only know as a fact, there are a lot of empty cabins on the river that used to be occupied.

An otter chirps. I recognize the sound, but do not see it. *Maybe Josh is right. Why don't we enjoy nature as it is given to us.* Why do we feel so compelled to change it to have grass we need to mow and trees that need trimming and care? Don't we have better things to do? What is wrong with the plants as they are? Can't we just enjoy that? There are plenty of wild edibles here. Mad Jay picked five gallons of blueberries in an hour yesterday. When I lived on the Kantishna River, I had enough rose hips for a year's worth of tea, and plenty of dry mushrooms, berries of all kinds.

I still like the garden with potatoes, turnips, beets. My canoe paddle dips in the water and scares the otter I heard. I see a smooth glide into the water, and glimpse of dark fur. No one has been here in many years I am sure, because the water level would not allow even a canoe here except at flood stage. I come back to the cabin and hide the canoe. Later in the year, or other years, when the water is too low to get the river boat parked here at the cabin, I can walk in and have access to the canoe when no one else can get in or knows there is a canoe to use.

I only use a gallon of gas in a round trip here. Something good comes of my conflict with Foil. Because I'll probably lose the Kantishna land as a getaway place, I had to search for someplace else. It has been an ongoing issue that it cost so much and takes so long to get to the Kantishna. There could be more moose there than around here. I hope I can still go there and hang out to get my yearly moose.

Where did all our moose meat go? We have shared a lot with others who never got a moose, but love this wild game. I was sure the meat would last two years! I see now we will not have enough. Moose season is around the corner. It is state fair time again. As always, rain!

Locals joke, "We know when fair time is, because it is raining." Actually it has been raining for a month. Not heavy rain, but rain drizzle and overcast every day. I'm glad I am not setting up at the fair this year. Rain means leaks, damage to goods, fewer customers. Still, there are aspects to miss.

---

"MILES, how did you move that barrel of waste oil the neighbor brought over with the fork lift?" Mad Jay told me I could not move it, or transfer it, and what an idiot I am to say I could. I did not know how yet, when I said I could handle it. The oil is too thick to siphon. I can use it in my waste oil drip heaters in the greenhouse and shop. The barrel weighs 400 pounds. I had to think on it to come up with a solution. I had to get some or most of the oil out first, just to move it. I explain.

"A barrel has two caps, a small and a large. I drilled a small hole in the small cap and soldered a truck tire air value in the cap hole. I drilled a larger hole in the large cap and install a long hose that goes to the bottom of the barrel and comes out with enough length to reach a can. I used an air compressor to put pressure on the liquid surface, which forces the thick oil through the hose." I add, "I have these caps that can now thread in and seal any other barrel to do the same thing." Mad will not admit I was clever, or acknowledge I was right, I could deal with it if I thought about it. It's the kind of problem solving learned from necessity. I have two other barrels of this thick oil available for free. I keep a gallon jug of such oil by all nine of my wood stoves. It is now easy to start a fire in a wood stove. A cup of waste oil ignites the most stubborn wood, without kindling or waiting.

---

"IT'S FAIR TIME!" I smile. Iris and I are headed in on senior day, when seniors get in free. I miss setting up and selling, but this is outweighed by things I do not miss, and various issues venders have been dealing with the past few years. There are only four hand crafters in the craft tent. There used to be thirty or more in a huge tent. The craft people all agree the fair is not treating crafters well. I repeat what I was told by management, "Why should I lease a space to a crafter when I can get much more money from a food vender?" Now I notice a lot of food vendors who are not happy. Sixty food vendors are listed, when there used to be twenty. Too much competition, not enough of a crowd. Food prices have gone up by a third this year. We spend thirty dollars for simple burgers and fries we are used to spending ten dollars on. Next time, we'll bring or own food.

"Honey, I do not mind spending money if I am having a good time. I just do not like getting ripped off."

I see a lot of people who know me, both vendors and customers. At the boat and snow machine booth I see a fund raiser poster for Sturgeon.

"Remember Iris, that guy who was in some remote area hunting and Fish and Game stopped him from going further saying his airboat was illegal." He'd been using it most of his life. They would not let him run it, not even to return home. State says it's legal, they control the waterways. Feds now saying no. He had to go back in winter to haul it out. Big battle over jurisdiction and whose laws apply in this situation. His legal fees through the courts has amounted to $800,000. So that is what it takes to fight the feds. The boat shop is putting on a raffle and fund raiser. Most of this equipment I see being sold is for weekend warriors. Nothing I could put to good use.

Iris and I enjoy the animal barn. We look at ducks, turkeys, sheep, pigs, and comment on each one for size, interesting markings and information on who raised it. The photography and art section is always fun. I like to see what the children have done. Sandy still has some pendants of mine she uses in her beadwork necklaces.

"Miles, I want to thank you again for donating so much of your art to me to help me out when my inventory got stolen!" I had forgotten about that. "You are the only one who did anything. I'll never forget that!" There are a few pieces she wants to keep for herself. I inquire how her husband is. He is getting old, with serious health issues. It's been costing them a lot of money. I have known the family a long time.September was just a toddler, she's now 27 years old, Miles!" Yes, mom used to bring the baby to the fair and I had the child crawling under my table. Good memories! I show her my latest work, but she can't afford anything right now. I want to give her one piece she likes, but tell her I can sell it for ten dollars, about a third of what I'd usually get. I have lots of inventory and I am doing ok financially, so like to help out those who deserve a helping hand.

Christine who used to set up next to me shows up. "I am checking out the fair, Miles, out to see if I want to come back, but there are only four crafters in the tent! Where is everyone?" She is the one who took up a collection for me when my homestead burned down years ago, when I was set up at the fair. Vendors chipped in and raised $500 to help me. It is nice to be able to simply enjoy the fair! I leave with good memories, but a troubled mind over the changes.

Something else is changing.

One aspect of being from a small community is that I personally know everyone on the city council and in the audience. No strangers or unknowns. The budget comes up, who we owe, how much money the city has to spend. It's a hot topic. With no direct answer. The barge line is pulling out of Nenana. This will be a loss of 100 grand a year to the community in docking taxes. It's a big hurt. After the meeting, some of us engage in a discussion on how to survive as a community. I mention, "Several of the biggest bills we have are owed to organizations not complaining." I have the view it is not so much anyone else's business, as long as those who are owed are happy. There are trades of favors, lending of equipment in lieu of hard to come by cash in a strapped community. It's not exactly breaking the rules, it's simply necessary agreements to survive.

"That's not the way to run things, Miles!" I agree. It is not how I'd run my personal life. I'd learn to live within my means.

I add, "As a community, we go to Fairbanks and enjoy the rewards they offer without paying. We dump our garbage in the Fairbanks transfer stations. We enjoy the good roads that are plowed, paved, and lit. We enjoy the library, theatre, and bingo. Then we come back to Nenana where the taxes are low." I shrug my shoulders to make my point. One of the communities reputations is as moochers.

"How do we motivate people, Miles? Get them to care about and support Nenana ?"

When I was the head of the Chamber of Commerce I pondered this and cared, struggled, worked at this very question. After thirty years I have a conclusion and answer. Summed up, "We do not." Say to the community, "You do not want to pay, fine, then you get what you pay for, minimum."

"Problem is, Miles, that is not legal. There are standards we have to legally meet. Services by law we must provide, affording it is not required!"

"That is the bottom line of the problem. Someone from outside is telling us how to run things."

The details of the rest of the conversation slip my mind. The thin brew ferments to paste as the heat of the topic is applied. A lot of people here feel someone is going to mysteriously disappear with the thought no one will recall knowing anything about it. Or not mysteriously, but found shot, apparent suicide.

I remind everyone at the table I'm on the committee to help resolve issues at the gift shop by coming up with some rules, guidelines, policies, so everyone knows their job and who is in charge of what. I'm trying to offer up something positive, legal, working within the system to fix things. We all agree it is too bad we have to write down some rules because workers can't work it out amongst themselves at the gift shop. I suppose that is where rules started in the first place.

"Though shalt not steal, and expect to keep your job," seems pretty obvious! Enforcing the rule is the issue here. Partly a cultural problem. The natives in particular do not trust the legal system. We would rather work problems out ourselves within the community and not deal with 'outsiders'. There is a tribal level of government in Nenana. With things like 'family court'. Maybe it is hard to seriously punish family?

IT IS September

The subject of 'rules' has me thinking of my own interpretation concerning my activities. Specifically fossil hunting. On my mind because I sometimes in the past made a late fossil hunting trip at the end of summer, like right after moose hunting season when the rivers are free of travelers. I have Neil's book as reference about his experiences as a fossil hunter. I think of advice and hunting techniques I might learn from the book.

I remember Neil discussing in his book not using poly rope, saying it will not hold a tusk well. He describes methods of fetching tusks I find educational and interesting.

I put some notes in my diary quoting Neil.

**Diary Sept. 2016**

Corruption: "Served some papers but never did charge us with anything, just confiscated our goods as evidence."

"The feds are asking about you Neil" Being careful who you talk to, what you say. Feds answer is always no. Never talk to them.

It's not the money the feds disagree so much about, it is the freedom and independence. Plea bargain- no choice. Confiscated tusk ends up on display in a government building.

Some of us just have our sails trimmed to another tack.

There is a lot of money to be made in the anti-fur business. PETA made half a billion dollars in 2011. Over half was used to invest in the oil business.

**Diary ends**

Hmm. I was not aware there is so much money in fighting the fur industry. Neil has similar thoughts I have come up with on my own. Some of the environmental protection groups might have more to do with making money than standing up for moral issues. I recall what I considered facts, brought up many years ago when a background check was done on the top five environmental groups. Money invested in making soap, fossil fuels, industries known to be hard on the environment. The answer when asked was, "It makes us profit so we can keep fighting." *But fighting what?* A conservation group getting its money from the oil industry seems like an oxymoron.

I'm focused on whether this is a good time to go on a mammoth hunt or not. My issues with Foil have me decide not to fossil hunt, but deal with supplies I had at the Kantishna homestead I removed to stop Foil from destroying. I have the old cabin closer to Nenana I am restoring and making livable. These Kantishna supplies can go there. Before the water level drops I'd like to boat a lot in to this place hidden up a side slough. A heavy wood stove is a priority on the list to haul in.

I have a hard time getting this cast iron stove on the bow of the boat and a harder time getting it off at the cabin site! I salvaged it from a remote cabin hanging over the river cut-bank with the stove teetering on the edge about to go in the river. Getting the stove in the boat was not hard; I just lower it into the boat. Here at the shed site, I roll it end for end; push, pull and manhandle it up to the door steps, then spend several hours trying to lift the stove up three feet to the floor level to get it into the entryway. I'd think it would be easy enough, just four more inches to go! *How hard can it be!* I cannot balance it once I get it higher than two feet off the ground. I have the stove on a handyman jack, but it falls off. I cannot quite roll it up. I am exhausted and no closer to getting the stove into the cabin.

I get an idea for the come along winch and cut a hole in the back wall of the entry way with the chain saw, then tie a rope to a tree against this back wall. I now have a solid pull point that should snatch the stove up into the entry way and in front of the door. This works, but now I have a big hole in the entry wall! Dang! But I am happy I have a stove in the cabin. Each day I make a trip and can get away from home chores for two hours to get out and work on my get-away cabin. In this way I have a secure cabin to store supplies and set up to be able to stop in winter when out cutting firewood in this area.

A door is made, a hasp and lock installed, bear proofing added, a window put in, spray foam applied. Slowly this becomes livable and secure. Many things I still buy at garage sales now have a home! Extra tents, sleeping bags, bed pads, lanterns, dry food, and all the items I get deals on out of a lifelong habit. Iris is happy I am making more room in the garage at home. She points out a news article while I am on a break.

"The Russians are accused of hacking into the voting machines, Miles!"

I am not surprised, but feel if it is admitted the Russians can hack in, then anyone else with the money and ability can as well. "Like our own politicians!" I am convinced the people did not vote Obama in, that he hacked in and gave himself votes. The Acorn scandal was just the tip of the iceberg. My views land on deaf ears.

Just an interesting admission and accusation, 'the Russians did it!' The perfect scapegoat. *But, hey, we are far from Washington, what do we know?* [1]

Foil is still around. I thought he had short times off from his work in Tucson. How can he afford to be here in Nenana so long? I can't help but notice he has made just two short trips to the homestead all summer. He has not talked to me about his plans. He has not made an effort to see me about payment he is a year behind on now. I'd think he would want to pay off and have title as soon as possible, since he is building on the shared land. I am interested in a new book about who owns any land! Well... in a way, this is the subject.

*"Windows to the Land,"* an Alaska native story by Judy Ferguson. Judy is a local writer I know through the state fair where she sells her books. She lives in Delta, where my houseboat was built.

It is interesting reading about Native history, as told through the eyes of the Natives. This is not the history I learned in school. The reality of the land history from the Natives perspective has a big effect on my life—anyone's life who wishes to occupy or use remote Alaska lands. Even land anyplace anyone thinks is 'wilderness' and 'available'. I was a white man coming to a Native dominated area, with no understanding of what I was getting into. I was interested, and learned, but it takes time. Many of the Natives who were instrumental in unifying the Natives, getting their views heard, respected, taken into consideration, are people I know, heard of, talked to, discussed some of these very issues with over the years, without knowing their role and importance. I knew them simply as 'people with a view' who I respect, and am proud to know. We often talked around camp fires in the wilds, at fish camps, on trap lines, in gathering places of subsistence people.

I was taught in school that the life of the Alaska Native before white man was like the aborigines in Australia—horrible and depraved. Natives were few in number because life was so harsh. They were lucky white man came along and saved them. The story I was told is the same for all minorities who lived awful lives in the jungle until white man came along and saved them. Any who complain are just ungrateful.

"Get a job!" The white man scornfully yells back. Yet I read, "In 1867, there were an estimated 35,000 Alaska Natives living in the territory, half of the 74,000 estimated in 1740."

"After 1880 whites in Juneau began to feel uncomfortable as a minority in a land with no civil government. They pushed for control over Alaska's resources and private land ownership. Russia had excluded Natives from having any rights when Alaska was sold to the US in 1867. Alaska was under military rule until 1884.

The first Organic Act of 1884 provided basic protection of lands for the Alaska Native, but allowed for title controversy by leaving indigenous land title for, "future legislation by congress." Section 8. This Organic Act was the basis for Alaska Native Land Claims.

SOME OF THE issues I see going on in Nenana, go all the way back to decisions made a hundred years ago. I considered myself well read on Indian history when I arrived in Alaska. After all, I had read every Indian book I could get my hands on. *All written by White people.* Even the romantic books and stories that favored the Indian, had an interesting White man slant I was not aware of. The 'Noble Savage' view of Longfellow in his Hiawatha for example. There were battles to read about, broken treaties. Taking advantage of the poor ignorant inferior simple people I could feel sorry for, and sympathize with. There were no stories of smart Indians who fought and won in a white man world on White man terms. In the beginning, it was in fact, the Indian who felt sorry for White man, who could not survive in the beginning without Indian sympathy and kindness.

"In 1886 Native employees got 12 cents a day, a noon meal of fish soup and rye bread, and three times a day a gill of vodka. The Russians took 100,000 sea otter and the same number of fur seals every year, till 1911 when there were too few to hunt."

In later years an elder recalls, "Under the Alaska Native allotment act of 1906 with its focus on the individual family, we could file for 160 acres Native allotment. But in 1971 that ended with the Alaska Native Claims Settlement (ANICSA). Land then got titled over to tribes and corporations. Individuals with quarter blood became shareholders."

SO I ENTER the scene in 1972. A pilot who first flew me to the wilds was one of the Wrights, a native (one quarter), mentioned in this book on Native Land Claims. He was instrumental in the Indian land movement. A neighbor on the Yukon my first year was at his fish camp near me. We traded meat for fish. He later became head of the multimillion dollar Native owned Doyon Corporation. I was right in the middle of the very issue at stake here. White man moving into Indian country. Could I somehow acquire their land by the very act of having been here, and claiming use? This exact subject was being discussed as I built my first cabin and began trapping on land no one yet had a legal claim to. I'm sure all this had to do with why my food was stolen and I needed to be rescued back in book one. I knew even then there were factors I was not understanding, but did not know what. No one explained it to me, maybe few knew themselves, beyond 'problems!' Anyone interested in the wilderness asks who has rights?

"How can I go build a cabin, homestead, be accepted, get off the grid?"

---

DUE TO THE impending oil pipeline, ANCSA was pushed through very quickly. We could select 69,000 federal acres abutting our original village site, or allowed three times that amount somewhere else. This exposed a hole in the federal protection to pristine areas. This contributed to the Alaska National Interest Land Conservation Act, ANILCA in 1980.

So between 1970 and 1980 I was in my prime and peak of nomadic subsistence life, living off the land where ever I happened to stop with my houseboat. So again, "Who has rights to what land? How do we protect the wilderness from being exploited? Who is exploiting? Who owns it?"

I was in fact only thirty miles downriver on the Yukon as the pipeline was being built across it. I could hear roadwork and blasting in the distance. Whose land was it? To a twenty-two year old civilized city slicker, it all looked like an untouched wild wilderness. The pipeline moved fast, before permission to cross lands was granted. I had personal knowledge from both land owners, and pipeline employees. Like a best friend Will, who drove bulldozer for the oil company. Moving quickly, the Natives did what they could to get their rights and lands protected from this invasion. At the time, I wanted to avoid the news, politics, permits, and go back to 'how it was' in the 'good old days'. But the good old days was about exploiting Indians! *But how could that be if I never saw anyone to exploit? Surely, I, personally, was not involved.*

---

"NOT ONLY WERE we not prepared for corporation life in the modern business world, but we had to rush to pick our land." By contrast, the state is still selecting some of its land fifty years after statehood. The original Native corporate heads were not necessarily the most skilled, but rather those who didn't already have a steady job, and could take the new risky position. Many of the corporations put up their land as collateral to get loans.

"We could lose it all - including culture." 1976-78 saw the first corporation bankruptcies. Senator Ted Stevens made it possible for corporations to buy debt to loosen the burden owed the IRS, cheating the tax payers, while bailing out Native corporations.

I knew Senator Ted Stevens who put in a great many years in our state politics. He relied a lot on the Indian vote. We spoke briefly about subsistence. Being powerful with a lot of money, I got the impression he knew a lot, had secrets, had the goods on people, maybe got kickbacks to get so rich and powerful. I still liked him, as he cared about our state, and stood up against the federal takeover. He died in a private plane crash. Many, including me, feel he was assassinated. Mostly because of the timing and what has transpired since. I read a book by Ferguson off and on when I have time.

FOIL IS GETTING ready to leave and stops me on the street to give me some final words before he goes. He makes a huge big deal about me taking my generator away from the homestead.

"Miles! I depended on the generator to charge my cell phone! You left me in a bind without communication! I should pound the crap out of you right here and now!"

There is a partial valid point. We agreed to share everything on the homestead. He depends on my supplies and, though he expects me to share what I bring, he does not share what he brings. I have to take many of my supplies home to fix and repair, including the generator I now cannot trust. I can say, "I plan to bring it back after it is fixed!" This could take a long time if it is not a priority, like if I am not inclined to have Foil ruin it again. Foil mentions a cell phone. I spent twenty-five years there with no communication. I do not consider either a cell phone or electric a necessity. Foil may forget he is 100 miles in the wilderness. I did not deliberately deprive him of something he feels his life depends on.

Foil put spikes going up a tree so he could climb for cell reception. A tree I am no longer agile enough to climb. Foil knows it. If he was looking out for me, taking care of me in my old age as he said he would, as part of our agreement, and he feels cell communication is a necessity, why didn't he build a ladder I could climb and share? I am not as filled with guilt as he'd like me to be.

Foil considered the removal of my generator an act of war, a reason to 'pay me back'. Kick me off the land, board up my cabin. Foil keeps his things under lock and key giving me no access.

At first I tried to understand; be in his shoes. *He is just getting started, has little money, and is in need.* He is not in a position right now to help out much, but given time, he can kick in, start replacing things, make up for wearing out my stuff. So I was patient and forgiving - as friends should be... for a while. I even felt sorry for him wanting something he probably will never have, that I do. Knowledge money can't buy, that takes decades to acquire.

"Miles, you left me in the lurch, so I cannot trust you anymore! You screwed me over, so you are no longer welcome on the property. If you step on the property I will have you arrested!" He makes a fist and leans forward. "What I feel like doing is beating you up right here and now! I should. You deserve it! But I'm a nice guy so will not." He reminds me, "You know I have a temper, so do not cross me or you will be sorry, Miles!" Followed by, "We could have been good friends. It is all your fault we are not! I could have helped you on the river as your partner. It's all because of that woman of yours, twisting your mind around. We were getting along until she stepped in!"

I am not saying anything. Just taking in what he is saying, so I understand his position, and where he is coming from. Giving him a chance to explain himself to see if it makes sense, is reasonable, and if I have been in error due to a misunderstanding. I feel he is not interested in listening to my side, nor asking. It's also true I am not good at expressing anything negative I feel.

"The Kantishna land is mine, Miles. You are no longer welcome. There is nothing in the contract about a payment plan. The contract says the land becomes mine when you die. I do not have to pay you a dime; just wait for you to die." This of course makes things quite clear, we are not friends. His view is, it is all my fault for screwing him over by removing my generator that he depended on, therefore I deserve whatever bad thing I get. However, I know he did haul in his own generator, never told me, nor wished to share it with me. It appears, he wants to use mine, so he would not wear his out. The agreement states we share all equipment, if I bring it, or he does.

Many aspects of the agreement were written loosely. It was a simple agreement between friends. There was no payment plan, to make it easier on Foil - if he had payment problems, we could work it out. The agreement is more of a partnership, where we share the land and I have use of it the rest of my life. In return for this, I share my wisdom with him. It was assumed when written, I could be a guest, visiting the land and using it, for the rest of my life, in the way Foil was my guest when he spent a month. However, as time passes I am certain once he has the deed, me as a guest on the land is not going to happen. So a delay in him getting title is not hurting me. I value the use of the land more than the money. I did not want to sell, but was willing to share. Foil talked me into things in writing more to his advantage than mine. I understood this to be a rough draft and we'd rewrite the arrangement as needed, as the situation might change. Meanwhile I hold the title free and clear.

Another loose part of the agreement is nothing specific is named in terms of what is to be shared, and what becomes Foils when the land is paid for.

Foil's answer to this is, "Miles, all your stuff is second hand garbage not worth fixing. It's throw away, use it up and get another one, type stuff!" So this is how he justifies his behavior. It makes sense in his mind.

However, it is not what he said when he needed to borrow it in the first place. He was as grateful and complimentary as can be! So I'm thinking, *If my stuff is useless junk, why is he so concerned if I remove it because I value it, need it, want it in good repair, while he does not?* If it is so useless, why not replace it with new and better, so we may both share in some nice stuff? Also, why would I share tools with someone when I value the tools, and the borrower does not? This gives credence to my view of how my chain saw was handled. Run by someone who does not value it, so runs it into the ground.

As Foil says, "Your $50 used piece of junk saw! I went and got a new good one!"

If Foil bought a new saw, and brought it to the homestead, then it is ours to share. He never told me, never shared it and removed it without telling me. I assume that means bringing and removing items we buy and do not share, is ok. Good. I am glad! I agree the 'sharing thing' is not a good idea, and not working out between us! However, it is a clear breach of our agreement. We need a new realistic agreement written up.

My original thought was expressed. Well Foil, if you are building a cabin and spending time here, there are tons of supplies here necessary for the work and lifestyle, a lifetime accumulation. I hate to see you buy all your own stuff when what you need is here! So use mine, keep it fixed up, replace it as you can over time." I assumed as a friend he'd end up buying 'us' a new canoe, generator, saw, as I would do. I'd be grateful in the future to have use of tools I did not have to buy or haul in. This was not part of any land sale. I was in fact not selling the land, but writing an agreement to share it. Or, that is how we spoke, and if the 'written' was vague and unclear, it was in legal terms not vague or unclear. Foil had a lawyer helping him without my knowledge.

It would be in Foil's best interest to honor the spoken agreement... I think. I mean... I have more to offer him, then he has to offer me. It is Foil breaking the agreement! I should be able then to get out of it under breach of contract. Glitter, the legal beagle, told me. However, little of the conversations with Foil are provable. No one but us even knows what is out on the homestead. If Foil says, "What is Miles talking about? What saw? What threats?" I took pictures of the wrecked saw, but can I even prove it is my saw, or that it was ever on the homestead, or used by Foil? I never saw him use it. He is saying my wrecked used saw did not work, so he bought his own. I'm pretty sure he built his entire cabin using my saw. He may have bought a new one that he can take good care of and not wear out. I notice he takes good care of his own equipment and is very fastidious. He likes it all new, in top working order, spare no cost, have professionals fix it if it fails. I spot his new good saw. The brand is cheaper - half the price of mine - and considered inferior to the brand I depend on. More of a toy for light brush work. Not for logs and building cabins. This confirms he does not know saws, or is not really seeking a more dependable better saw as he says.

*One thing for sure though, Foil is correct, I have drastically changed how I treat him, and not given much explanation.* It is true, I smile and nod a lot. He and others may take that to mean I agree. What it means is, "I hear you and understand." I may or... may not agree. I often wait to pass judgment rather than be quick to speak my mind, and later change it because I misunderstood, and live to regret it. I also feel there are some people it is not a good idea to explain anything to. Their objective is not to be reasonable, or try to come to a mutual understanding and work things out.

Before Foil goes, he sees me one more time, and is overheard on the senior center steps cussing me out. No one can hear exact words, but it is known he is showing a temper. It is understood I am trying to get away from him and he is cornering me. Once again telling me to keep off the land, that my cabin is boarded up. He hands me a few items later, saying this is all that is mine, so I have no more reason to trespass.

"Miles, I can get you in a lot of trouble. I know you own a gun! I will turn you in!" He hints of other knowledge he feels he has, concerning what happened to certain animal parts that mysteriously disappeared. "You do not want to make problems with me, Miles!"

I do not agree to comply, or acknowledge the land is his to control. Foil in fact has little on me. He has never seen me with a gun, "I'll need legal advice, Foil, and will get back to you." is my reply. I do not want him to be able to say in court, I agreed. I recognize the logic of an abuser. I ask myself, *How did I ever get involved?* I recognize a pattern from the past. 'This hurts me more than it does you.' 'It's for your own good.' 'If only you had not disobeyed me all would be well.' Is it old habits? Something familiar? I know I begin feeling sorry for angry abusive people. I believe I am stronger than they are. Seeking a resolution, a conquering of an earlier situation when I could not defend myself. Sucked in to someone else's game.

Foil leaves Alaska. I now need to get to the homestead and see what damage he has done. I wish to take pictures, because otherwise all I have is words that would not be evidence. He hints he might kill me. This is at least an empty threat until he returns. I can hunt for moose at the homestead and relax, get rid of some of this blood pressure. I can enter the property because I am the one with the title.

This is a place where I am usually successful hunting since I know the area well, and it is remote. If there is time, I can get further upriver to fossil hunt. I can pack up more of my belongings and move them at least temporarily to the old shack I am fixing up on Poggy Slough. I now call it, 'The Poggy cabin.' The issue of ownership crosses my mind again .

**Future Flash**
    Daily News Miner May 28th 2017
    Front page headline:

**Beaver woman tasked with looking into embezzlement sentenced for embezzling.**

Summed up, a Native woman in the village of Beaver discovered the theft of $900,000 in village corporation money. This is part of the new Indian money as a settlement with the White man. She was hired to clean up the mess, get details, find out how many shareholders there are, and 'fix things'. She gets caught up in the drama and partakes of the easy money rolling in. Doyon distributes funds to Native corporations and after that, it is up to the village corporations to determine what they're doing with the funds. Beaver has 191 enrolled members, and received $2,236,673. "Doyon, between 1987 and 2012 according to the document has never issued any dividends to its shareholders."

Part of the issue was explained by the accused...

"We were out of our realm, you know. There are still very few people in my village who understand anything about laws, corporations, businesses, administrative things. The people living there are still hunting and trapping."

I read this and understand, in relation to the book I read on the subject. In the Native Settlement agreement, each Indian gets something like a million dollars. As I read, the Native leaders did not pass the dividends of this vast amount of money on to its people. Meaning to me, 'Indians' do not have an exclusive on being, 'at one with Nature, in harmony with each other, kind, generous, fair.' No. In any race there will be greed, abuse, stealing. Yet here is an interesting part of the story, simply not knowing what to do or how. Money rolls in, it's easy to spend. It's hard to comprehend having earned it, why some have more and some have less. The amount you have has nothing to do with work or anything you did. The government simply says, "Congratulations, you now have a million dollars!" Then turns it over to tribal leaders to manage. The White Man says, " Because your ancestors were treated unfairly by us. You are still being treated unfairly. We will make it right!" The Indian says, "How can we own land? We did not make it, it's Gods land!"

Yet shrewd Natives know full well what 'ownership' is. When I was in Manley, a Beaver woman was run out for coming in and trying to find a fishing spot. She was considered an invader and would have been killed if she did not leave.

**Future flash ends.**

I quickly finish up Poggy cabin by making sure I have a canoe there to get in when water drops, spray foam insulation around the new window, fix air leaks around a door I built, and make sure the place is squirrel, bear and people proof. A wood shutter is hinged over the window so it can be closed and locked. No one can look in, and no bear can break in. I build a sheet of wood with nails sticking up as a bear proof in front of the door. The new wood stove is tested. All the stove pipe sections fit together and are the right items to fit out a wall chimney exit with an elbow fitting. I prefer a wall chimney exit. The single most important part of a structure in preserving it is the roof. A hole in the roof for a chimney is a weak spot. Ninety percent of the time when a cabin goes bonkers, it is the roof failing, and usually leaking around the chimney. A solid roof with no hole in it is my preference on an inexpensive maintenance-free structure I will be gone from a lot.

Iris makes sure we have our fishing and moose hunting tags. I can look for a moose with a bow as I did last year, but I got lucky last year. However next year I must take a class in Fairbanks to qualify.

"Iris, I saw moose tracks and places moose sleep at Poggy. I could bring you out in the evening or early morning, and while I work on the cabin, you could go out in the canoe and shoot a moose for us. I can come butcher it." This would give her a chance to see the cabin I have been talking about. I keep thinking she might want to come spend a night and enjoy the scenery. A nice change of view, something to do.

"But you do not have an outhouse yet, Miles!"

I can work on that, I have not had time yet. I use a seat on a five gallon bucket, and consider it a luxury.

Glitter calls and says he is going nuts at the Denali shop, and needs to get away for a day. Maybe we can get out together on the river, "Do some fishing Miles!"

I took him and his son once as a favor, something I can offer. Meant to be a onetime experience to see what it is like. I do not enjoy being out with others. The joy is theirs. My joy is to be alone. I'm not interested in this being a habit. I have explained this politely to my friends. Glitter ignores what I want. I'd think he'd have some friends he goes out and hangs with. Glitter is not my friend - he owes me fifty grand and I think he is never going to pay. But ok, I can offer him a break.

"Maybe moose hunting later, Miles, we could get out hunting together!"

"No Glitter, I cannot be in a boat where there is a gun I have access to. Nor can I act as a guide." He spoke of paying me.

"Glitter, I might be able to drop you off and pick you up again a day or so later, or you call if you get a moose." He could have his gun locked up in a gun case and only take it out after I leave. This would be how to be legal. Glitter shows up for his getaway boat ride. The issue we have is, this has been a very rainy summer. We have almost flooded. In fact the river level came up underground and into the rows of the garden. Within six inches of flooding the house. Some Nenana roads went underwater.

"Glitter, this is where I launched my canoe and paddled across this road, and went all the way to behind the school in the road!" I was testing out a used canoe I bought for $200 that needed repair. "Yes, Glitter, just yesterday!" This is the canoe I decided to take to Poggy. "Water is dropping fast Glitter, so hope the boat landing is ok. It was pretty muddy last I looked." I have the boat out of the river and on a trailer because of the unpredictable water the past two weeks.

The landing does not look great. We get stuck. I bend the trailer getting back out. We manage to get the boat in the river after two hours of hard work. We almost sink the boat trying. This is just one reason I do not like to partner up. On my own I would have taken one look, and decided it is not a good day to launch, and gone back home to do something else. But no, I have my buddy with me. He came all this way to go boating. Hate to tell him no. Got to give it my best shot. Now I have a bent trailer I can't fix. There goes about $800 damage.

Glitter has a great time. We can boat back into the swamps where I had already been going. Tons of ducks! We do not catch any fish, but a bull moose waits at the side of the slough and crosses behind us as we go by. The season is just a week away.

"Pretty big bull!"

I do not think so, but nod that it is certainly legal to hunt. It is late in the afternoon by the time we get back. Glitter had a wonderful time, but to me it was a wasted day. Glad I could help out, but I did not get done what I wanted to. I feel behind on the garden harvest and wrapping up the greenhouse. Glitter was going to help, but of course he needed a break, not arrive to do work. Partly my fault, as I do not work well with others. Or, I do well doing some aspect of a bigger job on my own. I'm not good at letting others know what it is I want done.

"I forgot, Miles, I'll have that paperwork ready when you come down next!" He was supposed to bring whatever paperwork he has on my art on consignment. Part of the fifty grand deal he owes for.

"I'm still messed up from that partner last year, he really screwed up the bookwork! Actually I saw some numbers the partner did wrong, and he did a wrong percent on consignment, so his numbers were too high. I will end up making it good though, I'm good to my word and you will make out ok, don't worry."

I'm not worried. I just want my money. I consider simply taking much of my consignment items back, since I own them. I point to a bucket of fossils.

"Oh that's sold, but the guy hasn't paid me yet." He hands me cash, and I am happy to get it, but not told what that cash is for, with a vague reply, "No time for the paperwork, busy, but this is part of what I owe you."

We will straighten out the details later. Always later. I view these as irregularities, causing me to want to see where we are, in writing. It does not have to be a complete list. Just total numbers we agree to. I believe I have a lot of goods with him, maybe 100 grand. I want him to acknowledge where we are, and agree on at least a ballpark to the nearest 10 grand. That is a lot of money to be unsure about. I want paperwork that reads, 'I owe Miles $100,000 for consignment goods.' At the end of each season, 'I now owe Miles $95,000,' or whatever.

I do not necessarily need any exact details. Each time he hands me cash with no comment, I deduct it from the 100 grand. At the end of each season I say. "My numbers tell me you paid a total of this amount, and it comes off the bill of such and such amount." We should have a starting number for next season and know where we are. We, in fact, began long ago with a starting set of numbers, just no total. Pages of numbers like "25 stone pendants at $35, 55 at $75" like that.

Glitter had asked me earlier how much it would take to buy me out, so he owned everything. I had said, "In the thirty grand range." Figuring a consignment long term 100 grand range, but cash today no receipt, for less than a third would be agreeable to me, and would allow him to make more profit if he had the cash to invest and buy me out now. *Money makes money.* Yeah, beats interest in the bank.

"I was thinking more like $5,000, Miles." Yeah, right! I simply say, 'no way'. Nothing more to discuss. There is not going to be any negotiating; offering five grand for 100 grand of product. Part of our issue is understandable. It's insulting, but I know the game. This is a typical offer. This is what Natives and country bumpkins, off grid types, get offered by city slicker shops and galleries. That's the deal. Not from crooks, bad people, but the normal pillars of the community, high end reputable shops. 'Rare' is to be treated better than that. One reason shop owners can give is, 'I can buy this stuff from China for a dollar each, I do not have to buy from you.' There is a counter argument, but that opinion is not coming from an organized group with clout. The answer would be, "That's nice, bye!"

Customers want 'local' and are usually willing to pay more if they are certain it's really local, and if 'provenance' is offered - a word I take from 'Antique Roadshow' on TV. I offer that authenticity… with magazine stories about me, pictures on displays of me getting the material locally in my boat, as well as me creating the art by hand in my 1916 log cabin shop.

I see something else in Glitter's behavior considered the norm. Most subsistence artists believe $100 is a lot of money, and $1,000 is supposed to put me in heaven… in La la land. I see this a lot. Dangle a thousand dollars in front of a druggie, dumpster diver, Indian, black, trapper, welfare subsistence person, minority poor of any kind, and in most cases they are happy and thankful. It is more money than they see at once… maybe in their lifetime. The assumption is, we are dealing with the mentality of the low life.

In general, artists come and go, are a dime a dozen, have to wait in line for crumbs tossed to them. There is Wess the Mess, Rooster, Mark, just off the top of my head. Knowing this is normal, I try to stack the cards more in my favor. Like products few or no one else has, so if you want it, I'm the man. You want it, it can't come from China. Try to get wolf claws originating in China. That of course was more my previous business. Rocks and wood is a harder sell. There are far more pieces of wood and rocks in the world than animal products. I'm also not the expert of wood and rocks that I was on animal parts.

Glitter and I had a reason to be vague in the beginning. I say, "Not illegal" But pause. Neither of us trusts the government. I could be Jesus H Christ selling salvation, and the government would call it a crime.

"Where's your permit, you vagrant!"

Jesus was tortured and killed by the government. There is no legal business I could possibly conduct and make good money to where the government would be satisfied with me keeping half of it. I assumed Glitter understood, and was on the same page. So he told me. Not me the sucker, and him the smart con artist who pulls something over on me. More like two business guys with a common problem who will work together so both make some money. I'm not wording it, "To put the screws to the government, cut out the 3$^{rd}$ party." More like, "So, we are calling the shots at the same time, giving our legal share of what is owed. Because otherwise, the government will steal more than its share."

There was some old outdated inventory between us.

"Before D day, to be exact." I assumed later, we could begin to get more accurate with the goings on between us. I believed he understood, and in time we'd figure it all out and make it right. After the dots could no longer be connected. As two smart business people who want a long term relation.

My take from his viewpoint when we met was, he needs a front of sorts. new location, dealing with Princess tours. They do not know him. He promised Princess, as part of the contract, "Yes, dealing in 'Alaska made' art and materials! You bet!" Princess has not seen his inventory yet. He told them… and me… it is held up in shipping. He says to me, "Overseas." Held up overseas implies customs issues. I believe, but was never told, partly he is going to pass off overseas slave labor worked goods as local hand crafted high end art. Almost every other shop is doing this. I assumed I was his cover. Mix in some truly local work, suggest there may be other local suppliers coming aboard, mix it all in a blend, a little of this, a little of that. Talk fast, shuffle the goods, and paperwork, switch tags. 'Poof' a completed shell game. Now you see it, now you do not. He needs my good name as a genuine Alaska hand crafter to hide behind. I'm the only one who turns out the volume required to make him serious money.

His new wife is from one of these places that fake imports come from. He flies over there. I assume, but do not know, he and her relatives manage custom orders done by Bali workers, and moves it into a primo tourist trap gift shop in Alaska. Five hundred percent profits. This may be why Glitter offers so little for my art. These are prices he is used to paying in third world counties, and to uncivilized Americans put in the same category.

"This is how it is Miles, what's the problem?"

Imported crafts and materials get, 'from Alaska' and 'hand made in Alaska,' 'authentic Native made' stickers. I'd rather not associate my name with this. Good luck with that, with a 95% problem. I'm not going to stay in business dealing with just 5% of the sellers. With Glitter I have a one time, move it in, move it out, single batch of goodies involved, because I'm in a bind. Also this is the game, normal, how it's done. I'm either on the train, or left behind. I do not know many honest shop owners. They can't stay in business. I can think of a few, like Taylor. Ten times my prices, to cover all the hands that are out in the process of being approved by the authorities. Their customers are people I do not understand very well. Alaska House was another honest shop I did understand, but this was long ago. Same with the Alaska land shops.

What I get out of the arrangement with Glitter is 100 grand worth of goods I can't legally claim ownership to 'for now' is made to disappear in the mix without a trace. Legally the items can belong to Iris. Or can be part of an old inventory of long ago before I had problems. Old consignment still owed for. I get paid a total, without needing to know what sold, with vague unclear paperwork, which is normal. This is inventory the Feds saw, but did not pick up in their confiscation, they left it behind. How bad do they want it? They may not... I'm just being careful.

"I do not really know what the guy owes me for, I have no itemized list, just a bunch of stuff." Or, "Actually, he owes Iris." I'd need a lawyer. Mine, Iris, Glitter, who has what. The goods, the money, moves around. How hard would the Feds try to get to the bottom of this, and why? I'm truly out of that business. There's just some old inventory floating around, getting sold, resold I am not in control of. When it's gone, it's done. legal antlers, fossils, teeth, claws, "Legal, just not for me." Or "yet." Or, "It is better not to ask." Or "Maybe legal, but I do not trust the government." I do wonder if it is legal to tell someone they can't sell such and such, a legal item for everyone else? ☠

I sigh. When I think about the issues, being angry is easy. The test of strength is in the ability to forgive. Or, is that hogwash? I follow the thought. The wonderful world of getting even, karma, and payback time. Who would be on the list? Sadly it's a long list. If I left a trail of devastation, what does that make me, right? Feared? A better planet left in my back trail? What I notice is, the other guy feels the same way and stands by the rightness of his convictions. Is this what I want? I fail to see how all this is my fault and I deserve it. However, the mysteries of the universe are beyond me. I suppose I need to simply have faith; faith everything will work out in the big picture. All my life the argument has been, "Look at them, where they are in life! Would I trade places? Are they happy? Respected?" [2]

Glitter explains. "Remember the custom build-it-yourself knife deal last year that went so well?" Sure I recall, Glitter handed me some orders to fill, and had me supply blank blades I could acid etch your name on. "Well the shop next door to me, that low life dirty &*%! brought in some cheap imports he sells as local made, undercutting my prices by half. I'm out of business now with custom knives!"

*Well what goes around comes around, huh?* I nod with appropriate concern. I'm not going to tell him how to run his business. But I see how he blends my stuff into the imports, then makes it have this over all 'authentic' look. I know he is not making all this inventory in his shop as he says. He has not got the time, nor the skill. I likewise suspect who supplies Glitter, as I meet these dealers in Tucson, and know what country produces what kind of 'look.' I can recognize China, India, Russia and Bali. I know one main guy who takes designs, and has it done in sweat shops to your specs with any signature or tag you want. It's not illegal as long as it's disclosed. Forgetting to do that is a very minor offense with small fine, and easy to claim that an honest mistake was made. The end user can easily remove the 'Made in China' sticker. It can accidentally fall off. Federal customs is not likely to show up in remote Alaska to check out a one man operation, or make a problem for the entire tourist gift industry affecting fifty shops.

Glitter does not admit he is part of this, and I do not put him directly on the spot. In court, I'd want no direct knowledge or proof. One problem I see with business as he runs it, is competition. The next guy gets it from the same overseas source, and settles for less mark up, then manages to make money undercutting your prices. *Why get mad? Get even. Control your product. Be smart, not tough. Free enterprise controls this problem.* That control in my opinion, as a business guy, is control your source. Treat your exclusive source right. Agree on the split and stick with it. Be honest. Simple. 'Problems' is being unethical.

I suspect Glitter thinks he has me fooled, and I'm a sucker. I am not an equal, not a gifted artist he wants to hang on to, not someone with good business sense he respects. Perhaps for good reason? After all, it was me who was in a bind and had to dump a lot of stuff in a hurry; hide it so to speak. Typical move of a druggie low life who can't pay rent and suddenly his stuff has to be stored someplace safe. The car, the stereo, etc. In most cases such a person has 'screw me' tattooed to their forehead. It's open season. They are in most cases lucky to get a penny on the dollar when they say, "Hey, Man, I need a favor!" *Try borrowing money from a bank when you are broke.* I had it all safely stored with Foil till buddy Foil decides to extort high storage and rent. It was not totally my idea to depend on either Foil or Glitter. Iris suggested, against my better judgment, "Trust, depend on, work with, others, Miles! When people want to help, do not turn them away!"

Like Wess the Mess when he lost his place to stay, and got kicked out of the basement of the store where he lived in a storage bin for several years. Inventory and tools went in storage, but Mess could not pay the storage, so his goods got auctioned off. Crafty heard of the public auction, got the stuff for a penny, a song, and dance. Mess got zero for it and never recovered; lost even his tools. Crafty would not sell Mess his tools back so he could make a living even though he knew Mess as kind of a friend and supplier for twenty years. You trip and fall, you get rolled. That's life.

A decade later Mess is a street person passed out in the gutter - if he is even still alive. None of us have seen Mess in several years now. This is typical, normal, what can be expected of life. It's the story of 'The Jungle' by Sinclair, written so long ago about the meat market in Chicago before unions. Crafty did not start out treating Mess like this! I recall now when thinking back, it was Crafty who helped Wess the Mess when no one else would. Gave Mess a place to stay, let him work in the basement to earn money. Crafty gave Mess orders to fill, gave him business. Even bought his stuff. Mess could have been lifted and changed, begun an upward trend, grateful for the help. But no. Mess brought drugs to the shop. Chased off customers, badmouthed Crafty, was stuck in the 60s playing bongos and wearing tie-dye and sandals, with, 'Cool man!' His only vocabulary. It got old.

There is no use getting mad. This is standard treatment I receive. The police would not care. Prove what? Where is the paperwork? Probably drug money involved somewhere along the line, so, "Good riddance!" Maybe I can be rehabilitated. Glitter has the outlook, "Consider yourself lucky!" Indeed.

"Miles, most of what I have of yours is on consignment, that means you own it, that is correct."

"So if I own it, I can remove it. I have a buyer. I can set the selling price."

"Well, Miles, I am used to marking prices up as I see fit, when I feel something can fetch more money. I give the supplier the percent of the amount they said they expected it to sell for! As long as you get the money you ask, what's it to you?"

"I priced it low, so it would sell fast. I hoped all of it would be gone in one season, and I'd be done with it." I explain that if Glitter holds out for a higher price, the goods are not moving. The bottom line really is, 'Am I... are we both... making money?' If money rolled in faster than I could count it, I may not care if he jacked up the price to the tourists. Yes, as long as I got the money I expected. However, the money is coming in slower than through a small Nenana gift shop. I can do better set up on the sidewalk in Nenana any day I want. The concept, the reason Glitter is involved, is moving volume - fast! I optimistically, maybe unrealistically, expected maybe forty grand in a couple of years.

He could keep the inventory on the market for years. There is no penalty to Glitter for waiting, holding back, waiting for more money, as he has no investment to pay back. It's someone else's product. My time and money is invested, not his. One argument shop owners give is, "It is my precious selling space I pay dearly for being occupied with your stuff!" This implies 'There is plenty more inventory where yours comes from.' In this particular case, Glitter needed inventory, and needed what I have, authentic hand crafted local.

Witty asks me what attracted me to Glitter, "Why did you get involved in the first place!" The subject came up because she is an artist on the side, making her hand made custom glass beads, and has an outlet on the strip a few doors down from Glitter. She asked where I am selling, so I tell her about Glitter and possible issues. What did you expect Miles, leaving so much inventory with someone whose name you forget." Put like that I see her point.

"Well Witty, I liked his daring innovative ideas. I liked his confidence, and the fact he had a specific plan in place, with big goal he believed he could accomplish. He's flamboyant, outgoing, friendly and filled with flattery. He bought me dinner." In truth I was not interested except Iris wanted this.

"Get rid of that useless junk that might get you in trouble" If she had her way it would have all gone to the dump in the first place. *She's just looking out for me.*

"If I do not like the situation, just go get my stuff and take it back." Witty has no comment and simply can't believe anyone would be like me. She is truly street wise. Witty used to drink, and I get the impression she knows every trick in the book. I remind her, there are not a lot of people in a position to make $100,000 deals, who can move that much inventory, or come up with fifty grand to buy it outright.

The plan I would have stuck with would be to just sit on the inventory, do nothing. In storage, safe, not really needed, a potential retirement way down the road. I had no need to be in a hurry. Even with Foil issues, I had options for places to store stuff.

Witty understands. We both agree, "The high end shops and galleries could pull it off, but only with high end well-dressed well known in the top ten artists, who often have agents, contracts, and lawyers to represent them." I might have tried that route if I had easy access to civilization over the past years. I had been lucky finding Alaska House all those years ago. But the 70s were different times.

Again, I repeat, there was a reason I chose initially, a life alone in the wilds.

Able to boat right up to a wolf swimming the river. Did not appear to be afraid.

# CHAPTER TEN

## WOLVES, BEARS, BEES

Valerie stops me in front of the senior center and expresses concern for my ongoing issues with Foil. She has looked up something on the senior center computer and printed it for me to consider.

> "**Adverse possession:** Obtaining property ownership through squatters rights."

Underlined in yellow for me to note.

> "Essentially, if a true owner does nothing to assert their ownership interest in a particular piece of property within a certain period of time, they forfeit the right to do so later."

I take this to mean an aggressive bully who intimidates, threatens, makes life miserable can 'chase off' a land owner, and eventually get clear title, without a direct confrontation, agreement, contract, court order, or deed.

I read further, not highlighted:

> "The adverse possessor cannot occupy the land jointly with the titled owner, or share possession in common. An adverse possessor must have been the only person to treat the land in the manner of an owner."

I see further:

"Cannot adversely possess if one is engaged in the permissive use of someone else's land."

So, while I appreciate the concern and fear on Valerie's part that I may face this threat, I do not think this is going to be one of my issues with Foil. Still, good to know what is potentially possible. Iris tells me she thinks I might qualify for free legal service by being both a senior and poor. Valerie is helping Iris, getting the forms on line to fill out to see if I qualify. There is a law office in Fairbanks that specializes in land disputes. I am concerned because Foil tells me he has a lawyer, and has had one from the beginning. By not knowing the laws I may inadvertently lose my rights through some loophole Foil knows about and I do not.

### Senior Voice Sept 2016 by Kenneth Kirk
Sorry no joint tenancy allowed in Alaska

In every state but ours you can hold real-estate as "joint tenants with right of survivorship." When two people hold real estate as tenants in common, there is no survivorship right.

Valerie finds this article in a senior paper, and passes it on to Iris. I read this and believe it is not possible for Foil to get title to the homestead just by my passing away. I'm guessing from this, there would have to be a separate special agreement legally drawn up as is explained to me, like what happens between husbands and wives in Alaska. I really need a legal interpretation concerning all these bits of information, and how it applies in my case. Then I can decide what to do.

THE SUBJECT IS CHANGED and Iris points out an article in the paper concerning the park wolves. It is hard to make an informed decision, on such an emotional topic as wolves.

In news under community perspective Dick Bishop writes an article. He has been around a while. I know him from Lake Minchumina on the edge of Denali Park, as a result, admire him and his opinion.

### The myth of the ancient Toklat wolf pack
There have been a lot of articles in the paper about saving the wolves. Not so many facts, or support for the state's decision to allow the trapping of wolves. A dozen articles in the past year discuss a need to save these park wolves. The tourists want to see wolves, expect to, and it means a lot to tourism. "Wolves are the symbol of the Alaska

wilds and what a shame to wipe them out." Asking why the state allows such disrespect for our wildlife.

Here are some facts in this article I find interesting.

"The myth that the Denali National Parks, Toklat Wolf Pack has a linage going back to the wolf pack studied by Adolph Murie in 1939 persists." This notion of the pack going back over 60 years is appealing, but not true. DNA disproves it.

It is pointed out that 60 years ago there was widespread wolf control, with bounties offered. No hunting buffer zones. Wolves were trapped, shot, poisoned. The population remained healthy.

Referring to a reputable research study, 'The wolves of Denali' 1998. A book by Dave Mech. Authors noted that during their nine year study period, wolves killed by humans accounted for 1% of the Denali wolf population per year; and the greatest wolf mortality factor was other wolves. "Wolf numbers are low because there is not enough prey to eat. A major prey species, the un-hunted Denali caribou herd is almost gone."

It is mentioned there may be political motivations for intervening to save wolves. I like reading something that contains references and facts we can look up and verify. I question the validity of a tourist expecting to see a real wolf in the wild on their Alaska trip as a 'must see'. That affects their trip's experience if denied. Wolves tend to be night time animals and very shy. After over forty years wandering the wilderness and traveling the wild and scenic rivers, I have only seen wolves in the wild a dozen times .

I tell Iris, "Of course I love wolves, and it is a shame their numbers seem to be on the decline, but wolves and mankind do not get along well. Wolves tend to feed off things people desire such as pets and livestock." I point out something I recall from maybe the 1980s. There were complaints about the decline of the wolf population, and facing extinction in many former wolf roaming grounds through the country. Alaska offered to ship wolves to any state that wanted them. We'd cover the cost. Not one state said thank you, and took up the offer. That to me sums up the bottom line. Talk is cheap. Stand in my shoes; someone wants something, you offer it to them and they panic and stutter, 'No', what would you think?

"You want to save them? Here they are, we'll give them to you." I also recall someone feeling sorry for captive wolves in a zoo someplace, and turning one loose. The wolf killed a dog and was shot the next day. Some other wild animals seem to be adapting better, even predators. Fox and coyotes for example, appear to be on the increase, turning up in suburbs. Falcons nest on sky scraper ledges. Mom has wild bobcats living and raising young under her trailer in downtown Tucson, Arizona.

Mom told me she was out on a walk and saw a coyote walking on the other side of the road. They both looked at each other, and kept on going. Mom did not feel afraid, neither did the coyote.

"Minding his own business, looking for ground squirrels along the path." I have a personal belief now that wild animals are smart enough to learn to leave things people value alone if they choose to. They understand ownership, territory, power, and who is in charge. They understand survival, pecking order.

Way back when I was a kid, I wanted to get into animal behavioral science. Figure out how to get animals to voluntarily do things we want to help save them. Such as leaving an area we are flooding when a dam goes in. Or crossing roads in safe areas we could set up for this purpose. This was a new science. Few people believed when I had the idea my goals could even be possible to accomplish. It was cutting edge ideas at the time.

"Iris, I just noticed wasp nests in places we walk by every day! Why didn't we notice earlier? Why didn't we get stung? One huge nest is in the entry way to the house!" My head comes within a foot of the nest every day. I had honey bees for a couple of seasons. The bees got to know me and my routine. I could run the lawn mower near the hive! The guard would come out and stare at me, decide this is normal, not a threat, go back inside, and all is well in the world. A stranger tried to get close, and the bees are nervous. I was not sure why, only it was not me, and the bees knew it. They got agitated.

Leave bees and wasps alone and they might adapt; leave us alone. If we do normal things maybe the wasps in the entryway get used to the routine. Maybe there is no reason to simply go poison the hornets for no other reason than we see a nest. A stranger hammers on the door, tries to break in, maybe the hornets would defend their territory? *Imagine, hornets being my friend! What a concept.*

I learned to leave bears alone over the years. It was possible for bears and I to share space, have our own routines, get to know each other and accept each other. Not friends - more like we understood we are each part of an environment we share and love. Tolerance is preferable to risky confrontation.

I forgot to tell Iris. "A bear visited Poggy. He went in the cabin, took my sleeping bag, drug it across the river and slept on it. I was able to retrieve my sleeping bag intact, not a mark on it." I found that odd. I interpreted this behavior as not being angry with me for being in his territory. He was not trying to teach me a lesson, or be aggressive. He was curious, wanted my sleeping bag. As if, "Yes, good idea. A great thing to sleep on, thanks, I'll borrow it."

My reply is, "Sort of ok, but I am not pleased you came in the cabin! We are not close enough friends for that. Please do not come in again. The inside of my cabin is my personal space, that's too close!"

I put board with nails in it out front. I will not shoot the bear on sight. If he leaves my things alone from now on, I will leave him be. *Enjoy the wilds, stay out of my cabin.* I will not even tell hunters he is here for the harvesting. I enjoy having the bear around. I view the woods as a healthy place that has a bear in it. Even as a trapper hunter, I am pro wolf, and bear. Or, believe in a balance, and working things out, if possible. If not, *Oh well*. But I'll give it a try. I have had other encounters with bears where the message was rage! A bear destroys on purpose that which he feels I value. It looks like a bomb went off when he is done. I no longer think this is simply normal bear behavior. Bears do not expend such energy in their usual routine. I have no problem killing a bear that's ordering me to leave, or else. The choice is yours to make."

"What's on your mind, Miles?"

"The Glitter situation, but not just that. I suppose Foil, and in general, how it is we can get what we want out of life." How is it relations can turn into such a mess? Most artists love to create! Most I meet, dream of someone else doing the selling. I ponder it is true in many ways of life! Inventors, designers of all kinds. These are not the ones getting rich off the work. However, when someone else does the selling the seller usually makes the money, not the artist. Most writers love the fun part of creating word pictures and would love for someone else to handle the marketing, selling, managing. Publishers and marketers tend to make the money, not the writer. The seller might be more the business person, while the artist has a different creative mind. Artists tend to think the seller over rates their role in the partnership. I say, "So you'd have nothing to sell without me, the artist and creator!"

Sellers say, "Creators are a dime a dozen; the line is long of potential to be tapped into! It is I, who promotes and brings into the light the undiscovered potential that matters most!"

I reply, "No, sellers are a dime a dozen, the line is long of people wanting to sell. Want a shop on eBay! Most any smart person can learn how to be a manager and sell in school. Creators are born and there are only so many." A classic Mexican standoff. But no, it is not. The sellers hold the cards in an economic business based society. Not all my relationships are based on business arrangements, but most of my relation problems are.

Glitter's wife, Sparkle, expressed the long term goal of the family. "We want a manager and employees to do the work. So Glitter has more time to be with his family!" *Good luck on that one.* I believe many books are written on how to run a successful business. There may be no one formula. But most successful people have one thing in common. They put in a lot of hours. Mrs. Sparkle describes the life of the rich person of leisure. You have to be born into that, be a genius, it's illegal, or you are very, very lucky, or work ten times harder than most.

As I contemplate how the world is, I am reminded of two women in my life. I have told the stories before, but the incidents had profound impact on me. One woman told me no one ever gave her flowers! I was puzzled, because I brought her flowers every day all summer.

"But that doesn't count! They are wild flowers!" I continued to bring her bluebells, wild roses, daisies, and she continued to cry, "No one ever gives me flowers, not in my whole life!" How could anyone buy her flowers at a florist? We live 100 miles in the wilderness where there are no flower shops. She has chosen to live subsistence, a life with few frills, and few can be afforded.

Another woman told me, "I have no transportation!" I am deeply moved, love her, so tell her, "Here, you can have my truck, here are the keys."

"I have no transportation!"

I repeat, and tell her I will give her the title, sign my truck over to her.

"I have no transportation!" I shrugged my shoulders. This truck was my reliable transportation for twenty more years. I did not drive it much, others drove me in my truck. But I used it to put my boat on a trailer and launch the boat, hard on the truck in mud and steep river banks. I ended up selling it, and the truck is, thirty years later, still reliable transportation with 100,000 more miles on it than when I offered it to this woman free. I know what she meant. She meant 'that meets her standards', as in, "You call that transportation?" I assume she found it to be ugly, and would not be caught dead in it. Meanwhile, who has reliable transportation and who is walking? Who lovingly gave flowers every day, and who did not recognize being loved?

We see what we want to see, and create our own reality, based on truth and facts we pick and choose from as we define the world. Or, someone else creates a reality for us by convincing us black is white. If enough people tell you that you are poor, have nothing, are unloved, unappreciated, only make mistakes, do not deserve anything, then this can become your reality. No matter how much money you have, what material goods are around you, or how much you are loved. It may never be enough, or noticed. A senior tells me they are poor and worried. Out of concern I tell them how there is money to be made! Not interested. They have an iPod with cool features they tell me cost them $150 a month. I just walked away from $50 a month because I could not afford it. I say I am rich, they say they are poor.

I think it is possible to see wealth, beauty, love all around you in everything you look at. *I choose door number two.*

One of my heroes and people I learn from through his experiences is Thomas

Edison. There is a special on Edison on TV! I think enough of this to make notes in my diary. Notes of inspiration. Or as I say, "Life choices, rewards, and prices to pay."

**Diary**:
  **Edison**
  Known as America's first inventor. Over 1,000 patents. Did not make a lot of money. His partners ripped him off for 500 grand back in the 1800s when that was a lot of money. He would not accept alternating current as the way to go, so General Electric took over where he left off, leaving Edison in the dust.
  He spent a decade sinking a fortune into an iron ore hole that went bust and ended up with a huge hole costing 10 million. His friends said, "Sure a big expensive hole!"
  Edison replied, "I had a lot of fun making it!" He hated to lose and was vindictive, would not compliment his competitors. Knew how to advertise his name. Was not a generous person. Acted like he did not care what others thought, and was eccentric.
  He showed up often at meetings in his lab coat covered in dust, hair a mess. Too busy to bother getting prepared. Loved best tinkering in the lab doing the basics himself. At 64, his factory burned to the ground. All he said was, "Tomorrow I'll start over."
  He used the word 'work' like many use the word 'prayer'. After retiring, he traveled with friends like Ford and Firestone. He loved the outdoors and often camped in the woods for weeks. He was not a good husband, was gone and preoccupied. His wife died young, basically could not handle this odd guy.
  Reporters followed him asking questions on every subject and quoted his responses. He enjoyed inventing for the joy of inventing. He did ok financially, but he lost way more than he ever made. Edison died at 84, well honored.
  **Diary ends**

I feel 'rocks and wood' will earn me chump change, never good money. I work with them because I enjoy it. If I cannot make money, "Oh well!" 😁 :)
Interestingly that could change. Making money is, in fact, part of the game, just not the first priority. After years of working wood and rocks I am getting the hang of it. My time working it speeds up and I make fewer mistakes. The rocks and wood are selected with a more critical and professional eye. I do not waste so much time polishing rocks that will never be pretty, nor stabilizing wood that no one will want. Like Edison, discovering all the wonderful ways it will not work! He may or may not discover the material that will become light bulb filament without burning up.
Leap out of bed and rush to the shop to see what greets us today! I once happily described casting metal work as being about winding up the casting machine and letting her rip! The look of molten metal spinning, the sound of it hissing in protest

as it quenches in the water! Does it matter so much what the designs are inside, or if they come out? First I grin, "Wow!" Then, "Oh yeah! I wonder how it turned out." It's true. There is a lot of lost money with that outlook in life. In time, I could tell by the sound of the hiss in the quench what kind of metal it is. I could tell if the metal filled the mold. I know the sounds of 'happy metal.' That is not science, it is art. Only then can I be a success using unknown metals, melting down scrap. Only art shows me when a metal is ready to sling using a torch, by the color and how it swirls.

The first mammoth ivory I restored and dyed I could not even give away! I was excited. No one else was. I tried again, and again, but others said I was on the wrong track. I began again from scratch with new methods. I tried PEG, acetone based methods, expensive epoxy. I tried dying separate, then stabilizing after. Years go by! Now I may be on to something. Only maybe. I arrived at some spot, where what I do works. It's gorgeous. I'm someplace no one else has been. If it sells, that's nice. But not necessary. Normally, unusual that is gorgeous, is worth money. But it may not be me who gets paid. I've had others take my ideas and run with them, and make big money.

I feel like the words in a 60s song. "This is major Tom to base control. I'm stepping out the door."

---

I HAVE a 100 pounds of stabilized dyed wood blocks I can neither sell nor give away. Yet, when I use this in my own knives, suddenly I am asked where I got such fine wood! Years go by! I am producing a product that's hard to duplicate. It's looking good. I might be able to sell it. I have my time and costs way down compared to the beginning. Meaning I have skills. I know how to make money. This takes me back to the Chinese proverb concerning teaching someone to fish.

Glitter knows how to take fish from fisherman. But what happens when his source dries up? Word gets around among the better fisherman. I know how to catch fish, so will always eat. Yes, it is a bummer I get talked out of my fish. However, I have plenty to eat and share. I am always hopeful the Glitters and the Foils of the world might change, might see the light. As an alcoholic goes to AA, and now and then gets sober with the help of the rest of us. It's like a disease.

Foil, Glitter, Crafty, the women I speak of, prison, and society in general, tries to convince me I'm an idiot, criminal, nuts, teller of tall tales, poor, stupid. Have I left anything negative out? That is hard to overcome, with nothing but my own sense of purpose behind me. Am I simply incredibly egotistical? Filled with my own sense of self-worth? As my Ex pointed out, "Have you noticed the only people who admire you are fans who do not know you?" Not true, but convincing, coming from someone you love, who is your wife and should know you better than anyone else. It's hard for this not to have an impact. *Edison's wife left him*

The good news is, many others have told me the same thing about their own similar struggles. *Like the Edison story.* I think of the mountain men I admired. Not one in a hundred ended up owning property. I keep pointing that out. They discover the wilds, Well... really the wilds had already been discovered by the Indian, but apparently that does not count. Except... before the Indian was the Viking. But the Indian wiped out the Viking. The Mountain Man who accepted the Indian, took on the Indian ways. He cut the path to the wilds, tamed it, learned to live with and get along with it. Then some city slicker comes along who files on it, kicks the mountain man off.

The Mountain Man moves on further west. This describes 99.999 percent of the interactions. Yet, who got remembered? Not the city slicker who had all his paperwork in order and knew how to read, write, get permission, file papers! Would it have helped a mountain man to be mad, get even, be filled with hate? Most were upset, but it gave them all the more reason to keep going west, even more motivated to keep to themselves, and happy to do so, perhaps mingling with the kinder indigenous savages if they wanted socializing. Ultimately fulfilling their purpose, of preparing the wilds for invasion by civilization.

Jim Bowie and Davey Crocket, gave her lives at the Alamo fighting for our country. Fringe leather jackets and coon skin caps are now insulted.

I am not that illiterate mountain man. Some explorer types went west for adventure. Audubon could be described as a mountain man, who went off alone for adventure and to paint. He kept up his civilized connections, but had issues. He died broke, not well recognized in his time. He called trappers and hunters his tribe. History repeats itself. I believe I can follow personality types through time and learn something.

While the likes of Glitter and Foil can pull a fast one on me, and it sets me back some, all is still well in my world. I'd say to them and their likes. "Ya know, we could have done business and made money together. You could have gone out boating with me, enjoyed the wilds with me. Now you will not, and that is more your loss." I do not hate them. Glitter and I are still 'friends'.

Helm and his wife for example, enjoy boating with me as friends. We treat each other well, and have a proven positive history together, trusting each other, and enjoying each other's company. They are due to leave Alaska back to Germany soon, on their annual cycle. We had made a time to get out boating, but Helm has a work commitment at the same time. Weather was perfect! We thought there would be plenty of nice days! Nothing but rain and high water follow! Only now, weather gets good again! Helm and I make plans for a picnic, and time on the river! It will overlap moose season.

"I plan to go to the Kantishna later on for moose hunting, Helm!" Getting a moose is not always about a plan. It's not like planning to go to the store for food. Iris needs to get out to sight in the rifle she will use if she is the one who needs to use a rifle, because depending on me with the bow may not pan out. She is not as experienced a hunter as I am! I take her to Poggy. She has some targets to set up on the river bank.

"This will duplicate situations where you would shoot at a moose. Try shooting from the canoe at targets from different positions!"

Iris is dubious about my methods, and repeats, "If a job is worth doing…"

I finish the line, "Yes, it is worth doing right, I know."

"Then why don't you follow that?!" She is looking at the bent boat trailer. I moved the boat guides so they steer the boat cattywampus to accommodate the bend. Now the boat sets lined up with the bent tongue.

"Miles you said you do not like to do it right, you prefer to do it your way! Then you wonder why it breaks again or does not work!" She is trying to be helpful, getting me to see the light. Be a better person, more to her liking.

"What you quote is not what I said, or if so, has been interpreted wrong." I explain how I feel.

"Lots of things do not turn out right the first time! These things are still worth doing!" Mistakes are not the end of the world! Anyhow, it's not a mistake, it's just an experience to learn from, not a matter of getting it right or wrong! Being perfect is not my goal! Like getting a moose is not about perfect. There is no right or wrong, "So long as we end up with meat on the table!" I grin. I forgot to mention 'right' is very often about having money, tools, knowledge. We may lack one of those things and not try! Like my greenhouse story! But I do not mention this. I have been on my soap box too long already.

I leave Iris with the canoe. She unlocks the rifle case after I leave. I go back with the river boat to the Poggy cabin to do some more work while she sights in the rifle. We do not know if she will hunt for a moose with the rifle, or I hunt on the Kantishna with the bow later, as I did last year. I hear the rifle go off in the distance and pay it no mind. I do not hear any more test shots, and wonder why.

Later, I head over with the big boat to pick her up as we arranged. The water is calm, fall colors reflect in the reddish surface of the creek, sometimes referred to as Red Creek. I get to where I dropped Iris off, but do not see her right away. She has shot a moose! It is opening day of moose season, and instead of the sights on the rifle being on the paper target, it was on a moose! Not even the same one Glitter and I saw earlier. It is late in the day. I can only get the moose gutted and partly disassembled.

"I will have to finish up in the morning, Honey."

"But tomorrow is our boat trip with Helm and Anita!"

"The weather is warm and the moose cannot sit here in the field a whole day, besides, that resident bear will snaffle it!" Nights are cold, so I know the gutted cooling moose should be fine overnight, if the bear does not drag it off.

I call Helm, explain and suggest we picnic in the area, show them the Poggy cabin, and maybe he and I can finish up the moose, or I can go take care of that while the rest picnic. This sounds ok. When I was younger I could have simply gone to fetch the moose at 4:00 am and been home by 9:00. Been ready for a picnic by 10:00. I know how it is now. I could choose one or the other, but not both. If I came home tomorrow at 9:00 am after butchering a moose, I'd be in bed the rest of the day.

Helm and Anita show up from Fairbanks, an hour drive away. Weather is good. We see the Poggy cabin. No one says, 'Nice!' Or, 'What a fantastic location!' No comment beyond Anita's, "Nice stove, that would look nice in town, it's worth some money." I take everyone to the moose site. I assume they will be excited, and want to see the moose! It is a big deal when living subsistence to get a moose and know you will eat well for a year! A time to rejoice! Even cut off a steak and add it to our picnic! *Yum, fresh moose steak!* But no. No one is interested but me. No one even wants to see the moose, or help. Getting a moose interrupts everyone else's plans for a good time.

To me, this would be like a civilized person saving their money, and being able to buy a year's worth of food for the freezer. Get a deal on beef and want to show a visiting friend what the latest excitement is all about. "Look! A freezer filled with 1,000 pounds of steaks! It's like heaven!" I'd want to look, even out of politeness, to smile and nod and congratulate them on their bargain find. Add, "Now you can eat for a couple of years with no food worries." We all smile. Isn't that understandable?

So I take everyone back out to the main river where we find a good place to build a campfire on a sandbar and have our picnic. I have to leave by myself to take care of the moose and bring the butchered moose back. We will eat, and when we leave at the end of the day, get the moose home. Not a great plan for keeping the meat well preserved. But there is no way I am giving the moose my last priority and letting it spoil! For the sake of a picnic! *Everyone else can be as upset as they wish.*

Butchering goes well, but already flies have begun to lay eggs. I had a mosquito net on the boat I had used to cover the meat, but apparently there was a fly size opening here and there. I had been in hopes dropping night temperature would keep the flies at bay. Perhaps the heat of the fresh kill wakened the flies even in the coolness. If I had followed my plan, the moose would be safe in the shop where it is cool, and away from flies before the sun was up giving warmth to the day.

I get to the picnic fire. The moose is on the bow covered in a tarp. Everyone has been waiting for me. No one asks how the butchering went, or wants to see the moose, nor roast a nice fresh steak. I am only late. We eat some good Fred Meyers steaks, homemade salad, and bread. We have a good time around the fire, collecting rocks and drift wood. These are good memories we have built up over the years. Helm has bits of driftwood and rocks in his home in memory of these outings. I am forever grateful for Helm taking me in to his home when I visited him in Germany. We are relaxed, and all feel like taking an afternoon nap when done! Meanwhile, the same hot sun is beating down on the moose meat. Weather is perfect for our enjoyment. We get home after a great day of picnicking. I need to deal with the moose now.

Flies have laid eggs, and the meat is hot. Maggots have already hatched! Geez! I spend a lot of time cutting and scraping maggots off the meat as I hang it. The meat will survive, but there are hours more work. I'll have to cut or grind it sooner than expected, and get it in the freezer. There is always an issue with moose meat when getting it during legal season, especially early on, like now. Weather is often too warm, so it is not possible to keep the meat unless you have electricity and own a freezer. I have had meat turn green, and be full of maggots. I had to eat it anyway, resentful of civilization that makes such laws that I had to live this way and eat such fare.

It is a blessing now to own two big freezers! One for the garden, one for the moose and fish! These new energy efficient appliances are great. We hardly notice the power drain. Still, the best tasting meat, is meat that gets hung in a cool place for a month before processing. The quality of this moose I got will now be cut in half, and will be lived with for the next two years. If Iris sees even one maggot it is possible she will not eat any of the meat, possibly never eat moose meat or cook it again, ever, as happened with bear meat. So it's a big deal.

We are all getting older now. Anita had knee surgery, and has trouble walking. Helm commented on my lack of a working windshield on the passenger side when riding in the boat. "There is a windshield that folds down, Helm, but the gas tank is in the way. The tank is full, and I am not strong enough to move it." I can tell he thinks it is a dumb idea to have it fold down. Truth is, I hardly ever have a passenger and if I do, the weather is usually nice. I tend to go slow at such times so we can all enjoy the scenery without a windshield.

"I save a mile an hour on speed with no passenger windshield." That sounds pretty lame. But that is an hour of travel time gained at the end of a ten hour day. But yes, I should care about my passengers more. I get it. It already took a day to clean the boat out enough to go on a picnic in. I smile, as there was a time none of this would have mattered. We'd take off in a crowded dirty boat and never notice! We are getting soft these days? Or my passengers would have noticed back then, but I was too occupied to see their discomfort! I'm in heaven and assume they have to be as well, *how could they not be?*

Even so, I decide I will do some work on the boat and try to improve it for the comfort of passengers. I might have a window that slides up and down. Take the window and track out of the side door of a junk old car maybe. *Turn the car door crank, and have the window roll up and down!*

😀 :)

Not even Iris is excited we got our moose. She shows no interest in eating a fresh steak. Never comments about, "Let's eat some good meat!" She waits until two weeks later, when it is frozen so no longer fresh, and treats it the same as last year's frozen meat. We eat a lot of Fred Meyer sausage instead. I cook up fresh moose for my lunch and enjoy it alone. I give away a lot to those who appreciate fresh meat. I may as well give a lot away, as we do not eat it so much like I used to. The meat has a smell Iris comments on when she goes in the shop where it hangs. She feels the meat has gone bad perhaps. Slaughter houses do not smell great, nor do meat lockers. I am reminded once again, that fresh wild foods can be the best there is of all choices of foods! But no guarantee. It is also possible for wild to be as nasty as any food can be! I recognize one beauty of Fred Meyer. We can expect a consistent standard to be met. Not the best, and not the worse. Certainly the Freddy steak will absolutely not have the possibility of a maggot in it! Nor ever be green! Ha! Iris carefully cuts out any veins, rinses all the blood out, trims any fat, sinew, muscle. Searches carefully for a single hair left on the meat. This describes the average person. Heck, the average person might not even try it! Might throw up at the site or thought!

I have a vivid memory I wrote about when in my twenties. Grabbing a 25 pound chunk of wild roast in my paws, slamming my face in, clamping down teeth, shaking my head, and eating what came off. Yes, I miss those times. But of course such a life does not involve life with anyone else. I now have the rewards of a partner, and civilization. Being wild becomes a private secret.

"You look deep in thought, Miles!" I smile "Hmmm," and discuss something civilized. "Did you wash your hands?"

I forgot. Such a child. I have to go wash my hands. My gray Einstein hair sticking out haphazard a foot to each side in every direction except on top where I am bald. I drop something on the floor and have to go down on my knees to pick it up. The life of the elderly! I do not call it 'hurting'… heck, I expect to live another two decades.

I head to the work shop to fool with stabilizing wood and fossils. I have been playing at this a lot of years now. I realize I am running out of room, so need to clean up the shop, toss out things I am never going to use. I come across a lot of cut wood I put dye in, At the time I was experimenting with dying first, stabilizing after, instead of doing both at the same time. It sort of works, sometimes gives a desirable effect under specific conditions. I never got to stabilizing this box of blocks. Upon inspection, I decide the quality is not up to present standards. I will burn it in the wood stove. So hard for me to do because I hate waste! Yet, I know stabilizing is expensive! I would not get my money back out of this wood if I went through the effort. Sometimes material has to be considered experimental, part of a learning process that can't be sold, that simply helped give me a lesson.

I HAVE an email from my longtime friend Will...

> I've been watching, 'The Last Alaskans'. Pretty good show that reminds me of things I used to do, wonder why you ain't in the series I think you would be a good one. Been watching more shows about Alaska than I ever have. Learn more about Alaska than I did when I lived there. The one guy on this series got cancer - 64 years old. Had to leave his Homestead. Good to hear from you again. I'm just working on walking. Glad to hear you're on the river, must be nice! Collins girls sell books on Amazon. Wow! Book 7 sounds pretty good. Wouldn't mind hearing the story myself, only got bits and pieces. Have a question for you, how come you didn't get a hundred and sixty acre Homestead in 1972? Was that the year you got there? Supposed to have been the last year of homesteading and yeah I'm hanging in there doing okay despite what's going on. Well, I've got to go. LOL. I don't know much.

I am glad to hear from him and remember the old days.

**Hello Will!**
Glad you are still with us and in ok spirits. Homestead in 72? Mostly I did not know enough. I never heard there was homesteading, did not know this was the last at the time. I had no money, no skills, no supplies. I did not even know where Fairbanks was. The only reason I got a flight into the wilds is because I worked Piper's fish camp in trade for the flying. He showed me how to start an outboard. Till then I had never even seen one. I was a city kid, spoiled upper class from a university family.

Even today there are homesteads for sale, easier than proving up maybe. If you know people that is. Some land, the state will let you take over when people default. I ended up with four what I call homesteads.

Yes, I have been asked to be on these various Alaska shows. They do not pay much. Now that I am older, less interest in me and what I do. A young friend is the guy in 'Life Below Zero', Jade. I work with him and his crew, and could get in the series, but he gets ripped off by them all the time. Has to get a lawyer to sue them to get paid at all. Plus, I now know how to do stuff, so it is not an adventure, just routine. Jade goes out two miles in a boat and it's an adventure, breaks down because he never checked the lower unit oil on an old engine since last summer. No spare, no paddle. The public and camera crew eats that up! I go on 800 mile trips and simply go and come back. Boring. Part of an adventure is a mistake. Like our early years.

Do you read? All my books are on Amazon and as kindle downloads cheap, they can be read off a computer. Book seven might be out in a few months. My friend who gets it ready for Amazon, works cheap, but at his own pace and has a job now so can only work on my project on weekends. I'm not in a huge hurry.

Yes, most of the elders I know had to leave the bush life when they got older. It's a young person's game for the most part. The twins are close to a small village, have help, and more money than they let on. They have family money when they need it. Like rescuing horses with a plane. They still deserve the respect, just that they are in an enviable position. Anyhow, dinner time. Take care. So can you walk a little and working on that? **Miles**

I am reminded Will has MS and not doing well. Unsure if he can drive any more. He said he got talked into assisted living. Told he would get taken care of. However, this required he turn over everything to 'them'. He could not own anything or have any money. Once they got control, he is given a $20 a week allowance or some small amount like this, and lost control over his life. His life is however he is told it will be. So much for the elderly. For now, I can take care of myself.

A LOCAL CAME by to pay me for something he saw at the bazaar. "Excuse me, Miles, but I thought you said this is $120! Now you want $150?"

I reply, " I said $120 is with a cash discount, no receipt. Now you want a receipt, and bring in a 3rd partner, our Uncle Sam. That disqualifies you for the discount."

"Receipts and partnering up with Uncle Sam are required by law. Are you suggesting we do something illegal?"

"If you feel cash, with no receipt for less money is illegal, you need to pay the extra $30. But it's not coming out of my pocket. I offered a discount, stated the conditions. If you do not accept the conditions you do not get the discount." This is a common conversation.

I learned about business way back as a trapper selling furs to Don. He had the furs of mine he just bought. My money is on the table. I'm twenty-three years old. He asked me if I had to have a receipt. I have my money, he has his furs, why do I need a receipt? Prove what? To who? I lived at the time in a six by six foot cabin. Where am I going to keep receipts? Who will ever see them? Who will care? I am living on $2,000 a year. I did not even need to file taxes, did not earn enough. Later as we got to know each other, I saw the receipt book on the table after one of our deals. I took the receipt book, signed my name at the bottom, and handed his book back. The rest of the page is blank. He said nothing, and put another $20 in my pile. I assumed we were both busy. I assumed he'd fill in the paperwork legally and properly later. I'd be shocked if he put an untruth down on paper without my knowledge. Paperwork can get complicated! It moves around here, to there, and back. In this pile and that pile, written, rewritten, counted, recounted, added, subtracted, and at some point it all needs to balance somehow, which is about impossible.

Or should I add, "For me anyhow!?" I can't get a phone number written down correctly, and it takes three tries to dial it correct. I can't get numbers to balance. I do not even like 'balance'. I prefer 'off balance'. In the way I do not do straight lines or geometric shapes in my art.

I can do quadratic equations with plus or minus rooms for error. I understand Einstein. Not adding columns of numbers. And I hate it. Because of this, I assume it's the same for everyone? So I say stuff like... "If you run a small business, who can afford an accountant and secretary? Who has the proper training to be an accountant among our many talents? A person could look forever for a missing receipt, an incorrect number, an imbalance someplace in a year's worth of transactions." It's good to say, "Ah yes! Here it is!" This missing amount being held over our head shows up! Imagine! "I knew I misplaced it someplace."

Everyone is happy, everything is in balance now. There is a truth to the fact, out remote it is often not possible to come up with paper and pen to write anything down when doing deals at fish camps, on traplines, in villages. Likewise, 'Was that a trade?' 'Hey, the government can't balance its budget either. Has to fill in the blanks, make up charts and graphs and trends. Come up with magic numbers.' Then tell us how many fish are in the river and how many we can have. How many dollars we are in debt.

I've explained before about assigning values, but the subject keeps coming up. "I hate paperwork, and it takes a lot of time to keep exact track of every single detail and get all those details to balance. All that time is worth money." I explain, "If I have a receipt, I must keep track of it, and be able to find it again." It is legal to simply have totals that are correct, and honestly answer, 'How much did you make this year?' "I challenge anyone to come up with drastically different numbers that show I have been cheating. Investigate to your heart's content!" I may not have this, may not have that, but there are as many expenses mixed together as bits of income.

The bottom line will be an income I faithfully report, and pay taxes on. I do so with the least mess and fuss. Proof? Go talk to Miss Lanious, my secretary. *Miscellaneous*. Look at what I have. It matches what I claim. "If you wish to try to prove something else, go for it." There is an assumption, if you fool with, misplace, lose, do not keep receipts, never had one, it is because you are hiding something, and up to no good. What other reason could there be? What if the answer is simple.

"I hate paperwork. Here is your cut, it's accurate, go away, leave me alone. You need more? Here, take it, now go away!"

The concept applies to 'boat rides'. (?) I can take you out fishing for $30. Between friends, you cover my costs. That's legal. If there are questions, such as, "Where is my lifejacket? Do you have insurance? Are you Coast Guard certified? Do you have a six pack license?" This same fishing trip now cost you $300, and I can't be the one to take you. If I am out on the road with my thumb out and a car stops to give me a ride, I do not ask if you have insurance, and if you are a commercial driver. Isn't that rude? If I were the driver I might say, "Call a cab," take off and leave you in the dust.

Iris points out all the money I have been owed.

"Like that guy who was supposed to buy your Bearpaw land and never paid you the twenty grand, Miles!" Oh yeah. I forgot to record the loss. So what would the IRS think or say? The reason for being, and doing, is not for money as the primary goal. When I trapped, it was common to head out the cabin door, hook up the sled dogs, and forget my traps! I rarely went back.

"Oh, well!" I loved being out… it was about being with the sled dogs under the northern lights. From a practical standpoint I needed some amount of money for basics. If it was this amount or that amount, I shrugged my shoulders and made it work. I've gone hunting and forgot the weapon. At some point I need to find food, but it does not have to be today. If one of the prices is, I do not have as much money, sometimes standing up for what we believe involves hard choices.

Iris says, "Well that is not what you said the other day! What about that picture of you counting all that money, honey! What is that all about?" So what is the truth?

"I may feel like ice-cream today. Yesterday I was in the mood for cake. Is one a lie?" Sometimes I trapped and took catching an animal seriously, and that animal was not getting away! Other times it's not important. One day the cat watches the mouse go by unconcerned. Tomorrow he may play with it and let it go. Yet, another time eat the mouse. Was any of it a lie?

"Money is not a mouse, Dear!"

I prefer to be vague and evasive when it comes to 'truth and reality.' The bottom line is, I am in control of my life. It's on my terms. It's my situation to deal with as I choose. No one told me to go get the mouse, or else! No one told me to put it down after I caught it, stop playing with it, or give it to me after I have it, or spit that out if I eat it. I didn't need permission, a permit, nor proof. If having that freedom requires I show little profit, so be it. In theory, I could bring in a million dollars, and find a way to spend it on the business, and live in poverty.

There is also an assumption if you fudge with reality and truth you are hiding something, and that something is bad. Such a person can't be trusted, is a teller of tall tales for the purpose of running a con. Life can be more complicated than that.

There was a time when telling the truth could have had me taken from my parents, and my parents put in prison. A time I was part of something I was not responsible for, did not understand.

"By the way, I got the electric bill today! It's up to $160 from $120!" I'm guessing Iris is used to being poor, and keeping track of every penny, accounting for each dollar. In her world, there is no such thing as easy money, or money rolling in, or having more than you need, being anything but stressful, hard back breaking work you hate. How do I reply?

"Well, yes, I expected a high bill. I've had the exhaust fan, the kiln, the grinder, and the vacuum pump, all running at the same time, almost all day long most days."

"I'm just saying!" Yes. But high electric is a good thing! It means I am in the shop, running equipment, being productive! It is something to rejoice about! The bad news would be a very low electric bill because it would mean I did not create anything that month." I show her some of the wood blocks I produced with that electric use. "Just this one small box is $1,000." I estimate I produced $10,000 worth of goods this month. Am I going to stress over an extra $40 electric cost?

It is hard to prove my production. Some return will not be seen for a year or more. There will be wholesale deals, trades, waste, theft, costs. Some high end retail sales, but with advertising and show fees. I could say, "Here is a calculator, add up these numbers." Or, "Here is what I produced, it is priced, add it up yourself." I have said before, what we believe, often has little to do with facts, proof, and truth. If we believe and focus on money being hard to get, and work, then it is, and always will be. If money is a game, fun, exciting, then that is what it is. I honestly cannot promise an exact amount I made last month to be matched against expenses.

"Have we ever been broke? Not been able to pay a bill, had to borrow money? Ever?" "Everything we need, we have paid for with cash, and seem to not have a wish list that has not been met."

As a game, money making needs to be played well to win. It is not good to spend what you do not have, bet into other people's games, have tastes we can't afford, etc. I was in the mood once to take a photo of a big pile of money just to prove it exists 'sometimes' because I realized at the time, few people ever see such a pile. It's usually numbers on a paper statement. I like to be a person of mystery and magic.

"But, Miles, where did you get such a pile of money?" Puzzled question. A practiced shrug, a wink, a rolling of the hands as a wave, the tide comes in, the tide goes out. Today yin, tomorrow yang. Me with a big grin, the listener, usually walking away, still puzzled with unanswered questions, confused. Some walk away with an all-consuming hatred.

*Crab Apple blossom and Bee.*

# CHAPTER ELEVEN

## LAWYER IN LAND DISPUTE

I like to think I have a feel for how things are going financially based on over forty years of experience. I have confidence I have a good product that will sell. Eventually. Yes, I gamble. I like to think, not foolishly. I always have the immediate future taken care of. Maybe two years ahead. I could have zero income for two years and survive at the same standard of living as now. Within that two years I believe I could liquidate assets not vital to survival and last another five years. Within that five years I am confident I could find other work, or adjust in some way to change my financial situation. These beliefs allow me to fool around. I get a dubious, "Well I hope so!" reply. I assume from general response, this is a novel unheard of view about financial strategy. This may not be what you'd learn in economics. But I wouldn't know, never took any money class.

I assume this would be a normal reaction from a majority of people. I live in my own world. It looks like a shell game, how could it possibly work! When I just lost a minimum of sixty grand to Glitter. If I said, "More like 200 grand." No one would believe me. Who believes even sixty grand? Not even my wife. Therefore, it did not happen. Glitter may count on that.

Anyhow, if I had the 200 grand I'd owe taxes on it if I could not find a way to invest it in the business. That kind of invested money might draw attention. I'd rather Glitter have it than the government. I do not need to be in another tax bracket! I do not need to give up subsistence.

The IRS might ask where the 200 grand went. I say, "stolen!" Then I am told this is a lie, and why didn't I report it to the police? What sort of answer would, 'Because I do not trust the government' be? Who would believe me? Where is my proof, my receipts? If I was believed, I'd be locked up. Who needs that? It's much better the 200 grand has no paper trail. It never came in, and it never went out. I had the fun of making stuff. Someone else got the money, while I got the fun. Is that so horrible? Like Edison said, "Tomorrow, I'll start over!"

---

I LISTEN to what my unconscious has been telling me and reply, 'It is said there are only two true things, death and taxes. And we have proven that wrong!' To those who hang their head and say, 'There is no way to beat the system.' I hold my head up and reply, "Watch! And behold!" We always have a choice! If I am, in fact, full of crap and there is no missing sixty grand and I never had it, and I am off in la la land, there is a fact that cannot be ignored. I do not pay much in taxes, I'm happy, and I have about everything I need or want. The rest is just details.

I mean, I like attention. Yes, it would be nice if there was even one person saying, "Sixty grand? Makes sense. You are gifted, smart, so this does not surprise me." But there is no such person. Not one.

Last time we went to town I had to pull money from the bank. There was getting to be too much there. I didn't think it looked good. Once money is cash, I'm in control of it. I know that is only a feeling, and not the truth. Or, it is my truth. Cash in my hands makes me happy, abstract numbers on the computer do not. I point out to Iris, "Every few months the bank sends out a ream of paper telling me it is a policy change to help my banking experience better." She nods, I need say no more. A month ago Iris spotted a five dollar deduction. She asks the manager of the bank what this is for. We are friends with the manager. Well, that is a fee for keeping under $2,500 in the bank." We didn't read the latest fine print. I tossed it in the trash with the other junk mail. Without Iris I would not know, nor care. I do not want to know or feel all the slashes that make up the '1,000 cuts of death.' I only need to know one fact. Banks are not my friend. I do not have to prove it. I know what a stick going up…

But we need banks, Miles!"

I give my usual reply. "Watch, and behold! Be amazed. Now you see it, now you do not! Magic!" *Yes Masta, weez poor, masta; struggling as you wish, masta; yoo slave masta. Take it all from me, masta, I no complain, nothing I can do about it, masta.* I frown and wave my hand in front of my face, making my unconscious go away. We get the picture.

"Trump says the whole system is corrupt and rigged. Did you hear him the other day in his campaign speech, Miles?" The news media reassures the public, "Rigging an election is impossible," and has experts tell us why. At the same time, telling us the Russians have hacked into the most secure of our systems and is trying to affect the outcome of the election.

"Life is good!" *Life is good!* "In what other country could I write such a book?" *We are eating, growing our own food, hunting, fishing, own our own home, controlling our transportation, choosing our occupation with no boss!* Downright spoiled! *Not everyone has it so good, so be grateful!* Yes, have you noticed it is spoiled people who complain the most? *Yes. I noticed.* Well, actually, everyone complains. The poor, the rich, the needy. *It can always be so much better than it is, but also so much worse.* Anyhow, it is wise to blend in, and blending in means complaining along with everyone else. If we do not complain? The public gets suspicious. Wants to know what your angle is. Turn you in as probably up to no good.

'Please report any suspicious behavior' is a directive. If you are happy, that has to be a secret. Some unhappy person will want to clean your clock because... after all, what right do you have to be happy when so few are? How selfish is that! It's hard to argue with that logic.

I NOTICE pictures in the paper that people sent in; a nice positive section I like to look at. It proves the entire paper is not about bad news! Every day there are pictures of pictures. The section titled 'Spotlight' is where something positive is said about someone in the community who has accomplished something good. It's not the front page, nor lengthily, but it exists. The best part I turn to first is the comics.

"What do you hear from your son, Miles?" As Iris and I read the morning paper, she brings this up. I assumed I'd hear from Mitch by the end of the summer to hear how his summer went.

"I emailed and asked how he is, filled him in on what I am up to. Have not heard. Odd, not gotten a reply from his mother either." I add, "Probably busy, you know how kids are these days." But it is not what I believe. I think something is up they do not want to clue me in on. Mad about something. Agreed to give up on me during his visit with her. There appears to be a resistance against getting to know me, or me getting to know my son. Over the years I had hoped to overcome that.

The 'Ex' would say she did all she could to get me interested in sharing parenthood! True. On her terms. Me the baby sitter. Her the one dictating how the relationship would be. What we could do, who his friends could be, where we could go, when. Participating is not the same as doing what you are told. No follow her rules, no get to see son.

Perhaps as Mitch gets older, and understands life, and how situations can get, he may change his mind. He wrote long ago that he flew to a comic book convention, paid to travel there, so assumed a subject of interest. I also assume if he wrote his life story, his relationship with his father would be explained through his eyes. Not the same story I tell. I deserve being ignored 'because'. Followed by something unforgivable I have done. *Be glad you are alive at all, rejoice! No one owes you anything, son. The truth is, he'd never have been born if I'd had my way. His mother and I did not get along well enough to have a child. She promised she was taking her pills...*

"One of my interests when your age was comic books. It got me into reading. I had a collection, would be worth a lot of money today if I had kept it!" I expected a follow up of, "Cool, so who were your favorite characters?" Opening up a meaningful bonding dialog. I might start a comic collection to have something in common with my son. Look for comics at garage sales for his collection. Nope.

Mitch used some money I sent to pay for violin lessons, so assumed this interested him. Because the money was sent at a time I heard how poor he and his mother were. I assume then 'music' had value to him, if he chose music lessons over gas money to get to school. Sure I love music, would be interested in sharing his progress and share music! Maybe exchange music tapes. I wrote poetry years ago, some of which was meant as musical scores. I had hopes of running into a situation where someone knew how to write music, or could play by ear. Maybe we could share those ideas, as he tells me his own music ideas. Music is a good way to know someone! But, no more talk of music. Heck I have a big connection to music I forgot about.

"Mick Jagger of the Rolling Stones, has my mammoth ivory on the knobs of his guitar. A guitar builder named De Temple buys material from me and supplies many of the rock stars with custom guitars."

"That guy in Aerosmith?" Iris reminds me,

"Yeah, Joe Perry? He wears a piece of my jewelry, saw it in a poster picture of him! One of my Japanese friends- customer Kenji sold it to him. I think he travels in rich cultured circles. I forgot till now, I met the Steppenwolf band. They came to Crafty's when they did a concert at the state fair. They all bought wolf claws from me and complimented my work!" You'd think a son interested in music would reply in some way. 'I like his music, saw him in concert.' Or, 'No, I like more folk type music...' Opening up a line of communication.

A sure way to get him to shut up about a subject is asking about it. For a short time we once shared an interest in photography. He forwarded pictures he took and edited. I believe I responded positively. I thought he had a lot of potential, very innovative pictures. It's almost as if, when I show an interest he quits for the very reason we might share or have something in common. Yuck! However, I felt a little like this towards my father! I did not want to be compared to him, compete with him. We share the same name, as my son does. I did not want to be a chip off the old block. I wanted to be 'me' in my own right. Dad was so good at what he did, so talented, how could I ever be noticed under his shadow? Except to be his opposite!

Yet oddly I believed we had so much in common, and Dad would never admit that. Both of us, dedicated to what we do, good at it, focused, independent.

"Not long ago, guess a year ago, Mitch asked for my advice! It was something he had never done, so I wanted to help!"

"How did that go?" Well, he wanted advice on where to go, stay in Hawaii for the summer, or head back to California where his mother and some friends are." He explained a little how he did not have enough money to come back if he left, so could end up in a bind, or have to find work in California to return to Hawaii for school. If he stays in Hawaii, he might get school work done, but have to find another place outside the dorm to live, and pay rent someplace, compared to staying with his mom in California free.

Hard to help here. I asked a series of questions I thought he should consider, to open up a dialog. He had written, "Well here is our chance to do the father son bonding thing, I'll give you a chance." But it is not that simple. Giving advice means knowing someone. Knowing someone requires ongoing communication. Hawaii would offer independence. But that is me, it's up to him. I do not know how strong his bond with his mother is, what that means, or how many friends he has in California. I do not know what his situation in Hawaii is. He may have a girlfriend. I need information. I forget, I may have given an opinion, "If it were me." Followed by, "But you are the one who needs to weigh the advantages and disadvantages, maybe write them down, let me know more about the situation." I never heard back. I assume, blew my chance to bond and offer fatherly advice. I got from this, "Well I gave you a chance! Obviously you have nothing of interest to say." Not even a, "Thanks for trying." He arrived with his question, pessimistic and dubious. Maybe his mother put him up to it. That's the impression I got. Just like

"Don't forget to thank your father!" An obligation. I wonder how he'd respond if I wrote, "Well here is your chance to be my son and do the bonding thing!"

I wonder if my ways are just too far out there to be understood or listened to. I can picture his mother saying, "Well you know how your father is, I told you!" She once said Mitch and her are bonded, close, share everything, have no secrets. I'm glad he has someone to share with, be close to.

"As long as he is happy, that is what matters!" If that requires me out of the picture, that is life sometimes. Sure, it would be nice to have a son! But you cannot make or demand someone care, respect you, want to get to know you.

"Maybe he wanted money, Miles, to go to California." I hadn't thought of that at the time. But I do not think he is like that.

"Iris, I sent him an email that I put the Tek homestead land in his name." He did not thank me or respond. Worth at least forty grand. I did not send the title. I was concerned if I just put it in my will there could be an issue when I die. I agreed with his mother to give it to him. I want her to know I have been good to my word. She seemed worried about it. If it's in his name, when I pass on, he can get sent the title. I still have hopes I might get up there to use the land, since it is only twenty miles off the beaten path. I remind Iris the trip requires a jet lower unit on the boat that I do not have. I'd have to buy another engine, and another boat to go. But maybe I can find something used. I keep thinking that.

This land area has changed since Mitch and his mother were there so many years ago when he was a baby. More to their liking now, less to mine. Crowded. Being so close to civilization it is a prime recreation area. Hardly any game, lots of airboats and loud noise. But high property value, compared to what I paid for it.

Hate to have Mitch sell it while I am alive. I do still consider getting the boat needed and putting a shack up on it. I figure after I am gone he can sell and do something useful with the money. It's going up in value faster than interest in the bank.

"I doubt he will ever see the land he was raised on his first year. I would not expect him to come up when I die. It might be hard to sell if he is not here. It needs pictures. It would help to show where the corner posts are. He'd have no way to get to it. So if he does not want it, and it is too much trouble to sell it, why did I bother giving it? His mother wanted it this way. Mitch never said if he'd like me to try to sell it, and give him the money.

"I should have listened to his mother and given it back to the state! What a waste of effort, time, and investment. I made payments on it when I had a hard time doing that. I think Mitch does not need the money. Not from me anyhow. He's mad at me, I assume." But doesn't want to talk about it or straighten it out. He'd prefer to be distant. He has never indicated he'd like a father, misses me, has any regrets.I think I told you? I'd get visitation, and she'd call him back home when she did not like how I was raising him. She had full custody. I do not even know how she got that." Whatever. If there was a court date concerning the subject I was probably out of reach in the wilds. Who knows what she told the court? I wanted her to be happy. That's what she wanted, a son. Yeah, there were some early days with Mitch I reflect on, when we were two peas in a pod. No use bringing it up. It only gets me upset. I ramble, and no one wants to hear it. Personal stuff is like that.

"No use beating a dead horse" The Ex used to say. I'm over all that now. "See! I can stop talking about it."

Iris sympathizes. I change my thoughts to good ones. Iris gave me a shirt out of the blue. Saw it at the store and figured I might like it. I do. It is nice to think of me, look out for me. I have a lot of shirts, and wish she'd look for things she needs! The other day we got a colander for the seniors. A huge one, sixteen quart, hard to find. Iris looked all over on line and everyone wanted $75 to $100 shipping! Forget that! For a $30 colander?" We decided to look at a restaurant supply store in Fairbanks and found one! I bought it for the seniors because they need a new one. Just for something nice to do. It is a sign of a good person - to do something like this, think of giving the seniors something. Iris can be thoughtful of others.

"Iris, Dodger sent an email in reply to mine. I have not heard from him in a long time. He shot a huge grizzly bear and forwarded a picture. He sure loves to hunt! Good at it too!" He is an example of a friend who trophy hunts. We talk guns, skinning, and such topics. I do not say much about me not trophy hunting or believing in it. I'm not bothered that he does. He says he got his tickets to Tucson too, and looks forward to seeing us there. Mike emailed. He is going to be there as well. He is headed for Kodiak to deer hunt. There is an art show there at the same time." I think he is able to write the trip off on business doing the art show. Hunting deer is providing food. Mike is a hunting guide. I'm not sure how that is for him financially. I think I saw him with a client, and Dall sheep in a picture he showed me in Tucson. You could not find a kinder more mild mannered guy than Mike. Certainly not the stereotype hunter guy civilized people scorn.

With the fossil show coming up in a couple of months I have been getting inventory ready. I still have to sand the twelve foot tusk. I did some major repairs and am letting the tusk dry out more. It hangs from a rope in the shop so I can spin it around and work on it. It is too heavy to lift and easily move. Alaska Air Cargo will transport it for me, or I think so.

"It has to fit in the underbelly of the plane, the door opening is small." The clerk told me it is best not to put the tusk in a box so two people can feed it through the door.

"Just wrap it in bubble wrap." That makes it easy. I had to sign up as an approved shipper months in advance, answering some basic business questions. No big deal, mostly because I knew the air cargo people from sometime in the past, or more like, they remember me from places like the state fair. I'm more likely to trust them if they know me by name.

I have so much 'stuff.' I am unsure what to do at this point. I have a complete truckload of fossil bones and scraps. Some big specimens, but mostly broken pieces. I have partial tusks, broken scraps, good color and garbage ivory. First I clean out the fossil room, that I had taken over temporarily as the wood drying room. I did not trust the Feds when they told me my fossils are no problem for them. Then I had an empty room I could use for wood. Fossils were a major part of what I needed Foil's help with, needing a storage place, then ended up in Glitters hands. I could show my probation officer, "See! Wood! I'm the wood guy now, nothing so cool as wood, check it out!" A whole room full of wood, my new passion. I have a nice snow machine for winter trails. I enjoy going out to places no one can reach in summer. I can scout firewood areas, look for small game to eat, and look for odd wood growths worth cutting and selling. Spalted birch works well, burls of all kinds, but poplar seems to work out best.

Since we need firewood resupplied about once a week, this gives me a good opportunity to get out, explore, have fresh air, greet the world, and of course, explore for interesting wood.

"I'm headed out across the river, Honey" Iris reminds me to be careful. "You know me, 'Careful' is my middle name!' " She snorts and I'm out the door. I wear my snowsuit, Native mukluk winter boots, Marten hat, my old Otter mittens. The temperature is thirty five below zero. Coolish but not cold considering I have seen seventy below. *But did not want to travel until it warmed to sixty below.*

I run into Dim and we just each raise a hand by way of greeting. It is me who pulls over to give him room to get by. I see he also has his marten hat on as I see the tail flying behind his head. He's running his twenty year old Arctic cat machine. The same machine that broke down and I had to come help him haul in the heavy load of beaver in his sled. Today he is getting wood like I am. The chain saw is sticking out of his sled. Dim built a special sled box that fits the carrier box on the back of my machine. He does meticulous carpentry and did a great job. A bracket on the side of this storage box holds the chain saw. It is just a slot where the bar drops down in and the rest of the saw sticks up above the box for easy reach. This way my saw does not get full of snow or snag on brush growing on the side of the trail.

At this temperature all brush is frozen so hard it snaps and breaks when hit. Crystals bind snowflakes together so it clumps up and does not easily fall off the trees, even when bumped. Our exhaled breath freezes as we exhale and sounds like bacon frying as it passes by our ears. It usually takes colder temperatures than this but even at thirty-five below, steel is not happy and begins to crystalize. I'm aware of this, so do not allow my steel skis to be jarred or slam into anything. I'm more careful going over logs, side swiping thick brush or dropping off creek-banks. Mostly I just go slower.

At noon the sun is just peeking over the horizon with a red glow. Even God feels the cold and the horses pulling the chariot the sun rides in are slow this time of year. I feel the chill, so am slow to wave at the horses and God. In an hour the sun will be down again. I will be coming home in the dark. There is only an hour of sunrise-sunset light to get firewood and look for burls. Across the river, the gas exploration people are bulldozing dozens of miles of trials, knocking down more trees than the whole village burns in a decade of heating. The news media never mentions this - only that we all need to sacrifice and stop cutting precious trees for heat.

My hope is to look over what the bulldozer pushed for any salvageable firewood. By agreement they are supposed to cut and stack the wood alongside the road so locals can utilize it. It's not being done. Who will make them? Felon Miles? Most of the trees are green, bigger then I like to cut up, and the bark is impregnated with gravel my saw would not deal with well. I find enough dead trees to get a sled-load though. Green trees are dozed over by the hundreds of thousands. Among these piles, I spot two polar burls I want. I tie them on top of my firewood and ratchet strap the load down. As I am tying my load, the mayor comes by on his machine. He stops to say hi. I tell him, "You're a little late, it will be dark in fifteen minutes!"

He laughs. "I can use the light from the snow machine. I have a spot I am headed for to get wood." He admires my burls, has said before, how he admires the fact I make a living collecting our garbage. I notice he carries a rifle over his back, and think he is looking for a moose. This is not exactly the legal season. The rules are simply vague on the subject. He has pointed out what I already know, "The train kills an average of 800 moose a winter going through our area. That never makes the news." Yet, let someone kill one to eat and all hell breaks loose in civilization. So the answer is, to not let civilization know, since locals are more concerned with biology, and conservation than laws made by outsiders. "How did civilization like it when the laws came far away from England?" No one likes to be ruled by outsiders. I have no direct knowledge our mayor is hunting moose out of season.

Someone I do not know goes by. I recognize the snow machine as someone who gets wood to sell to the locals. I see the ad up at the post office.

"$200 a cord delivered." It seems like this could be good money for about any local without work. Not hard to cut a cord in a day. $150 profit. It requires the snow machine, a sled and saw knowledge - something all locals have.

I heard the Native Council buys the wood and gives it away to the poor in the community, maybe only poor Natives, but it's a good program and anyone in the community can get free wood. No one is allowed to go cold. I do not think we have any homeless either. Even the poorest have a place to stay - cabins built specifically for the poor until you get back on your feet again. Natives have their own way to deal with social issues. Is it better than big city ways? Worse? My own view would be simply 'different' and should be respected, left to deal with our own world in our way.

It is dark by the time I get home in the ice-fog. Almost every home is a log cabin, and has woodsmoke coming out of chimneys. Every home has a woodpile in the yard. Few roads are plowed. The side roads are run with snow machines, or people walk. A few bundled up citizens are seen hauling wood in hand pulled sleds to their home. This is a lifestyle we all choose.

Iris hears me pull in and I know she is getting dressed to come out to split the wood I cut. I go in, "Too dark, Honey, you can bring wood in tomorrow."

"During the one hour of light?" I smile. There is no other time, so of course.

"I got a couple of nice burls today." In the dark, I drag them to the shop and get them in the back drying room that used to be my fossil room. Well it still is, just a lot fewer fossils right now. Mrs. Probation can see my promising wood business. These burls will have to dry a year before I split them into smaller sections, then dry them for another year. However I have wood leaving as new wood arrives. There is a big pile I have cut and dried for two years.

I get the wood stove going in the wood section of the shop so I can spend time doing final cuts of blocks on the bandsaw to sell. I'll do a rough sand and soak in cactus juice resown with dye, then bake it later. Eventually selling for over $50 a pound. The scrap wood left from making the blocks is used to heat the shop. *So why would I want to buy and use oil heat, even be forced to? Do what? Toss the wood slabs in the garbage and pay to have it hauled to a dump where it can be buried?*

Once I organize what I will keep, I can have the wood on the right side of the room. On the left is shelving for fossils. Now that I have the space, I sort my fossils. Knife materials go in a box. Specimen goes in another area. I grade the mammoth ivory. A, B, C - small medium and large, good color, and a box for potential stabilizing. That done, I know better what I have to work with. I'm upset so much of the wood I spent so much time on is no good to sell. This was a lot of time, work and money. I tell Iris, "This is part of education, a learning process. What would a college course on the subject cost?" She reminds me I have said this about a hundred times. *Yes, I so often seem to be defending myself and community.* Why?

"Is what I say so hard to understand? Is there another view I am ignoring, some reason what I say does not make sense?" I get no reply.

---

I HAVE BEEN MAKING finished knife scales to set aside and tentatively looking on the internet to see who is selling mammoth, and what the market looks like. I have been out of the loop four years now and see a bigger price spread than I am used to seeing. Knife scales from $20 a set on up to $900 a set! Much of the pricing appears to be like the opal situation when I was trying to learn about pricing them. The top dealer in the world tells me, "Because I say so." Big name dealers get the $900. It's nice material - on the large side, but not a lot different than $250 material others are selling. Interesting. I am not even interested in getting $900 a set. I might ask $300 for the same set. I see the low end has gotten lower. Some descent material that used to be $75 a set is now $50.

No one is offering dyed stabilized. It is hard to know how mine will be received. So far, a good response, but what will the real pros think? What kind of prices can I expect? I'm anxious to find out. Begin testing the market. I'm partly motivated by Iris commenting on her concern over a $160 electric bill.

"How will your fossils or wood pay that bill! I hope you are right!" I try to reassure her we will do well in Tucson, not to worry! I show her some prices on the internet. I have boxes of knife scales in the $50 range, no problem; material that looks as good as the $900 stuff. I'm too busy making them to count them, organize them, price them. That is not my interest. Money has never been a worry. It is an interest, a game, only because it impresses others. It does help me feel good about myself to make money statements that look good. Even Iris believes all I have is a vivid imagination, pulling big meaningless numbers out of a hat. So getting some cash coming in would be good. Unless the situation is like showing flowers to Karen, or a truck to the Ex.

I understand Iris has a past that offers insight into not trusting others. Raised poor, just stuff going on. Personal family stuff, but a reason not to trust what a provider says about bills, income, responsibility.

I show Iris a picture of a mammoth tooth selling on eBay. eBay does not allow the sale of ivory, but other parts of the mammoth can be sold. I will have to remember this. It's a mixed bag. Sometimes selling on eBay with its huge customer base has worked out. However, it is set up to protect the customer, and does not look out for the seller very well. It's all about satisfying the buyer and ensuring they have the best experience possible. I get the impression some buyers know this, and take advantage. Customers will, in general, complain at the drop of a hat. Holding good ratings over your head to give them a top experience. This means lowest price, free shipping, and replying immediately to any complaints, making it right. In a dispute the customer is always right. I only had one bad experience, but saw potential for a lot more. I dealt with more customers that I had to make it right with who scammed me.

One customer complained to eBay that a claw I sent them was not as big and nice as how it looked in the picture. I offered to refund the money, but the customer was not satisfied. Preferred to discredit me, warn others about me, etc. So we went to eBay court. I explained I gave accurate claw measurements. Taking pictures is about taking good pictures that help sell the product, making it look it's best. The picture was not doctored. I even took the picture next to a penny to show size reference. Yes, it looked big, it was a close up picture. It looked cool. It was impressive. eBay had a dissatisfied buyer and that is what was most important.

I forget the outcome, beyond the customer being right. Similar to the customer that wanted a seven inch wolf claw, when a record size 300 pound wolf might have a three inch claw. eBay would side with the customer, "Why isn't the claw seven inches like the customer wants?" My bad. I ended up not selling lynx claws on eBay.

"Iris, this looks like a tooth I sold to Glitter." I had a buyer for $800, but Glitter wanted it now, and I caved in and sold it for $500. Listed on eBay for $3,500, I think by Glitter. I had told both Glitter and Iris at the Tucson show it could fetch two grand. Neither one believed me. Even now, showing Iris the tooth. Oh, this one looks bigger, Miles." *Yours could not possibly have been worth that much.* Partly because she has seen me sell cheap. She has not been around much when I sold in Tucson, before being a felon. Maybe did not pay attention to what I was selling, what I got, or the total amounts I saw. Much got reinvested in inventory.

"Inventory you do not need, Miles. You already have a lifetime supply of junk!" Yes. I imagine someone like me is hard to put up with. I should be more grateful. *I'd rather invest in junk then give it to the IRS!*

I notice not everyone feels the same about 'cash' as I do. Iris appears to notice numbers in bank accounts, steady regular money, like a person's salary. The Ex was like this for sure.

"So what are you worth financially?" Is the question.

"It depends," is not what these women are fishing for. They want to see pay stubs, bank account records. Flashing cash comes across to smart women as, "So you gambled and won the lottery once, when that's gone, then what?"

"It never stops!" is not the right answer.

So here I am, wondering if I should test the market before Tucson, because dang, I have a lot of stuff. No one I can think of would believe the numbers if I told them.

My book seven just arrived and I need to let the world know. I review my email list of friends and many are longtime customers. This reminds me they used to buy fossil material from me. Maybe they are still in that business and would want to buy my new material? As long as I am telling them about the book and what I have been up to, why not put feelers out for who is interested in mammoth ivory. Iris is not excited. It is part of my animal parts problem indirectly in that there is a question as to where the fossils come from, and the fact I was told there is no legal ivory of any kind, despite what the laws say. Iris does not want 'problems'.

"Just stay away from it!" We do not have the same personality type. *But who does have my personality type?* Or the issue might be, it is me who invested all the time, knowledge, love. It is my retirement out the window, not hers. It is easy to tell someone else to walk away.

I wonder as well, if she would as soon I not work at all. It would, in her view, allow more time to be together. I'm in the shop a lot, 'fooling', doing, 'who knows what'. Silly unimportant stuff she has no knowledge of, does not want to, and I should not either. Life could be good for us in our retirement, getting all the bennies of being poor, with assistance, free this, free that. Why work? Why do anything that creates bills. Worse, do things that attracts the eye of the beast. Why do things that create a price on my head?

I can see the logic. I wonder why she chose me to partner up with. There are plenty of guys with no ambition who'd love to sit around, play cards, watch football, go to bingo. Iris loves sports, knows the names of all the players on all the teams, football, baseball. There are guys that'd sell their soul for a woman they could talk football with, watch games with, who is interested and knowledgeable! Sit around, eat popcorn and chocolate, get fat together.

Iris may not want what I do to succeed. There is no joy for her to share, no triumph of a job well done, or solving a problem, after a decade of trying. All the joy for me is within. A fire that burns, whose warmth and glow only I know and feel. That's selfish.

So the question I am presented with is, how will the government react if I am known to be back in the fossil ivory business? Will it appear an affront? Getting uppity? After all, rehabilitation means knowing your place. Frin, my friend who is on the Subsistence game board explained, "The word Subsistence begins with sub, meaning beneath." A lifestyle beneath other lifestyles. Poor, society feels sorry for you, give privileges based on pity, lack of other opportunity. So I find a way to make good money and how will that look? What will happen? Should I try to keep more low profile? Sell only at shows, do not advertise, contact old customers, wholesale, work behind the scenes?

If I get back on my web site with mammoth ivory I could build up a business again. Mostly unload all this stuff I am creating. I need to get rid of it. I'm running out of room. I am on my way to having over 100 sets of knife scales. So part of the game is money.

If Iris worries about the electric bill, the answer is money. So I look around at present prices. Let's see, if I average $65 a set for knife scales, some being as high as $320 while some as low as $20... six grand. Hmm. Not that much money. Ok, the 100 pounds of descent raw material, average $100 a pound... Hmm. That's only ten grand. Oh, the tusk. I'm told up to thirty-five grand! But let's say I can get twenty grand. Add it up, it's at least forty grand. Then there is all the art, the custom knives, the wood. Is fifty to sixty grand a lot of money? People get robbed and killed for a lot less. If someone robs a bank for that much it makes front page news as a big haul. What do I use as a yardstick for doing well? What would cause Iris to say, "Good job! Well done! Good decision!?" Or, "Let's not worry about the $160 electric bill!"

I do not discuss income and wages with anyone. I assume I do not know anyone well enough. I know what minimum wage is. I do not know how many hours people work, or what comes out in taxes, or what their expenses are. I heard the mayor makes $45,000, which is considered ok money. Ok for what class? Ok in our area? Is this normal nationwide? Again, how many hours are being represented for this kind of income? I've never had such a discussion with anyone. I've never had a regular job.

Of course, until it sells what I have is just junk. *Depending on your view, right?* Yeah, like watching Antique Road Show - 'You have an American icon!' "Oh, gosh, I was using it for a door stop!" "I got it out of the trash." 'I got it at a garage sale for one dollar.' 'I got it out of a dumpster.' Until someone who knows tells you it has value, anything is so much junk. Get it out of here. I see the logic. I get it.

I admit that when I die all this stuff is nothing but a problem for surviving relatives and friends. They will have to pay someone to haul it to the dump. I know. I acquire goods now and then at estate sales. 'Animal parts' are commonly unwelcome by relatives who had someone die who happened to be a nut case hunter trapper. All these trophies, horns, guns, fur coats, rugs. Yuck! Get it out of my sight. And people like me oblige. Rocks! How many seniors have a pile of rocks they collected! I haul it off for the relatives who thank me. 'Oh that's turquoise? An opal? Whatever.' They do not know who to sell such things to. I already know it would not be worth my son even coming up to Alaska to deal with it. Someone like Crafty or Glitter would offer $5,000 for all of it. Iris or Mitch would happily smile and thank them for taking it away, glad to get a free $5,000. So! It is my personal treasures. What am I going to do? Every time I get motivated to make out a will I get depressed and stop. I want to just apologize there will be so much to take to the dump. The good news is, you only have to do it once! Nothing more to put up with. Nothing to discuss in a will. Ashes to ashes.

When I got this Nenana house I live in, the old lady had a fella around who stored his stuff and built the shop. The shop was and is worth more than the house. She did not want to talk about it.

"Junk!" Filled to the brim with stuff. I knew it had value, but not to me. Big 500 pound gears for heavy equipment packed in grease. I had no idea who buys such things. A guy come out from Fairbanks and bought it all as scrap metal, by the pound just to make room for my stuff. I'm guessing this fella would say a good hundred grand there. Who would I call to say, "I have a 500 pound gear in my shop."

"What's it go to?" I do not have a clue. As will be said about my stuff, "Some kind of animal part, who knows what. I'll pay you to haul it off." So this is on my mind trying to price my goods, and figuring out who my market is.

A reply email:

> **Hi Miles,** good to hear from you and getting back to mammoth, I'm still buying some here and there, send me a couple pics of what you have and the $$. **Thanks. Bill**

I do not reply right off, just seeing who might still be interested among past customers. Telling him about my new book, as I said, may as well inquire if mammoth still interests him.

> **Hey Miles.** How about some pics of the mammoth ivory ?? Thanks. **Fowley**

I am unsure how to reply. I ran into this issue from the beginning. Customers expect personal service. Pictures, discussions, descriptions, negotiations. Followed by a hefty discount because they are a regular customer and friend. If I say, "The more you need to talk to me, the more it costs." They are offended. I used to explain, "Four out of five inquiries end up with no sale. The average inquiry occupies a total of half an hour of my time. That's an average of over two hours of my time per sale. Followed by a discount. How does that work??"

☹ (:

I began the business thinking I could figure out how to run an internet garage sale, flea market atmosphere. I got tons of stuff. You browse around and bring to the cash register anything you find and I give you a price. Like Crafty runs his shop. The success of this is, I just pile it up and forget about it until you find it. In this way I can be the creator. I make it and keep making it and do not even know what I have, how much, or where it is. It does not matter! You do all that work! Like any flea market, it's half to a third the price to you, and half the work for me.

If I have to price, record, do paperwork, inventory, know where it is, keep it clean, hire helpers, pay workman's comp etc., I have three times as much time and money invested. Instead, I treat it as if it has no value. In truth I do not appreciate it enough. I am the creator. I make stuff. That is my role. Once it is done I have no more interest in it. Each and every art piece is one of a kind. I do not even keep much track of good designs that sell, that could make me money. I just seem to not have time, or interest. I'd rather start from scratch with some other idea. I have never in my life had a creative block.

So I see what will happen with Bill. I send a dozen pictures. "Got any more to show? What else? I want to see it all" Or, "Got any with more blue? Got any cheaper? Got any for 1911 grips? Anything thicker?" When I am all done he might spend $100. I'm supposed to be grateful. If I am not, he will not come back. This is how typical deals go if I do not prevent this from happening.

I know others who pull this off successfully. The guy getting $900 for a set of knife scales for example. He can give you all the time you want. He also knows where everything is, and has a limited supply. He keeps good records, with a picture of everything, all in an organized file. Probably hires a secretary. It's the difference between a flea market and a gallery. *Should I put in the headlines on the web site, "This is a flea market, not a gallery?"* But it is not entirely the truth, as I have something for everyone. Dollar items for kids, cheap stuff, old stuff, but also pretty high end quality. At shows I set up items range from one dollar up to $3,000. I label the areas, 'Under $10' and 'My best, $500 and up.' I do not feel the need to explain the difference and the rules or etiquette for each.

"Do you order a big Mac and hold up the line expecting to discuss who supplied the bun, meat, pickle, fries, condiments or napkins then want a discount? When you buy a $100 meal it's a different story. "Anyhow," sigh.

IN THE PHYSICAL world of my shop, you get to walk around and look through my boxes, while do something else. On the internet, it is all about organized pictures with descriptions, and the ability to match that picture and description with something in the physical world. I cannot say, "You find it, I can't" and make that work. Theoretically I could zap a few hundred pictures. Toss them up on the web. Now what?

"I like the blue one!" Ok, which blue one on what page out of the 500 pictures I put up? So I number them.

"Tell me about the blue 336 please!" Ok. Now I have to go find #336. How am I going to do that? I have nine buildings full of stuff packed to the ceiling. The beauty of 'flea market' is volume. It would take half an hour minimum if lucky to find #336. How can I then offer a deal on it? It's not even a promised sale, just a question.

"Just wondering, thanks, bye."

I was somewhat successful listing items and being able to find them. Only the tip of an iceberg of my inventory got listed, but if you do not see it, it is not for sale. No personal tours. When an item sells, I replace it with more stuff I can keep track of. It began getting complicated when I had sixty-five web pages. Over 1,000 items I knew where to find. That was still a fraction of the iceberg. I'm best at moving it in and moving it out! Raw hide! Rolling, rolling, rolling! "Remember that TV series, Honey?"

:)

So maybe the answer is to offer grab bags, or 'cheaper by the dozen.' Take a batch picture.

"For this price you measure it when you get it in your hands, I don't have time." It is what it is, here is a reference picture for size, next to a penny or a ruler. Most customer who get a hint that I have a hundred times this amount of stuff want to see it all.

I used to get fed up and reply, "My shop time is $100 an hour, how much time would you like to purchase?" In general it worked ok. Not great, but we got the train a rolling. A rhythm. Selling large amounts in batches seemed not to work as I had hoped. I still ponder and wonder.

Maybe I Need to appeal to stores who need inventory, or serious artists who need large amounts. How do I find them? Maybe focus on specific customers, chase off the rest. Suggest they shop with the people I sell to. Have a minimum order. Offer huge discounts for volume. That still is not the flea market I envision. I can sell mammoth knife scales others ask $900 for, for as low as $75.

"I'm the man! I'm the source! No added costs! I find it, do all the work, have low overhead in a remote village with low taxes and no permits." But things need to be understood. "It is not $900 because…" You got to put up with me. You do not get to chat. There is not a lot of discussion. I'll get to it as I can, shipping might take five days, not three. I'm a one man show. There is no Facebook, twitters, twatters, videos. It might not be perfectly square, but accurate to the nearest sixteenth. Polished satin, not mirror. "That's the deal, in or out?" It's the deal because we do it my way, not your way.

"If we do not get along, fine, go shop someplace else. Bye!" Go pay $900 someplace where the seller rips you off and in return smiles and kisses your A*&^. "I do not rip you off, do not waste my time." It's the same piece the other guy wants $900 for; in fact I may well have sold it to him for $75. Can a business be run like this? Usually it is me who gets ripped off! Geez! I had it working, close to a 100 grand a year, until I got stopped.

I wonder if this is a main issue with Knife, who no longer wants me in his shop? Years ago he suggested I raise my ivory prices to match his; in other words we engage in price fixing. I need to join the others, not undersell! He was angry I cut into his business with my low prices. Would he be angry enough to clean my clock? My answer was, "Knife, we have a different outlook, different customer base. You have high overhead with a shop in the city. You have employees and have to buy your product. You have access to a different high end customer. I am the source, out of reach, low overhead, no employees, so I offer the more raw rustic look at a lower price." I do not see the problem. Is a flea market in competition with a gallery to where they have to hate each other, try to put each other out of business?

This is an issue I come to understand better as I get older. I am asked, in stunning astonishment, out of respect, how I can produce so much, have so many talents, make so much art! Wow!

I have no answer. I would say, "Talent!" I have to this day never spent time with another artist, stepped into another's work area or been around anyone to see how their life goes. If I want my work to be different and stand out, I do not want to know how you do it.

I get up at 5:00 a.m. and create. I do not like to be disturbed. I do not wish to be interrupted for the phone, a fire, family… nothing. Not a wedding, nor a funeral. When shop time comes to an end, then I am available. "No, Bill. I do not have a clue if I have this in blue. You look, you find out."

I begin a new unit of time. Time stops, and starts again. I jokingly describe it to people, "Like my computer." I scan-disk and defrag at the end of every day. I wake up in the morning new, fresh, full of energy. I begin loading a blank hard-drive. I see the morning sun for the first time every morning, smile, load that, and say *"Sun!"* as a child would.

😀 :)

I open my emails, see the list and say 'friends!' And load that data. That is not especially good for customers or friends who expect me to remember them and know who they are and what the past is. I smile. I nod. I pretend I remember as they do. The train moves forward on the track. All the engineer sees is the view through the front window of the future.

Perhaps I was raised believing, 'If you build a better mousetrap, the world will make a path to your door.' Very New England, very 50s mentality when the sky was the limit, the economy good, we will be going to the moon. It is the industrial revolution in full swing. Since then I have learned if you build a better mousetrap, you have to explain it to the world. No one knows unless you tell them, and convince them. Mouse traps are a dime a dozen in general. Anyone says theirs is the best, the newest, and the one you have to have. Maybe more important is the cheapest. There are more mousetraps in the world than mice, or people who need them. Lots of people are perfectly happy with the old mouse trap. Like myself. So now what. How do I keep this a seller's market?

I had dreamed much of my life I might partner up. I am the creator, someone else will sell. A wife who likes to keep track of stuff and handles 'all that'? A manager, handler, agent? A gift shop owner, gallery? Never happened. Anyone interested expects ten cents on the dollar as commission. I keep reviewing my options. Being older now, I do not have the energy this takes. I am impressed I can still turn out product. Is volume more important to me than quality? Possibly. Not necessarily.

I think of a discussion about putting up food at our last WIN meeting. I am used to putting things up by the hundred pounds. I do not peel potatoes or inspect every berry to end up with an edible product I like. I prefer thirty jars of ok pickles, maybe oddly cut, short and long, number two grade rather than five jars of number one grade, (for the same time or price involved). I heard from others who disagree! I was surprised. No, they'd rather have best quality, less volume. I heard growing up, "If a job is worth doing, it is worth doing right!" That is usually interpreted in taking your time, doing your best. A perfectionist view. Having a showpiece product to be proud of.

I'd rather have a huge natural woods around me. Others prefer a smaller manicured yard. I'd rather have a huge bloody moose, others prefer a few quality steaks. I'd rather have a hundred books to read on a Hodge-podge of subject and quality. Others would rather have half a dozen quality books. I'd rather have a wardrobe of clothes from the Salvation Army, others would prefer one nice suit. I see all these differences in how to view the world.

"You explained it before, Miles; you repeat yourself in your old age!"

Now and then I like to take my time doing a fine work of art. I support that by catering to the tourists, or the average customer who is not rich. My bread and butter art. Many of the great artists of the past had patrons. They did not have to market their work. A rich person, family, or group who took care of them - ded them, gave them a place to sleep, clothed them, and in return, the artist painted things like the Sistine Chapel. A more modern wildlife artist, Audubon was greatly helped by a patron, otherwise he was broke! My mind drifts off.

"Miles, is that the new shirt I gave you?" I absent-mindedly look at the shirt I put on this morning.

"It's filthy, Miles!" I smile and nod happily, child-like, *indeed!* Replying, "That means I have been in the shop producing!" *And that's a good thing, right?*

"You ruin everything I give you, Miles! Why did I bother!?" *I shall put it away and never wear it until you tell me to.* I feel bad and guilty. I show no appreciation for gifts. I agree, if it makes her feel bad, she shouldn't do it! I treat gifts like they are mine. Well, truthfully I forget who gave me what, or where I got a shirt. It's a shirt. I grab a shirt, put it on, wear it. Isn't that what shirts are for? If I do not like it, I do not put it on. If I put it on, I get it dirty. Me and dirt have an affinity. We love each other. The good earth. Isn't that what 'down to earth' means? If you love the earth, what's wrong with getting some of it on you? I wonder what it would be like being with a farm girl! I bet farm girls understand dirt! The lovely smell of cow dung. Ya gotta love it.

"The smell of money!" It means you own at least one cow, or even a horse! That's good, right? Bad is no smell of cow. Very bad if you are a farmer.

But hey, what would farm girls know about editing my book, huh? No one is going to have everything you wish for right? But all I have to do is look at dirt, and it leaps upon me. I have no idea where it came from. I remember my poor mother! Every mud puddle I saw I had to go in the middle and jump up and down. My great goal in life was to get away from my mother and go jump in every puddle I saw without getting yelled at. Now look. I went from the frying pan into the fire. In my next life the first question I am going to ask is, "Do you like dirt?" *Maybe I could come back as a worm.*

♥

I change the subject. I haven't heard back from that lawyer lady, maybe I should email again?" Iris got me a free lawyer. *Because we are old and poor.* This legal service specializes in land disputes. I had to fill out paperwork showing income and state my need,

I had trouble summing it up.

**Attempt:** number four
I am in a property dispute.
I am the owner of a remote homestead. Bla bla for 4 pages a few hours of work.

Can they can help me? I bring in whatever paperwork I have with the Foil deal and the Kantishna land. A lawyer there says, "Well, our priority is disputes where someone would be left homeless."

I reply, "Well, what about left dead?" Foil says he will wait for me to die and get the land for free. He may not want to wait. He's threatened me. She looks over the agreement.

"Miles, you hold the title and it is not jointly held?"

"Correct. The title is in my name only."

"Then you own the land, Miles. Foil has no right to kick you off of it, or board up your home. Go get a restraining order. Go to the police."

*Yeah right, what planet are you from lady?* The truth is, I can deal with Foil better than I can the police. I know this is not the correct answer. I have been rehabilitated. *"I'z poo, Masta; pity me, protect me, call in the professionals to deal with this, masta. I is helpless."*

"If I call the police, Foil might retaliate and burn my house down or something. Past experience showed the police would neither care nor do anything to prevent this. I'd rather avoid head on conflict and handle this more diplomatically. Appeal to his sense of reason, if possible." Explaining, "All I want is the money he agreed to pay me and live up to our contract we shook hands on." *I'd rather not play the, 'I have you by the nuts' card unless it's a last resort.* "Why punch a bully in the nose?" *Why bring in the artillery, don't lawyers know how to delicately joust; avoid fisticuffs?* "Foil is now far away in Tucson, so a restraining order is not necessary in my view, as I hope to have this settled before he returns next summer."

"Miles, first thing for sure is to notify Foil there has been a breach of contract. That means there is no longer an agreement. Someone has a building on your land."

"Well, he has paid me money, but I might call it 'rent' for the use of the land." It sounds like legally, Foil and I might be back to zero. *Foil, I have a piece of land with a half-finished cabin on it, would you like to buy it?*

So now I am supposed to compose this breach of contract letter. I want it proper, fulfilling the legal definition and requirement of a breach of contract, without saying anything that legally puts me at risk, or is not to my advantage, is not clear, might be seen as threatening, does not cover all I need to cover, is not illegal. After all, Foil tells me he has had a lawyer and legal advice all along.

My legal help tells me, "I doubt this Foil has had any legal advice. He might be just trying to intimidate you. No lawyer would give this kind of advice justifying his behavior." Well that sounds encouraging. I am in general, not fond of our legal system.

In the paper today front page headlines—"Police Chief Resigns." Over the past few weeks there have been allegations he used his position to enhance another job and make money on the side. However, the name of the accuser is not forthcoming.

No evidence, or formal charges. None the less, the chief is on leave with a front page reprimand. Right at election time. The new mayor sees no reason to dismiss the chief; he is needed during the Indian AFN meeting, a time the community has a lot of new Natives in from the villages, who are not used to rules in the city. The police chief is brought back, takes charge in a time of need. Then, when AFN ends he is dismissed again. Now the latest, he quits. Iris comments.

"I saw this coming." No formal charges, no one named. And this is the head man, the top dog! If he can't expect justice from the system, who can?

I used to get beat up as a child for drawing pictures of birds and flowers and handing them out as gifts. "So what are you, huh? Some kind of short shit faggot artist?" Shoved against the wall. Surrounded by a gang. Interestingly, I do not recall ever actually getting beat up or losing a fight.

I think of friends; people I admire, who seem to have found personal answers, have what they want, are happy, successful. All of them are generous and kind. I have friends who are not kind! Are not happy. I feel I got ahead more by cooperation. Like this house I have. Without Helm and Anita, I would never have been able to visit Germany. The rock guy... long ago... some customer I entertained who appreciated me taking my time to talk to him and help him in some way. I forget what I did for him. The rock guy sends me a heavy box of slabbed rocks worth more than I could afford at the time.

"Maybe you could try working with rocks?" I had never considered it before.

I cannot give up on humanity, believe in dog eat dog, kill or be killed, screw you before you screw me. I believe in pay it forward. The only way I can thank all these people is for them to know I, in turn, helped someone else in the same way. In the same way, the only way those I help can repay me, is to pass it on; be thankful, enjoy life, spread the idea, that being nice, sharing, works.

Obviously I can't randomly be nice to everyone. The solution is to figure out who to be nice to, who to leave be. Being nice is not easy. It is not something I was taught. I arrive at my place in life with baggage. Defined as 'stuff that gets in the way.' Deep rooted pain and even abuse. All I can do is try, and at least have as a goal, to be as ok a person as I can be. It is possible to rub me the wrong way and get a snarl and fangs. Usually some issue out of a troubled past. Related to just two to three topics. Permits, permission and authority issues.

It is possible Foil has friends, is happy, and we just rub each other the wrong way. Except his wife said he has no friends. She appears to be abused by him, and Iris agrees, is afraid of him. Well, I can feel bad, but again, not everyone's problem can become mine. There is so often a reason people are where they are in life with the things happening to them that ruin everything. Usually we get what we deserve. "Except you and I of course! We deserve better than the rest, we are the exception!" I answer myself "Right! Got it!"

Metal crane, front yard flowers. Are you seeing this in color? It's pretty in color!

# CHAPTER TWELVE

## BANNED FROM SHOP, COMPUTER COMPROMISED

I get an annoying email from Knife, accusing me of being a back stabbing low life, and let's stay away from each other. It took me a lot of attempts to come up with a reply I thought covered the situation from my viewpoint.

**Hey Knife**
I ran into someone. I think at the gun show, said he worked for you for a while. I asked "How did that go?" He said you were a good boss. He wondered why I looked puzzled. Bla bla long and involved song and dance.

I tend to be too wordy with my communications. Hard for me to keep things short and to the point. I could understand Knife's view, if we in fact were friends. Did he think we were? It is true, I am usually in a good mood and chatty when I visit. Iris likes to walk his dog, talk to the wife. I assumed it was understood there are sensitive subjects we avoid because we do not agree. In the same way I do not talk religion with some people I enjoy visiting.

Anyhow and whatever. I am reminded of the gun show where I met and talked to the guy who worked for Knife. I had hoped to find a deal on a black powder rifle. Everything seemed overpriced; as much as or more than new. I sold some wood scales to gun dealers for pistol grips. One dealer carries my books. I owed him a book. I was amazed how many people I know here! Almost all the dealers, as well as a lot of the customers. I ended up having to buy a black powder rifle at Sportsman's Warehouse where I had been looking at one for a long time.

I tell Dodger about it. "I figure this is the way to go. As a felon I can have black powder. The compound bow is ok, but I'm more used to a rifle and feel more comfortable with one."

Dodger knows about bow laws and how I'd need to take and pass a class to get a permit to use one. No thanks. "Miles, that fifty caliber is a powerful gun, you'll like it!"

"I tested it out, Dodger, and I notice Pyrodex is not the same as the old black powder I am used to from several decades ago. I recall the six foot of flame, cloud of smoke, and smell of rotten eggs! Black powder would draw into me and sound like dynamite going off! I miss all that!"

"Yes, black powder is hard to get, as it is more dangerous. A small spark can set it off; it is unstable compared to Pyrodex which can be shipped in the mail."

"It is good to know I have a simple weapon that can be adapted to use a variety of possibilities in an emergency!" *A time could arrive we can't buy bullets.*[1]

I comment about the big grizzly he shot and the email picture he sent to me.

"I'll tell you all about it when we meet up in Tucson!" He said. I forward pictures of the fifty caliber to Dodger. I cast new rod tube of copper and a custom front piece so the ramrod is going down a bear's throat. I carved a front piece, and the butt plate swivels. Underneath is the hollow gunstock I can use as the patch box like the old rifles had. With black powder you have to carry bullet parts, cleaning tools, a plug wrench, caps. It is good to have these actually in the rifle. I have enough material for five shots, and if I am out a while, the gun needs a quick cleaning done if fired. A copper bear is inlaid on the stock just to personalize it as I did with a previous one I had in Galena. I had named it, 'Holy Shit.' Mountain men named their rifles, same as naming a boat 'Old Bess' or some such. I was really in to being a Mountain Man back then.

I'll have to contact Will! He got a black powder like mine and really got into it. He killed a musk ox with his, and gave me some meat. I gave up black powder becausesi I decided it was not practical. Too often there was a misfire. The open canoe was not good for the gun as the nipple and cap were exposed to the elements and black powder and moisture do not get along. The rifle was long and heavy. It took too long to reload. In the cold, my exposed fingers would go numb before I could do it.

Modern changes. This new style is only six pounds, like a lightweight modern rifle. The modern shotgun primer is not exposed to the elements, uses more powder and goes off more reliably than original mountain man 1800s equipment. The barrel comes off easily and the finger tight breech plug comes out for easy cleaning or unloading. In the old days submerging in hot water was the way to clean it, and is still an option. Even the new rifle is so primitive, the rifle could be set off with a fuse or source of fire if there were no primers available. It is possible to make gunpowder from elements off the land. It is possible to make even a wood mold and cast lead slugs from old tire weights or battery lead. This new rifle is accurate and deadly to 400 yards. Only if I use all modern components, for about four dollars a shot. Acceptable if I am shooting only moose or bears. Five hundred grains of lead is three times the weight of my favorite 270 rifle. Moving the same speed. Imagine a car with three times the horsepower. It's the difference between a hammer and a sledge hammer. I'm all excited, and have it hanging on my wall. *I'm armed again.*

I notice the mayor has a shotgun in his office and rifle in his truck. Loves to shoot and is a hunter as many of us are in the village. It is common to see someone walking down the road carrying a rifle. Headed out some back road to look for rabbits, grouse, ducks, etc. for dinner. Even children. Seeing citizens with pistol side arms is common. I'm not sure if even a carry permit is needed now in Alaska. Bear protection is normal. Bears could show up anyplace, even downtown. I recall a grizzly was shot in front of the bar right downtown, for example. Bears are somewhat common at 10$^{th}$ Street, where I launch the boat. Josh has had to kill them in the dog yard over the years. Not that I worry, or that bears scare me, just that they are common and people with guns are normal. It's not weird to have a gun interest. There is still no law against discharging a firearm in city limits. I think this perspective is not the norm for the country.

Iris points out how many people were killed or mauled by bears in Alaska this year! There have even been a couple of wolf encounters that resulted in attacks and people trampled by moose! If there was a wounded bear on the loose in town, it is nice to know I am not helpless. I could carry my new black powder on the boat and camping too. I've spent most of my adult life being armed. I never realized what an important part of my life this is until the right was taken from me. *We call it a right, really it is a privilege, or we need to repeat that. Otherwise, we might be viewed as getting uppity and put in our place!*

NORTHERN LIGHTS HAVE BEEN WAVING ALMOST every night for two straight weeks. I take a picture of the whirly gig loon on the cabin roof in the foreground with flashing northern lights. Iris stands out in the yard with me. We marvel at the beauty together. She says, "This is why we put up with the cold and dark!" A comet zipped across the sky a few days ago when we were walking, and she missed it! It was just so fast! But so spectacular! I am glad we can see this together.

"I do not know what is going on with my son." It's often on my mind. We can't, or… shouldn't, control anyone else. I cannot make him care or communicate. He is free to decide how he feels; up to me to deal with it. Words to a song on the radio remind me. "I don't remember loving you." His mother. "They say I went screaming down the street calling out your name, but that does not sound like the sort of thing

I'd do. If you hand me my crayons I'll take down your name..." But no, this describes another situation better. A child I probably fathered. *We are not sure, but probably.* She was sleeping with others at the time. "I think I told you about this before? I do not recall. Assume, I mean I must have." It's a vague memory from the past. I go on,

**Past Flash**

"Mitch was with me helping out at the show in Tucson over a decade ago. He and I hardly ever get together and when we do, we rarely do anything together. So this was a big deal to me, and somewhat stressful as it was not going that great. Where we had once been two peas in a pod, now he's a bean, and I am the pea." I tell Iris, "This kid comes up to me at the show with some of his friends. He introduces himself." I pause to recall.

Kid says, "Hi, I'm your son!" Right there in front of customers, Mitch, and the whole world. I was flabbergasted, shocked.

I have to backtrack the story.

"I met this woman once." She seemed unhappy, so I got her an ice-cream cone. For no reason other than she looked sad, and thought on this hot day as she was working in the sun and could not get away, she might appreciate a kind gesture. She tells me her name, and invites me home for dinner. I was only slightly interested in going to her house. Not looking for a thanks for the ice cream. I did not really want to know her name. At the time there were only two things in the world that scared me. It was not bears. It was women and telephones.

This woman invites me home with her and proceeds to tell me how she is sad because she wants a child. So what is the problem? Find a good guy; you seem nice enough; settle down, get married, have kids. How hard is that?" For some reason she thinks she can't find a boyfriend; good guy. Or maybe does not want a partner, just a child, not a husband or father. It's the new modern thing involving women's rights.

"So ok, just sleep around until you get pregnant, that does not seem complicated." I mean to where she needs to be so sad and unfulfilled.

"I want to know who the father is!" Well this does complicate matters. But you can't have everything. "Will you help me out?" Hey I'm trying to be a nice guy, get you an ice-cream. I am not interested in being used and abused as a thank you!" She cries. She's heartbroken. No guy will sleep with her once they know the deal.

To make it short here, I end up helping her out. *What's a guy to do when a woman cries? It's heartbreaking.* We have an arrangement. If she is bound and determined to get pregnant, I'll have the fun. But that's it. She is grateful and forever in my debt. She swears there will be no responsibilities on my part, no obligations, no connection. I'm just a cheap sperm donor. *I mean what's the huge difference if I donate to a sperm bank, and she goes and pays big bucks she can't afford, to get sperm out of the bank? Cut out all the middle guys! Direct deposit.*

We will never hear of each other again. Her child will never know who the father is. That's the agreement. Just like any sperm bank deal. I lived up to my end. Bye! Have a nice life!" She was not grateful anymore.

The song changes to, "I got the mine, you got the shaft." Wanted my signature on the birth certificate. *Yeah right!* I might as well commit suicide! There would go about a hundred grand I do not have out the window. Tied to some woman I do not know for the rest of my life. There go all my life dreams. Why not just put a gun to my head and pull the trigger? That would be kinder. She was going to take me to court, prove the child was mine, bla, bla. But did not have the money; or for some reason spared me. I'd be hard to find. Living on a boat, traveling, no address, no job, no ID, no money. I disappeared.

**Past Flash ends**

*So here we are fifteen years later.*
I tell Iris, "The thing is, all this happens in front of Mitch. He's excited to have a brother which I think means more to him than having a father, based on what his mother tells me, and seeing how he behaves." I'm not interested in bringing a kid into my life, when I can't get along with the son I acknowledge. We are not going to be one big happy family together. Who is this kid? I have no idea. Now if the circumstances had been different, he had come up to me in private, and let me absorb the news, I might have heard him out. Decided yeah, we could tentatively see if we could get to know each other and all that. Before I broke the news to the world. He did not want any money or anything. Just wanted to know who his father is, and get to know him. His mother had pointed me out on the street one day when he was five years old. It was just a bad time, right now. His argument was, "Well, whatever deal you had with my mother is between you and her. Us getting together is another subject."

I was stressed to the max. I do not know my mother well, and am staying with her during the show, trying to get to know her. I have not seen her for over twenty years. I'm really going to say, "Hi, Mom, by the way, this is my child I never told you about." I'm already in trouble because I could not make my marriage work and here, "Yes, just like your father!" *What's wrong with you?* Mom now trying to get to know both Mitch and me. Even Mom suggesting Mitch not take after me.

"Maybe he has a chance if he is not like you," and such talk. "It's a nice thing his mother takes care of him, what a nice woman."

So I want this other kid out of my life. It was nothing I agreed to. I am sorry there is a child in the world without a father, and it's my fault. The mother told me she'd be happy and I believed her. Should I have said, "Go find another sucker!"

Have I got, 'kick me' written on my forehead? *Kid, be glad you are alive, no one owes you anything.* I have met several women in my life who feel a family is a woman and her child. A father is not necessary, single parenting is fine, workable, desired, because men are so evil. I do not agree. I think it is important for a woman and a man to get along, work together, and decide to commit to a life together before having a child. But who am I to force my opinion on other people? A woman believes something else? Go for it! See how it works! But do not come to me later agreeing with me that it doesn't work! Da! Now I'm not going to be part of the problem. I'm the one that said it is not going to work. If I find out later, 'It worked out great! We have a content happy woman, and a well-adjusted wonderful happy child in the world now, thanks to you!' I'd say Great!" With relief… and mean it. Glad I could help. I was wrong, fine, good for you. Lesson learned, now go have a good life! *I'm sorry you do not feel I could be part of it, as a family, a man, woman, and child. Who am I to say society is wrong, and men should be part of families.*

"Iris, I went to Mitch's Facebook recently to check it out, and here is this kid as one of his best friends."

So I assume from being ignored, I'm the bad guy. I can guess Mitch saying, "Yeah, he treats me like that too, irresponsible; must hate kids." Bla, bla.

My step brother is on the Facebook page as well. I hear from him about every two years at Christmas. He emailed me a year ago maybe, telling me what a great son I have. Apparently they visit each other. I wouldn't know.

I'm glad Mitch is a good person. It has nothing to do with me though. His mother raised him; go thank her. *Thank God I didn't raise him, huh.* He was deliberately kept away from me 'for the best'. So I accept what is best for everyone else, and, if it is not for my best, who am I? Nothing and no one. I do not deserve a child or the chance to be a parent. I have a lot of ideas about raising a child; ideas I think would work. I have not had that opportunity. I get the message, understand, accept this is others view.

I have to create my own world, another life, another identify, in a place where I am somebody, far away. *Doesn't that sound like a reasonable solution?* So yeah, the child I acknowledge whose mother married me does not want anything to do with me. The child that may not even be mine, wants to get close and know me! Go figure. But my guess, it is only temporary. Once he gets to know me, he's out of my life. I don't need that pain. I mean, what are the odds some random kid is going to get along with me? Odds are, we have zero in common. Or, "I just panicked. Felt trapped. I do not like forced relationships, obligations we are not in control of."

I'm sure we can never totally escape. The unconscious has a big job to do sorting information and defining what an acceptable reality is. Civilization in general, is not kind. I believe the road to happiness is to look around, count my blessings, see what I have, not dwell on what I do not. Take what I do have and make something meaningful of it. I learned long ago it is futile to beat my head against a wall that is not coming down. 'No use beating a dead horse' as the Ex put it. I tried, and all the effort was considered disgusting. "Put a sock in it! Shut up!" I find a path with obstacles, but no insurmountable wall! Learn my limitations. Know what it is I am good at, accept what I am not that good at. So I smile, looking up at the northern lights, with Iris.

🖤 ☺ :)

I do not drive, so Iris runs me around when it is time to go to town. She does not seem to have many places she wants to go so finds something to do wherever I want to go. It's all about me. That is not necessarily how I want it. I'd be fine, even happier, if she told me she needed to go someplace, left me someplace else. Told me this is not a good day, she is busy, has plans of her own. Partly she wants it to be about me. Unless I am not aware of being controlling. Or, the situation has changed in time. When we met I took care of her the first three years. She had no savings, no job, no money, and was ready to retire. I paid all the bills and supplied all the stuff. Under those conditions I felt I had more of a vote. A time came Iris could tap into her retirement fund, has a job, became eligible for Social Security, which is over twice my retirement.

My 'big bucks' slowed up when the business got shut down by the Feds. I did not make much money off the controversial items, but still, my entire business was affected. The change was not all at once, but at this point, I realize Iris's vote is worth more, so to speak. She has more of a vested interest than when we met. I had expected, assumed, anyone I was with would be part of my business! It is hard to imagine anyone with me who is not also passionate about the things I am! I envisioned someone being strong in areas I am not good at. Maybe handling the business or marketing end. Running the web site, and web store, advertising, lining up shows. Maybe doing prep work, or cleaning up the shop. Maybe taking pictures. Or like one lady who made beaded chains while I made pendants. That did not work, but in theory it should! Second choice would be the same personality in someone who had a passion for their own thing. An artist, writer, professional of some kind. Respected in her own right, 'busy' as I am, and happy to be so. Our paths cross at mealtime and in bed. Vacations together. Two independent people, alone, but with emotional support when needed, which is not often. A shared respect, trust born of not stepping on each other or invalidating each other. Over several decades, having this grow into a healing process, leading to other higher levels, ending up at what is called, 'Love.'

It is difficult to defend myself. No one sees me working. I am alone in my shop. No one knows my income, and I cannot show it, prove it to doubters. There are no pay stubs. I have time to be seen around here and there doing who knows what? Off on the river! Wink, wink. Dressed like a bum. Saying I am making a living. Wink, wink. Like most guys with toys that women have to put up with. "Who needs men around anyhow? It's like having another child!"

I hear it a lot from women. "If there were no men, there would be no more wars!" I agree. But if there were no women, there would also be no war. Men would have no strong need to compete for luxuries they acquire to impress women.

Just the other day, Jade came over. "Miles, can you make a special custom knife for my girlfriend, I'll pay you!" He gives me $200. I give him a deal on a $350 knife. He hands it to her and she smiles. He'll do anything for her. That's called **love**. She is supposed to come to run the puppy team, as he only has time to run the adults. She says she will, but does not show up. What else can he do to please her more? How much harder can he work? That's love. No one I recall disagrees. Most laugh and think that's funny.

I see it with animals. The male bird builds a nest, and if the female accepts the nest, that is his reward. That bird is in heaven. For the honor and privilege of breeding. To successfully build that nest he may have to go to war. The laughter is, the female slaps him around with her wings, kicks him out of the nest he built, and says, "Men! That's all they do is fight and go to war!" It's hilarious to watch. She takes over the home and territory he fought for and won… without a thanks. The male bird appears to accept life as it is and is happy, but who knows how a bird feels?

Iris and I are figuring it out together. Possibly it is simply time that bonds couples. We learn each other's boundaries. The subject is on my mind because I seem to be in a lot of social situations that end up as serious disagreements. Logic tells me it must be at least partly my fault in that I believe we are responsible for our own lives. That also means responsible for what happens to us. We are in control and life can be whatever we make of it. If we are not happy about it, it is because we are willing to accept how it is rather than change things.

I sometimes think of my prison time. It is hard to imagine anyone coming out the other end of that experience a better person. The experience effects individuals differently.

"Blame?" No, I am not talking about blame. If I install the batteries in my radio backwards and fry it, is the radio at fault? The electricity? I, who hooked it up wrong, simply goofed. There is no blame. I did not fry my radio on purpose. I just get a new radio. Not all radios are destroyed. Just damaged. Perhaps only certain stations come in now, not the full spectrum. It might be there is no more full volume.

Or no more stereo. Base and treble is tweaked. The life expectancy of the radio is shorter.

Our own lives when tweaked are the same. If the electric got hooked up backwards in our life is that our fault? Are we responsible? What we are responsible for, is what we do about it! Pretending we are not fried does not seem to help, nor hiding it. Nor assigning blame someplace. Nor using this as an excuse to give up. Make a joke, get on with life, focus on the good, *"Hey this one station sure comes in loud and clear now! At least I have a radio, lots of people do not, am I ever lucky. It's all a lesson!"*

"Miles? What are those marks on your arm? I noticed before and meant to ask." Iris is curious. I come out of my introspection and answer.

"Old Scars."

"Well geez, how did you get such odd scars, they look like burns?"

"Yes," I calmly say. I'm not avoiding the subject. It's just a personal experience I expect few to understand. A Zen thing. This is like discussing spiritual beliefs. She wants to know how I got them.

"Cigarette burns." Is all I say. She assumes someone burned me, and what sort of situation would that be? Someone belongs in jail! That's torture.

"No. I volunteered." So I explain. I spin into a...

**Past Flash**

I am nine years old. Just moved to a new school. Moves do not often go well, I had learned. I'm short, like art, appear to be sensitive, have mental problems, partly based on previous situations that resemble this. I come across as weak. Other kids move in on the weak like predators. Teachers and parents allow this as it builds character. Teaches children to be strong. Adults will not always be around. Life is like this. Get used to it, or perish. The weak fall by the wayside and end up wards of the state, in prison, mental institutions, or among the street people of the world. I am determined to be one of the survivors.

A group of peers have me behind the school, out of sight. "You are not like us Shorty." They are smoking cigarettes. Trying to look tough. Consider themselves tough. I believe, even at nine years old, that if you have to brag, intimidate others, and dwell on the subject a lot, you are not so tough as you say. I'm not afraid as much as disappointed. Hurt. This pain is greater than physical pain. [2]

These punk kids are telling me I'm a coward, need to be treated as an outcast. They review what should be done with me. Take my clothes off and send me into the school naked? Wouldn't that be fun? Other ideas were discussed. They are watching me. They want to carefully choose something to do to me that will unwire me, and freak me out. Shave my head? Force me to give them a blow job? I do not change my expression. They are getting disappointed. How can they best destroy me? They begin to get bored. One of them drops a cigarette on his hand, "Ouch!" I smirk, that this bully cannot take even this tiny bit of pain without being a pussy. I tell him so, in a polite voice.

His reply: "Yeah! We'll see who the pussy is!"

A contest of pain tolerance was set up. I suggest, "Put our arms together and drop a lit cigarette between our arms, and he who pulls away first is a pussy!" *Right out of Shakespeare! 'Let he who calls halt!' Such contests have gone on forever.* We do this, and the bully pulls away as soon as the cigarette touches.

He makes up excuses for why he pulled away. "The cigarette is red hot and hit my arm first!" Uh huh. I ask for the cigarette and lay it lit on top of my arm. I let it burn as we talk, then calmly remove it after half of it is burned and the smoke and smell of burning flesh drifts away. I quietly ask if there are any questions as to who the pussies are here.

I got left alone. They even wanted to be my friend. *You have got to be kidding!* But in the end this is all there is. These guys, or no one. About everyone else is afraid of me. I'm a freak. Or no. Parents tell their church going children I am not a Christian, so stay away. I'm considered a trouble maker, not getting good grades, so the non-religious cream of the class also avoids me. I'm not in any of the gangs or groups. Not always deliberately avoided, simply left out. I do not want to hurt people, or smoke and do drugs. There is no group that consist of anyone doing what I like to do. I'm on my own personal path. Everyone knows me… or of me. I'm perhaps respected, even admired by many for my artistic talent, sense of humor. Just not 'one of us.' I see myself as a lone comet burning across a night sky.

At first, I considered catching my tormenters alone, and caving their heads in with a brick one at a time. I'm sure I could handle it. My unconscious reads me the newspaper headlines. This became my method of thinking it through to its logical conclusion.

'Children found murdered. 'A psychopath on the loose! No one is safe!' What would be accomplished that was good, or solved anything? So I tried to be in their shoes. *I would not want to be them.* I looked at adults I felt held the same view on life, who were once these children. Such people tend to be their own undoing. I did not know what, 'Leave it in the hands of God' meant. It's just people being people.

**Past flash ends**

"That is weird, Miles." *Yes, well I suppose I should expect such an answer. It's why I do not mention it.*

"Why is it weird?" Now it is me who does not understand. "Don't most people go through this as part of life? Or participated, understand what goes on? As the runt puppy is not allowed to nurse, nor later eat at the dog dish. The weak cry to their mothers who ignore them until they are too weak to cry until dead. That's life." The baby bird gets shoved out of the nest unless it competes for the worm. Pigs raised alone do not do well as they require competition and struggle. Sometimes survivors end up with battle scars.

"True, Miles, but not self-inflicted."

"That is only true in our culture. There are other, usually primitive cultures, who have rituals and such." I think of people in mourning who cut their fingers off, slash arms, mutilate their hair. Those in great pain over a loss. *What was my great loss?* Oh, perhaps the chance to be accepted, to be human, to see hope for humanity, peace, kindness, salvation.

I think of the sun dance of the plains Indians. Putting pegs in your skin, tying yourself to a pole by a rope, walking around that pole staring in the sun as the pegs rip your flesh out. Done voluntarily. Other cultures use red ants on males being initiated into adulthood, after which they get a man's name.

"Miles, a disgusting cruel primitive ritual with no positive purpose whatever!" I do not agree. The ability to know what pain is, and tolerate it, is a useful survival skill. It makes us strong. If we can take, and are not afraid of pain, many doors open up. We can think rationally in a crisis. We are there for others when arrows and bullet fly. In war, in car accidents. *In prison.* It is hard to control us by threatening harm. It is not so easy to blackmail us. We accept dangerous challenges that push us in directions of accomplishment and discovery at great sacrifice. Not just in war; not just physical hardships. Being innovative artistically, being creative, a free thinker is to lose some of the protections of the group. There will be more disappointments in such a life, even as there are great personal rewards. Such a life is not for the faint hearted. It takes balls.

Speaking of balls, putting the ball in the woman's court, not all women who are brutally raped loose their minds, come totally unwired and useless. Not all can or do get therapy, or sympathy. As with prison, 'You probably deserved it.' And, 'If you complain, or fall apart, you'll get more of the same!' It is possible for no one to be on your side. Not your family, no one. "So shut up and take it like a woman; let it continue, or die." Some women choose to live.

"You do not find that understandable?"

"Yeah, but what woman in her right mind volunteers to be raped?"

"Chosen over death? I didn't volunteer to go to prison. But one inmate told me he did, to save his family. Many women choose to be raped to save family, children, need the bills paid. Partner up either temporarily or permanently with the man from Hell. Some eventually escape and accomplish great things. It is good they survived.

If one hopes to live as a Mountain Man, you better understand pain. There is not much for anesthetic, no halfway house, shrink, medication. If a tooth needs to come out, you will grab it with a pair of pliers and yank... or die. Welcome to the world of your ancestors. I can look at the red badge of courage on my arm and say, "I can do this!" Those Natives who volunteered and survived the Sundance are now special people.

"That's the one I want watching my back through life." With a high Jesus factor, who does not give up. The one with the scars and the calm serene face. As we - the lemming - civilization go off the cliff, I want to be among those comforting the scared and freaked out.

"It only hurts once when you hit bottom; then all will be well. No need to be afraid, everything is fine." Or another answer I often give when asked about my life, "Dying is easy! We are all going to die! Not all of us get to live! That can be the hard part, to live, to survive, to go on, even to be useful, happy. That is the true test of courage, character, faith, strength."

When people in my life come to me with a splinter in their finger crying, I sometimes smile, pull out my hunting knife and say, "I can fix that, no more hurt finger, I can remove the finger for you." Since I know that is not the correct response, sometimes I sigh and try to fake sympathy, wondering how this person has made it in life as far as they have and are still alive. Sometimes if I pull the knife out, they laugh and thank me. You are right, it is nothing, what am I thinking? Thanks for putting it in perspective! *Would you like to know how many times I get splinters, as in how many times a day or an hour?*

I turn around and show Iris my coat.

"What happened this time!" Half my winter coat is gone. Burned. Well my propane buddy heater works!" I grin. It was in a small cold space I needed to heat up to work. The coat is loose and I did not know it was close enough to the heater to catch fire. I assumed the heater is hot, but not hot enough for instant fire. The coat is a plastic material that when lit, melts and burns like, well… the petroleum product it is. It was a bit of an exciting time trying to put the fire out on my back. My pants got covered in melted burning plastic… also hard to put out.

Her reply is, "Mrs. Post Office told me, 'Who needs television with Miles around.'

Glad I can be so entertaining. We toss the jacket out and do not give it another thought. I could have burned up? And I could have died in my sleep. Lots of time can be spent on 'could'. I focus on 'did'. Accident prone? I'm a senior. I have arrived here with all my appendages, no limp, no injuries beyond the ordinary. My biggest health issue has very little to do with anything I did, but is hereditary. My indignant reply of, "I know what I am doing!" Gets an amused smile.

I told a 92 year old customer friend the other day. "I hope my last words are, "Here, watch this!"

She burst out laughing, "I'll have to remember that!"

She is having to go into a nursing home. Her younger companion is leaving state to have a more active life. They are not married. She resents being told what to do, how to do it, where to go. She appears to be still active and alert. I do not know what her issues might be that she needs to be put away. I'm guessing more of a family decision. Guessing she is a handful. She used to buy fossils from me over several decades period of time. Her companion is a fossil hunter. He's one of the guides taking the university fossil people to dinosaur remains up in the Brooks Range. The Colville River. He's described to me what layer the dinosaur bones can be found in, what to look for. If I have some basic knowledge, there is a chance I could stumble upon a new find when I am out. I have found a couple of dinosaur bones in the Nenana River that tumbled down from Denali Park where there are known discoveries. I became more alert when I heard dinosaurs were in the park, figuring remains could wash down. Nothing earth shattering, but still interesting. It is the same valley I live in, so if dinosaurs were in the park, they were on my property.

While fossils are still on my mind, I review an email that represents the ongoing internet selling issue I keep facing. The 'trying to be a flea market/garage sale on the web' issue. As a customer, you are not viewing all the stuff I have sitting around that you can get at discount prices.

I buy a box of 'who knows what' from Tusk, Dodger, or Crafty. 'A deal.' I could pass that deal along if a customer comes to my house and I say, "Go through the box, I have not even looked in it yet." I can sell what you find at ten cents on the dollar of normal retail. Or I find fossils, and end up with a truckload. You come look through my shop and grab up what interests you. No pictures, no description, no mess, no fuss, no time involved. So I tell an old customer I am tentatively getting back into the fossil business. I have a pile of knife scales done up, and is he interested?

The reply is,

"Send me pictures of your best!"

I reply,

Need to know best what? Best price, colors, size, most solid, my best seller, stabilized? Restored? Natural only?

What I really need to do is get a page up and going on my web site for people like Bill to review, even if it is just samples so I know what they call 'best.' I go ahead and send some sample pictures. I wonder if there is a way to make up a check sheet/questionnaire to fill out? I have tried over the years, but it gets too complicated, too many categories and choices. I'd need to offer samples or even pictures of samples for each category. While I can see the difference between twenty grades of ivory in the way a fur dealer can see that many grades of fox fur, the average customer may see three grades, or not know the difference between bone, ivory, or wood.

**Miles!**
You forgot me? I have not got a reply.

**Bill!**
I replied with some pictures and want to know if it is stuck in your spam box or something. Here are more pictures, but let me know, as I do not want to take the trouble to sort pictures and match that to inventory on hand, if you are not getting the emails. Let me know.

I get a reply

Monday, November 07, 2016
Hi. This is the email-send program at dns2.nenana.net.
I'm afraid I wasn't able to deliver your message to the following addresses. This is a permanent error....Connections will not be accepted because the ip is in Spamhaus 'list;.....

Bla bla and more…

In our continuing efforts to protect our users from unsolicited email, Yahoo Mail doesn't accept SMTP connections from:
Dynamic or residential IP addresses as determined by Spamhaus PBL
We recommend you review our bulk email standards and best practices
If you think that your IP address has been listed in error we suggest you contact your administrator so they can contact Spamhaus. Once your IP is delisted by Spamhaus, Yahoo Mail will automatically unblock your IP within 48 hours.

I follow the link and read:

The Spamhaus PBL is a DNSBL database of end-user IP address ranges which should not be delivering unauthenticated SMTP email to any internet mail server except those provided for specifically by an ISP for that customer's use. Server patterns are consistent with end-user IP space which typically contain high concentrations of "botnet zombies," a major source of spam.

I read an insert notice:

Use of Spamhaus's free public DNSBL service is restricted to low-volume non-commercial users only. To make sure you qualify for free use, please see the terms:

The below indicates my customer may have applied a block:

Before including this block list as part of your mail

I read a caution:

Caution: Because the PBL lists normal customer IP space, do not use PBL on smart hosts or SMTP AUTH outbound servers for your own customers (or you risk blocking your own customers if their dynamic IPs are in the PBL). Do not use PBL in filters that do any 'deep parsing' of received headers, or for other than checking IP addresses that hand off to your mail servers.

There is a lot to read, much is over my head. I follow a few links. Pages of this kind of stuff too boring to repeat, but possibly containing answers to why my customers are not getting all my emails. A few things are not making sense to me. Maybe my customer has deliberately in the past blocked me out. Perhaps once in the past he and others did not want to get messages from me, but forgot. Did not want their name associated with mine during my time on the federal hit list. I can only speculate, but probably never get to the bottom of what is going on. I may not be meant to ever know what or why! One possibility is the Feds did the block, not wanting me communicating with certain customers they think deal in animal products. Or giving me a hint, they are not pleased I am talking fossils.

But here I am trying to be legal, having 'strange' problems trying to run my business. It does seem odd that an ordinary internet customer should have to understand such deep levels of issues I'm told I have, just to send and receive email. Is this normal? Does everyone on the net have these issues? I have been told by trusted customers that for sure the government launched an investigation on them through their link to me. Questions were asked about me, their transactions with me and what else, who else, they buy from. Followed by threats.

Invention—my propane buddy heater is now a cooker as well. Designed a screen that folds up and locks.

# CHAPTER THIRTEEN

## LOSE WEB SITE, LOST EMAILS AND ORDER SHIPMENTS

I get another email concerning a customer I informed of the release of book seven.

> **Hi again, Miles-**
>
> I haven't pestered you for a while! Neither your website nor your Amazon page has any update re your # 7 book. I am guessing that your emails, phone, etc is still "monitored" by "Big Brother". But, you are not responsible for emails sent to you by someone else (like me!). It did occur to me that in the book, you may have offended 'BB' by saying some truthful but unflattering things about the time when you were 'a guest of the taxpayers'! And that may have delayed the release of the book! Anyway- whatever the reason-Here's wishing you well!
>
> **David**

It is easy to be paranoid. Some customers encourage me in this direction. I had book seven up and running on the web site. There should have been no reason a customer cannot access it. I am unsure how some aspects of the internet work. I have seen delays, or my view of a web site is not the same as your view of the same page. Even the content, or the pictures can be different.[1] A customer can call and tell me they are on my web site, telling me what picture they are viewing. This is not my view of the same web site. In this case, I asked Dave to go back again, gave him the link, and he was able to access. I rather doubt it was part of a government conspiracy against me. But here we are. Some people think so. It is certainly possible. There is a partial doubt of what the government is capable of, likely to do, really did do, could do, or will do. That fear, concern, or knowledge, of capability alone is a threat, and form of power and control. I do believe the government is happy people believe the false information, and encourage that belief and fear.

When I was interrogated the opening line was, "We know everything, we have satellite pictures of you on the river." I did not fall apart and wet my pants. I waited for the punch line. I heard, "You and your friends." I knew then they know very little about me. I do not have such friends. I travel alone, always. But there it was, wanting me to believe I am watched, recorded, controlled, everywhere I go. There is no escape.

My more major concern is, if the government does not want me to sell fossils, tell me. Why illegally block my business, play games with me, or secretly monitor, collect data, and wait for years before pouncing on me with charges. *As was done already*. If I had been told there is a problem, we might have talked, worked it out, and saved us both a lot of hoop-la. I suppose if someone was a drug dealer you would not tell them you know and ask them to please stop. You'd never get evidence to convict them! I would ask, "Is the objective to lock up as many criminals as possible? Or create a safe environment at the least cost with the least damage." If anyone had said, "You know, you are not really covered by subsistence laws as you believe. The government is not pleased!"

I am not a "Screw the government!" type person. I'm not stupid. I'd have asked, "I am trying to run a business, so how may I do so in a way you are happy and I still meet minimal survival income?" I could have been told, 'Hey you have skills! We need people who can identify fakes! Even identify what animal part this is!' I'd love to help. Have a job, get paid.

It is easy to blame someone else! It comes across better to say, "The government did it!" Get a reply, 'Those stupid bastards!'

"Yeah! You and I (bonded) against 'them'; the common enemy!" Much better than, 'Stupid me!' I notice privacy on the internet is not treated the same as privacy on the phone. Conversations used to be sacred and protected. Breaching this law was serious. A President got impeached over it. I believe tapping into emails, collecting, monitoring, is common, and not illegal. The internet is not considered private. I believe the government is profiling every single citizen. What I am unsure of is how, what, or when the information is used. There was serious talk, people who vote a certain way or register democrat, are more likely to get audited. I'm told, this has been proven. If so, what else is possible?[2] I know certain key words in email headings, maybe even in text, get intercepted.

Bomb, kidnap, along with racial slurs are found and understandably monitored. But what about words like ivory, furs, Trump, republican, abortion? I can go in my own computer and find any word I choose, wherever the word shows up in any document in the computer, and change it to another word anywhere it shows up. I can change 'republican,' to 'democrat.' I get a notice, "Found 500 incidents." I can click, 'change all.' Phrases can be found like 'Save the whale,' or 'Stop drilling.' Any phrase can be easily found and replaced, anywhere it turns up on the internet. Found, watched, altered, deleted, monitored, controlled. My question again is when, why, how?

"Iris, Trump is saying the entire system is rigged. Including the election ballots." Once selected, label a message spam. No internet communication allowed, using those words. There would be an outcry? Only if you knew, cared, disagreed, it was allowed.

I noticed over a long period of time—several years—a high percent of my mailed packages with animal parts were slow to arrive at the other end, never did arrive, or arrived severely damaged. Where this rarely happened to shipments of rocks, wood, or other materials! *How can that be?* How could my packages be selected? It's easy. The postal system is computerized.

My address shows up in a computer anywhere in the country, or world, and it gets flagged. Once flagged, anything is possible. More than once, customers wrote, "It looks like someone took a hammer to this package!" We at first assumed this is due to random carelessness in the postal system. Then assumed some postal employee who is an animal rights person is deliberately putting a monkey wrench in my gears. To say, "The government is deliberately targeting only my packages!" would be ridiculous, paranoid, and get no agreement this is possible from the sane portion of the public.

A situation arose with the government where I was told, "Do this and your problems with us will go away." This is off the record. I agreed, and suddenly my packages got through, on time, not damaged, as promised me. I only know about this case, I am not trying to spam anyone. I am very careful to discuss my business only with those who ask, and are for sure interested. I do not send anything out in mass. All my emails are one of a kind, written individually, to one person at a time. I'm even told this view is hurting my business, as in this day and age, selling is about reaching the masses, sending out automatic notices to thousands of people. I do not like it to be done to me, so I do not do it to others.

"Honey, I'm considering starting and running blogs. Offering a mailing list. Perhaps sending out bits of knowledge, 'how to' tips, excerpts from my books, whatever might be of interest." Iris does not know what a blog is. So far, I am not getting much interest. I started a whole new web site using a different domain on recommendation from my book helper I hire to get my books in the Amazon format and accepted. He owns a small publishing company and has knowledge I do not. So far, the site is not justifying costs. Oh, it began as 'free.' But to sell anything, there is a fee. Then a bigger fee if you want it to go through your system, like pay pal, or bank. To do it their way requires knowledge I am not understanding. Or I refuse to give them the routing number to my bank. I forget the issues now.

My own web site is now not paying for its costs. Even though I have recorded 200,000 visits since I got out of prison. Less than a dozen sales a year, all minor and often for outdated items I can't find. I have little use holding inventory for years, organized and unavailable to sell anywhere else for fear of not finding it if it sells on the net. It is hard to understand how this many visitors arrive, with so few inquiries or sales. This is not matching twenty years' experience where four out of five visitors bought something.

I lived a simple, subsistence life, living off the land without electricity for over twenty years! However, many remote people understand the internet is a way to reach people, either family, or to run a business they are in charge of.

Few people have indicated they could not order something from me. If emails were blocked I'd run into people at shows who told me they could not get my site to work, and how come I did not reply to an email. I suspect that my earlier animal product sold well. Maybe wood, metal, stones, are a dime a dozen in search engines. Lots to choose from. Computer geek friends say my web visits could be robots, not real people. This could be something new that did not exist in the beginning.

I'm told my web site looks twenty years out of date. My web program is outdated, with outdated slow code. No video, maybe not user friendly as it had been when I was on the cutting edge. I do not know. I first paid for Dream Weavers, the top of the line program. I could not understand it, and am told it is outdated and no longer in existence. It seems odd that this same type inventory on the net, sells at shows. Even sells on other people's sites when on consignment. I'm reluctant to start from scratch and rebuild my site from the ground up, learning a new program. It's sixty-five pages and a thousand pictures.

I now think much of the web store format is much more standardized than when I, and internet selling began. In the beginning my innovative ways to offer product were learned and explored by customers. In time, customers are in and out faster, so do not have time to figure out how my page works. I had no page navigator or search capability. Customers got used to the accepted standards, and expected this. The internet store scene is not new, fun, exciting anymore. Maybe buyers simply want in and out with the least fuss.

The Weebly site I'm exploring, uses modern code, is interactive, offering the latest of everything. *Should I make my own site go away and go to Weebly?* I hate to keep paying for both sites! I do not think my web program translates into what Weebly uses, to simply cut and paste pages. Someone could make it happen maybe, for a nice fee. I have had my site forever. Hate to dismantle it. Would it cost less to make it one page and a redirect? I sigh. I have a lot invested. A lot of heart and dreams. But it's the past, in that I cannot bring up that level of Rah rah again. I'm a senior. My rah rah energy is limited, followed by nap time.

I KEEP a diary in the computer. Sometimes the diary note ends up in the book. When I write it in the diary I have not decided if I will keep the entry or not. Sometimes the diary entry becomes more than a note. I move it over to the actual book. As my web frustration progresses I move this diary entry about the subject to the book.

**Friday, November 18, 2016** web internet issues continue

My web problem has become more complicated and is not an email and internet access only problem. First there was the frustration of extremely slow page uploads to where it became almost impossible to upload new items to my web site.

I investigated, and see a .5 meg upload speed. When I am paying for 10. For a year or more I have had slow speed and been unhappy. I can make a pot of coffee in the time it takes to download a page. Getting worse over time. Mail will sometimes time out and not download. I can't get updates. I sigh, and accept it. Most everyone I talk to is having internet connection issues. Few think their connection is fast. 'Faster,' is solved with a $100 a month satellite connection.

A lot of time is spent within my computer looking for errors. I do diagnostics, compatibility checks. Get reports that all is fine, no issues found. I wonder if my computer is now too old and outdated at seven years. I may need a new computer? Out of memory, who knows what? I run an on line test that says my computer memory is great, operating speed is above normal.[3]

I call MTA, my web provider. They tell me my connection is fine; they have a speed over 100 recorded as far as my connection. The issue has to be within my computer or with the web domain provider, another company, BBN which changed to United Utilities. The helpful lady at the desk is looking around, having me test different things. Disconnect, connect again and such. The secretary tries this and that and goes "Huh, I have not seen this before." She cannot access my IP address. As if it is hidden.

I just now put my fossils on the web page. Just now sold my first piece on the internet in four years. At least possibly I am not in control of the business. The Feds control maybe my computers, or my web site, or my internet connection, or all of it. Watching. Waiting. Approving some things, not allowing other things. Some selective emails are not getting through. Usually ones that have to do with the sale of fossils. Why only those emails? Does that sound like a technical error? I am asked by more than one friend, "Why do you think the Feds took your computers and then voluntarily gave them back?" Followed by "I guarantee they did not come back in the same condition." Followed by, "It is likely the Feds use your computer to monitor emails and your web site, as a way to find and get evidence against your customers. In the same way as a drug distribution house would get wired and monitored, then allowed to reopen."

No I can't believe that! Because what would the implication be? ☺

I end my web subject in the diary.

---

I HAVE thoughts now and then of Flower, the tribe, the life they live. How it is dangerous, but not complicated. No internet problems, social issues solved, even if it is with violence. I may have been happier back in my earlier years living much like the tribe. I a sigh.

It's Friday before Thanksgiving, the day the seniors serve Thanksgiving meals. I'm coming back for lunch! But have things to do in the shop. Another batch of wood and fossils coming out of the kiln after being stabilized in resin. *"These things sand better when still hot and the resin has not yet set up rock hard!"* I check the internet and have a connection for a short time, long enough to get some email and the address of the customer who bought the mammoth pieces. I get to pay pal, get to my account, and find the customer, but cannot open the page to print the shipping label. I can open some sites but not my own domain, or eBay. Google shows up but no sites.

Spending hours looking, going here and there. One good thing is, it helps me learn and understand my computer. How it works. I know a lot of places to look when I have issues. This is like a review. I can talk a tad of 'Geekeez' the language of the computer techs.

### Saturday, November 19, 2016 Diary

Web issue continues.

Our morning ritual is breakfast at the local café. The usual group is here. I never inquired, but think some of them are here every morning. Iris and I used to have this coffee ritual every day. This is a time to talk to her, relax, as we do things together, socialize. Only Saturdays now.

We began to hang out at the seniors instead. We get caught up on the news, visit with locals. Mad Jay hangs out here, and Dinner and his wife Cook. Known for having parties with good food several times a year. I have known them for years. I think I could say, they have had an impact on my life. Both are pillars of the community. Cook has been on several councils I was on - the library board and she does a lot for the community, mostly children. I value her advice. We used to be dance partners ages and ages ago! Maybe thirty years ago. Dinner is not as social as his wife.

He helped me get interested in bees. Like me, he is a custom knife maker, has a special forge and specializes in making Damascus steel. He used to work at the school, a computer expert, and she was the accountant for the city. Now they own homes to rent out and can afford to go on vacation overseas and to southeast Alaska to fish for good salmon.

Dinner and I show each other our knife work and exchange ideas on prices, 'how to' information and such. Today he shows me new steel, "This is what motorcycle chain looks like made into a knife blade." I'd heard of this, so am interested to see it. There are dots that were the chain rivets, which is a different kind of steel.

Dinner is friends with Knife, a subject I usually avoid. Knife has a club or class teaching or sharing knife skills. I have no ride, and do not want Dinner to feel obligated to take me. Knife and I are not on speaking terms, or so he says, or wants.

"Dinner, you still up on computer stuff, or is it outdated for you now?"

"I did not keep up with changes, not an expert anymore! Why?"

I go into the basics of my computer internet issues. Mad Jay jumps in, "It's the Feds again, got into both of your computers!"

Before Dinner can reply, the owner, Mrs. Carver gives her two cents worth, "Why didn't you just get rid of your computers!"

About everyone agrees the Feds did not give me my computers back in the same condition they took them! I agree, but reply, "The issue can as easily be with my domain web site, my email account, my server. My entire internet connection could be rerouted by the Feds. I could have an infected disk or thumb drive. A new computer is no guarantee I am rid of the Feds, if that is even the problem." Dinner asks me some questions about my symptoms. "Well, I was at the senior's and tried to access my email and crashed their computer with a warning of illegal activity!" This tells me the issue is not my computer.

"Is it all your internet access, Miles, or just email?"

"Everything; even using the laptop." We are all stumped. I joke, "Guess I can go crash the library computer trying to access my email!" We laugh. "Maybe, Dinner, we are at the end of a line and out on the boondocks with primitive make-do equipment. Hackers may discover us over in India or wherever and easily hack in as outlets for spam?"

No one answers, but think about this. I add, "I had issues a while back and it turned out to be my domain. My email was insecure due to being associated with a web site that was unstable. Because of this, my entire site got put on a blacklist as suspicious. Many sites would not allow a connection or accept me accessing them. I had sporadic service with returned emails and connection issues. Having one company supply my internet and another a web site is in itself suspicious. Computer robots that look around for anomalies, suspicious behavior, look at this and wonder if someone is trying to hide something. An expert had to go in and manually override the automatic system.

I pause, " It could be the same issue again. Only now no human intervention is possible, it's entirely automated now." I mention I once accidentally put a partition in the computer and the computer could not see it. Could the Feds do that? Is there a way to find out? I wonder if I took my computer to a geek in town could they tell me? I had the laptop looked at when it was in for repair and they did a scan and looked after I told them the Feds might be in it. They saw nothing. Can the Feds bypass anything and everyone undetected? Or would geeks be ordered not to tell? Under threat of who knows what? National security laws? In charges against me, it was written that illegal ivory trade supports wars against the country and international terrorism. The implication is, I am involved in, and supporting, international terrorism.

Dinner simply says, "I think the Feds are in your computer." He agrees I'll never prove it. Nor will they ever admit it.

I ponder again that my web site has had almost 200,000 visits with only five sales. I buy programs, pay internet fees and advertising, tweak my product, optimistically stockpile product because I believe. I wait days, weeks, months, years, while the Feds hit a button that castrates me, or secretly collects data about me and my customers as I wait for sales. I call the experts and providers; they all tell me "Everything is fine!" Do they then call the Feds and tell them I am asking questions and might be on to them, and collect a reward? So the turn of the screws gets tighter. If true or not, it is possible. Even plausible.

### Past flash

Hmmm. I buy my first computer. It's used. I acquire programs off the street on CD's with no labels. I did not appreciate Bill Gates telling me I am only renting the computer and do not own it. My usual control issues. It is a new concept to buy something and not own it. I have an installation code for Word. I suspect it was pirated, but have no actual proof. I did not get in this situation on purpose. I had paid a shop to custom build a computer at three times the cost of a standard one. I was not told it would not have Word. Word is $500.

The installation works. Down the road my computer suggests an update. I vote 'No'. I believe I control the computer; it does not control me. I suspect the issue involves having an outdated program, but am forced to get an update. I decide to try it, mostly I was outvoted by my computer. I am asked for my installation code and put that in. The computer stops and thinks. About fifteen seconds later the famous blank blue screen of death. My computer dies; I have to buy a new one. I'm convinced, indirectly, Bill Gates saw a pirated program. Rather than ask if I'd like to buy the full meal deal for $500, he destroyed my computer. No questions, trial, charges. Could I prove someone hit a button and destroyed my computer? Not likely.

"Buy my program, and no one else's, or else!" Can Bill Gates, anyone, the government, do that? Observe, and behold. "Oh. Sure, trying to upgrade a pirated program and my computer crashing at the same time is just a coincidence!" I understand all right.

**Past flash ends**

*Iris and I selling at the Tucson show.*

# CHAPTER FOURTEEN

## FEDERAL GOVERNMENT TAKES OVER STATE, REALITY TV COMES TO NENANA

Iris saw a movie when she was younger that she loves and wants me to see. She orders it on line for a dollar. "A Beautiful Mind." We watch it together. I assume a true story about a mathematician who gets the Nobel prize. He is schizoid; has friends no one else sees, experiences that never happened in the real world. Spends years under the care of a shrink. Drugs, shock treatment. Candidate for a lobotomy. A guy about to be turned into a dysfunctional zombie. He's going to be involuntarily committed, like for the rest of his life in a Looney bin. He, of course, does not want to go! He is lucky to have a wife willing to take care of him and deal with whatever there is to deal with. Someone to help him figure out what is real and what is not.

The turning point is his realizing a little girl he knows is not real because she never gets older. Few can understand his ideas. Kind of an Einstein story. He questions and explores what reality is. He wants to believe in only the provable. If you can't prove it, how can it be real? His wife helps him understand the concept of love and blind faith. By the luck of the draw he found such a partner. He could have easily been, and was for a while, just another locked up nut case. Instead, gets a Nobel prize. After all those years of getting laughed at, not fitting in well socially, being one track minded and what not. He becomes a respected university professor.

I wonder how many others are like this. People who do not accept things as they are, but have to tweak everything they touch. Thinking they can make it better? I want to know what makes it work. I wonder what else it might do. I'm curious if it can be adapted. Or there is a situation and no answer has been discovered or invented. I feel sure I could design and make something to fill that need. Like the arrow I made that shoots out of a shotgun .Because something must be able to come out of this gun that is more accurate, and goes further than this outdated lead ball!" I had a need for such a thing. An option that was not a modern rifle, an answer using old technology. I made the arrow, it worked. I had no money, no connections, no lab, no way to produce more arrows; I was simply satisfied it could be done, but not worth the work to build them one at a time. I moved on to other things. The one arrow went twice as far as a shotgun slug with deadly accuracy.

I am not some 'brain' who deserve special recognition or a prize, etc. That happens usually in dreams and movies of rare people, one time anomalies. I wonder how many people are simply different. Not really 'sick' needing to be shock treated, drugged, and locked up. People who simply see the world differently than the norm. In a harmless way. Yes, well, it could also be said there are not enough people identified as dangerous, who get left alone to be free to do their dastardly deeds! The drug age creates a lot of real wackos, that seems for sure.

---

"Time to get up, Honey." Iris wants to get up at exactly seven most mornings so we can get out the door and be at the senior's by eight. She lifts her head and groans. I have come in from the shop. I usually get up between four and six. Getting her up at exactly seven, not a minute one way or the other is a big deal to me, whose personal time needs to be accurate to the nearest week. But no excuses! I keep track of the time while I work and make sure I get in to wake my wife. To help her wake up I say, "Oh, look honey, a white bird! Loose in the house! Hovering over your head!" She is used to my shenanigans and practical jokes. In her half sleep state she wants to know what I am up to now, catching her while she is not fully alert. She assumes I am up to something. I take on a hurt tone. "No really, look, a bird." She opens one eye, and there is a bird! There it is, white, fluttering over her head in the dim light! She is ready to be a believer. A miracle has come about. She can't figure it out.

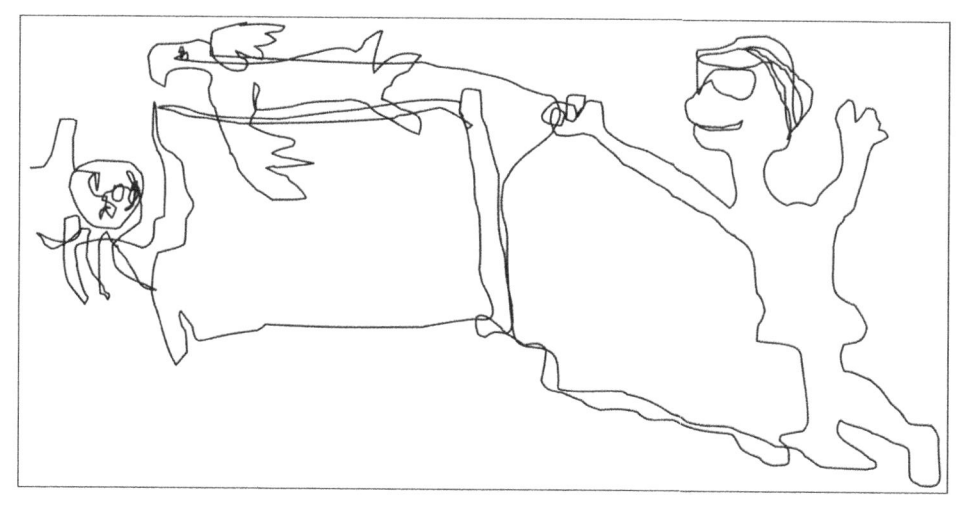

*Look! It's a bird in the house! Wake up.*

I have a cut out bird made of white cardboard on the end of a curtain rod she can't see in the dim light. I spent half an hour building it.

"It's magic, Honey, look! A dove from heaven!" She still can't figure it out, as I move it here, there, so she only gets a glimpse of the white bird. Is it part of the dream she was having as she wakes up? She is not sure. Then a shaft of the nightlight reflects off the curtain rod.

"That's you!" Puts her head back on the pillow to catch another minute of sleep.

"Good morning, Dear!"

I bend over and give her a kiss as I do every morning. I'm not good at showing affection. Call it baggage from the past. Somewhat like in the Beautiful Mind movie Iris and I just saw when the main character says, "So are you going to slap me now!" I tend to lack the social graces of normal intimacy. When his friends asks how he is in the movie, he happily tells of his ongoing work. His friend says, 'No I mean other things in your life.' The main character frowns, "What other things?" All he knows is his passion for what he does.

On days Iris does not need to get to the senior's and can sleep in, I try to get in the house from the shop and hang out in the house about the time I think she will wake up, so I am here when she opens her eyes. I do not want her to wake up in an empty house. I do not even know if she notices. It's just a silly thing. *Probably does not make up for my many inconsiderate acts.*

Because the Internet connection is getting slower, I have been reading books instead of using the internet as entertainment. Computer scans, tests, and what not has been occupying days of time. Like thirty hours into a full scan now and not done. I suspect the scan stops when the computer goes into sleep mode or energy save mode. This happens after about five minutes of nothing happening. I'm sure there is a place in my settings to ask the computer to stop that, to wake up and let the scan run! But I can't find that place to give the command, not trying hard, no time, so let the computer run as it does, and the scan will get done, eventually.

**Pilgrim's Wilderness. A true story of Faith and madness on the Alaska Frontier** by Tom Kizzia. I had heard of this book and somehow ended up with a copy in a pile of books I hope to get to one day when I have time. The subject came up recently, so I decide to find my copy and read it. A true story. I'm unsure what it is about, except a rough idea about this family with land problems, and social issues that get them in trouble.

This is a story that touches my life. I met the family I think. They played music someplace where I was dancing in my youth. One of them, Neil, has bought art from me for his lodge over the years. I know families similar to this; I understand parts of their story.

A wilderness family with about fifteen children. The story starts with them wanting to get away, be left alone. Eccentric, weird beliefs few would live by, understand, or accept. Persecuted, much like the whole reason our country was founded! The focus is on a land dispute with the government which got into the news and created interest. This ends up one of the first court cases over new Alaska land protection laws. The story gives me insight into my own complications over land. Anyone who realistically thinks of being 'off the grid' needs to understand what wilderness land they can go to and under what conditions.

In 1980 Congress enacted ANILCA—Alaska National Interest Lands and Conservation Act.[1] This was the biggest act of wilderness preservation in the world. 100 million acres. Not everyone at the time understood what this meant. The exact details of what this looks like had to be worked out. Some Alaskans maintain, (I quote this from the book,) *"Written in haste to cover the building of the oil pipeline."* The purpose was spelled out, *"Maintain unimpaired the scenic beauty and quality of the land."* Congress wrote in special provisions for Alaska, different from federal parks in other states, to protect the lingering frontier lifestyles of rural Alaskans. But what exactly does this mean?

I make a distinction between park and preserve. Park is more strict, preserve means to 'leave it as it is' which may well include homesteaders and small communities who depend on subsistence. Those in power trying to be fair and do the protecting may not have understood what it is they are trying to protect! Some—most—cared, and tried to figure it out. But some flat out were opposed to giving special rights to a bunch of undeserving hippies and Indians. *It is after all, discrimination, which we say we do not support!* The Pilgrim long hair hippie family, is a case in point. Part of this conservation act was to protect and preserve basic subsistence life in and around a federal preserve in small remote communities off the grid in the wilderness.

This is how the Pilgrim family read it and understood the situation to be. Exactly what they are looking for! As I did, and many others! *"Alone in a wilderness surrounded by a remoteness God created that will never become industrialized."* No fighting off civilization! It's written in the law! The Pilgrims legally buy a piece of land in the preserve that is grandfathered in, that was once a mine claim. There is an old, unused, impassable access road, assumed to be legal access to the land. The law states access to legally owned land will be made available. There are no park rangers around, government is not welcome here, and the locals use the land and manage it well enough. The access to the land has always been used.

Pilgrims put all their money, work, and dreams into the biblical prophecy come true now their home. *No more civilized problems!* The locals sort of accepted them, saving judgment until everyone got to know each other better. This is how wilderness communities can be, trying to be open minded, understanding they are all considered weird by the outside world! The park service states they sent an agent on a friendly visit to introduce themselves. The Pilgrims were not interested in being introduced or obligated to be friends with the government, who they have had past bad experiences with. *"If you are truly honorable and care and wish to be our friend, go. Leave us alone in peace!"* I sympathize and understand.

"I'm with the government, I'm here to help!" Yeah right! It's a standing joke! The best help you can be is to go away! It seems so much of what the government touches it screws up.

The government's view is, it is their job to manage this preserve. To manage it, they must know what is going on. To know what is going on, they have to stop and visit. If you live in a preserve, you need to understand there are certain obligations and rules. You can't just move into a preserve and do what you want, exploit the land all around as if it is your personal planet. When the agents explain this to the news media and civilization in a reasonable calm voice it makes sense, "Parks belong to all of us." Civilization nods and agrees it is reasonable to go visit the Pilgrims to see what they are up to. The Pilgrims call that trespassing. The right to privacy, as long as you are obeying the law. Agreed! So now the park service wants to find a legal reason to go visit with a warrant.

The governments assumption is, "The Pilgrims must be up to something! After all, why else would a reasonable legal person not want the government to stop in for coffee, ask some basic questions?" Rather than take a chance, the Pilgrims will scream harassment. The government tries hard to find laws that are being broken so they can get a warrant for a search, and if all goes well, the Pilgrims will show the true colors the government wants to see, unacceptable anger, hostility and damage to land that belongs to the people. This would justify the government's reason for imposing. I understand, because I have had it happen to me. The government wants to come out of this being right and looking good. Some officers want to win points by making this happen, to the extent they will intimidate, harass, do whatever it takes until they get a rise. Even a rabbit will bite if you poke it enough. There is a somewhat common personality type like the Pilgrims who fear the government and want to be isolated. You only need to say,"Boo!" in the dark and they come unglued, lose it, come out screaming and shooting in 'self-defense'.

The park focuses on the fact a road is being built in a park! This should outrage the public! Factions form. Each side has an agenda with something to prove for their own reasons. There are environmental protectionists who want no one on the land in a federal park doing anything at all.

"Leave it pristine; untouched in any way by man." Said of coarse by those totally unaffected, not even living in Alaska. There are those who favor doing away with all parks; stop creating them. Let's get the oil, timber, and other resources, instead of protecting resources as the needy starve!" There are rural subsistence people and those who wish to protect the mountain man life of freedom, being at one with nature that big government is trying to stop.

"Let subsistence people in, but keep industry out!" There are government employees justifying and protecting jobs as caretakers, enforcers. This Pilgrim case becomes a focal point, a line in the sand, encouraged by the news media! 'Another Ruby Ridge! Waco! Stay tuned to this channel!'

The homesteader protection faction says they are promised access to legal land, and this access the Pilgrims opened up is an established road which is a protected previously approved access. However, the road was approved before it was a park! Does it qualify as a historical trail worth protecting and making available as grand-fathered in? No one applied for such a permit! You are supposed to apply and have it approved! The Pilgrims opened the old road up with a bulldozer! I understand the public's view would be the word 'park' and 'bulldozer' do not belong in the same sentence! But everyone in the area is a miner with access to mining equipment. This is how roads are opened here. People come to town for groceries riding a bulldozer! The local lifestyle needs to be put into perspective. *No one complains the gas exploration people behind Nenana are bulldozing hundreds of miles, just so they can go out and blast seismic grids.*

Now the focus turns to the Pilgrims using land not actually within the boundaries of their purchased lot. "The horses are grazing on public grass!" The government sends helicopters out almost daily, then pays for a survey to prove the horses are indeed eating public grass. They clear the property right of way. Laws vary, but sometimes the original homestead or mine survey requires the surveyor to brush out only two of the four boundaries. The owner may or may not wish to make the boundary sides more clear, depending if they expect neighbors or not, or expect a dispute. It is easy to side with the Pilgrims, as I read along. Iris makes a case in point. She has friends she tells me about who just wrote us.

"Miles, they had serious house damage by hurricane Sandy. The government came in to help." I nod. I recall they wrote many years ago when the storm hit! FEMA paid to have everything but the main structure torn down. Contractors built a new house using the same framework. The government would not approve the house. I think because it was not on stilts above the flood plain. The house was torn down again… completely. A new house was started. Meanwhile, years are going by in which the family has no home and is living in a hotel. The family assumed they were going to have a house soon. They see a finished house for the second time.

What did they write Iris?"

Iris quotes from the card they sent, "The new finished home is eighteen inches too close to the sidewalk. So far there has been no solution. We can't get hold of the contractor and we are still living in a hotel." The entire home may have to get moved eighteen inches. Not exactly the same subject, but having to do with boundaries, rules, access, and homes connected with firsthand knowledge of government help. I also had similar FEMA issues after the Nenana flood. I would have been better off without them.

People flying over the Pilgrims place, point out the park service killed more trees, did more damage to the land, and made more of an eyesore cutting out the land boundary than the Pilgrims have done trying to live at peace with nature. In my own mind, I wonder how the park can claim that the Pilgrims damaged public land? At the same time, I can see why the public would not be happy with the Pilgrim family and their objectives. I have seen similar stories played out over and over, that run a similar course using the same arguments. I'm very familiar with what is at stake and what is going on. Not all stories make the headlines. The park is now requiring a permit for road access by the Pilgrims, that it never asked for before. The Pilgrims argue that acknowledging a permit is needed, means the possibility of having the permit denied, charging a high fee, or later taking the permit away. [2]

I understand. The permitting office is hundreds of miles away. The family, by definition is living off the land, poor, and may not be able to even get to this office. Like me, who does not drive. There is local outrage, demonstrations, acts of defiance, a big community meeting. The park is represented by armed troopers with flak jackets and rifles at the ready. This in itself is an intimidating act of authority that creates enemies. Yet many of the town people are also armed.

"If you can be armed, so can we!" Neither side feels safe. But only the safety of the public servants is recognized by society! Some hick town hippie freak gets shot by the swat team and good radiance. Or an accident, or justified due to fear for the officers life. An officer gets killed and its world war three on the community. Can a local resident also claim he was in fear of his life as self-defense? And it is all about what? Hundreds of thousands of dollars are getting spent for what? To accomplish what? The news media ready with the camera, "Here we go again!" Waiting for breaking news of who shoots first. Even Papa Pilgrim says, "Someone may die here." Troopers admit, and they are prepared for, the same. Each will stand to the death for what they believe. But it's not a Mexican standoff because we all know there can be only one outcome.

It's possible that way back in the beginning, if the Pilgrims had cooperated, been friendly, invited the rangers in, promised to be good, protect the park, follow any laws required, sucked up, kissed ass, bowed down, the rangers would have been satisfied. Left, and never come back. All would have been well. Possibly never a next time. The Pilgrims are the sort that do not get on their knees saying, "Yes, Masta. I obey, Masta!" And frankly, those types tend not to make it in the wilds. A lot of wilderness people would say to an officer, "Get out, do not come back unless you have a warrant!" To civilized people this is usually out of line. I wonder if this is just cultural. In the wilds people can simply be more gruff, rough round the edges, with a no nonsense view. After all, the entire reason to be here at all is to be left alone!

The plot thickens, otherwise the story would not be a book. The Pilgrims have a base in the community to store stuff and stay when they come in for supplies, as many bush people do. They leave a mess, by civilized standards. I joke, "You can tell when you are approaching a homestead in the wilds when you start seeing blue tarps, five gallon buckets, and wrecked snow machines along the wilderness path."

It is suspected the Pilgrims are stealing. This has never been acceptable by bush standards, but more common than we like. Being poor and wanting to survive, some people will do anything before giving up. Certainly the Pilgrims are running scams, taking advantage, bothering others. They take tourists out without a permit, taking business away from the local taxi and guides. At first locals felt this was acceptable until the Pilgrims get established. Guides and taxi accepted some loss of business to help out a new struggling family. Locals felt they gave an inch and the Pilgrims took a mile.

This is a common issue small villages face. Kind of a lack of agreement on what being polite and nice means. Handshake deals require a high level of understanding, compared to written contracts city people abide by. Smaller communities would prefer not to have the confinements of contracts and cost of lawyers. Aggressive newcomers see marks, move in and take over. If you have been around for a while you pick up on what is locally acceptable in agreements.

The Pilgrim children are not in school and have never been taught to read. The spiritual beliefs are more than weird, even more than fanatical, but abusive. It is suspected, the father is having sex with his daughters. *So this is why society wants to know why you are alone out in the pucker brush, and wants to know what you are up to.* It's not just about Mr. Pilgrim, but underage children, a wife, and infringements on the hospitality of poor locals who can't afford to support users.

The Pilgrims at first won the locals over when their children put out a fire at the local store! Then the family came over and did repairs! However, it was later suspected they started the fire in the first place. I have met people moving to smaller communities exactly like this. Create a problem, then solve it.

"Like buddy Foil, Miles! Steals your boat gas tank, offers you a solution to protect your stuff with the long term solution requiring paying Foil rent." *It happens in civilization as well. I suspected Norton antivirus creates the computer viruses we pay Norton to protect us from.*

While Papa Pilgrim uses the word, 'subsistence' and not bothering anyone, taking care of himself, his family collects the oil revenue money. Times fifteen children is a lot of money, as in fifteen grand, that comes out of everyone else's fund money. Likewise, they appear to be running a welfare scam. Then expect neighborly help, in the old pioneer spirit. Over time, it appears Papa knows how to manipulate a crowd, and put people against each other in a Charles Manson sort of way. The community gets divided, little can be proven, and here is pious Papa with mindful polite children. Kind forgiving reasonable people say, "Give the family the benefit of the doubt, base decisions on facts and proof." One reality I suspect is, some scammers are very good at what they do. There never will be any proof that could hold up in court. For what is being said about the Pilgrim family is also said about the government.

Possibly nine out of ten government agents will understand and work with the remote person. It only takes one to start a crisis. At least a few government employees get on board for the power it gives them. They enjoy hurting people, making others' lives miserable. Their job gives them a gun and the authority to act on their basic nature. Or the same damage done with a pen. There are other issues and agendas not in the land owners best interest. In this book, a high up official is quoted... *"I helped the park service high handed preference for secret land acquisitions and condemnations. I did not like this, so I quit."* How many others did not quit?

In one court case the judge is mentioned. *The same judge I had in my federal case!* The judge rules, *"That the guarantee of adequate and feasible access to in holdings in Alaska was subject to reasonable regulations."* A park official is quoted. *I met this same guy! One of the good guys.* "Started in National Park issues at Kantishna in the 1980s."

I was there and involved. I had a trapline and cabin at what was then, the free side of the Denali Park boundary. That became part of the park preserve extension of Jimmy Carter. The quote has to do with mining, not trapping, but similar issues with what can happen in a preserve. The public has more understanding of parks like Yellowstone, so cannot imagine anything like a bulldozer, mining, or trapping going on in a park! I understand! *But envision the headwaters of the Amazon, or Siberia.*

I saw a documentary where a big waterfalls was not protected. Some rich guy bought it and the surrounding land. He held parties where he'd have a burning log raft go off the falls so he and his drunk friends could watch the fire display. Logging was done for economics. Total disregard for what it did to the beauty and the future. This was during Roosevelt's time. The forming of the first National Parks with the word 'conservation' being coined. How lucky we are today that some of these pristine areas were protected from development! I and most people I know, support this; we love the land!

In the case of myself and those I know, we were on the back side of a park few but the likes of us ever got to. I spent twenty years cutting a trail you could not see from a plane. Hard to find or follow, even on the ground. It impacted the environment not one wit. Neither did my lifetime of trapping. There was as much game when I stopped as when I started. I saw maybe two outsiders in the area in all those years. A meager living was provided, where I took care of myself. It was argued when the preserve was established that part of the preservation would be a lifestyle for people like me!

Park ranger, Ray Kreig says, "*A third of the gold mining district was engulfed in 1980 by the preserve. The park set up visitor displays about Kantishna mining history once it had finished driving the last few placer miners out of business.*" Later he says, " *The Park Service is waging a heartless and fanatical war of intimidation against the Pilgrims with this investigation.*"

The government almost always wins. There are unlimited resources for investigation, punishment, and eviction, with hardly any dollars for prevention. That is what concerns me. As it happens, the Pilgrims were a pain to everyone. There were issues other than just land dispute. In the end, the park seemed reasonable with the family after Papa went to prison. Possibly all that negative park publicity did not look good, so this time and 'in this case' the land issues were resolved positively. I have observed situations that never made the news that ended in a shootout, or arrests.

There is more than one catch twenty-two involved in land preservation. Some people who will never see this land, never want to; it is their hope to simply know it exits someplace. A pristine wilds no one ever sees. This is a minority group. Most want a park they can visit. Like the Denali I know, where the equivalent of the state's population passes through each summer. Tens of thousands of people. Such visits require a road, restrooms, places to warm up, eat, visitor information, viewing spots, busses, garbage cans, bear proof containers, observations turn offs, roads, and on and on. Visitors protest, want something done, if they get mauled by park bears. Doing wildlife studies, taking pretty pictures, getting land surveys done requires a guide, services, employees. This has been the function of the locals. This land is usually remote with no jobs around. In my view this creates a series of tradeoffs or compromises. There is no big industry, but at the same time, there is some limited use of resources so locals can exist, have jobs in tourism. This is my idealist view of how we could all get the most of what we want, compromise, and work together.

"You the tourist get so see the beauty, because I, the local who knows the area, can take you there. Or maintain the road, pick up your trash."

Even the Pilgrim family took tourists out by horseback and put them up in remote cabins for tourist's life dream of a wilderness experience. An experience not possible for civilized people without help from wilderness people. This is part of what a preserve offers. The Pilgrims and many like them are friendly enough when they are appreciated, complimented, treated with respect, when it's on their terms, and getting paid. I have had many visitors whom I guided for including biologists, government employees, wildlife photographers, news media, tourists, friends, etc.

"But, Miles, you said yourself, this unregulated activity of taking people out in the wilds interferes with those who are licensed to do so!"

I agree. There are professionals who paid to go to school, pay for insurance, have equipment certified and up to code. Who am I, or the Pilgrims, to take their business away without paying dues? In my defense, I do not see the people I take out as among the sort who will pay ten times as much to have the same experience with someone who is certified. Likewise, those who pay the big bucks are those who have expectations I cannot meet! It's the difference between those who shop at thrift stores and those who get it new from someone reputable with class and insurance. In most cases, two different sorts of people. Flea markets are not in competition with galleries.

I'm in business too. I go to a big show and bring business cards, displays, paid for advertising, as a professional. I take notice of artists who set up on sidewalk, pay nothing, have no overhead, no business card, brochures or advertising. Horning in on the publicity someone else created, worked and paid for. These street vendors sell art on the sidewalk near my booth. Sometimes they offer a deal, sometimes it is a rip off. Most customers who come to my booth are not shopping with the street people. Few of the street vendors have either the quality or the variety I have. I understand both views. Because I started out as that street person selling with no overhead out of a shoebox on the street. Today is different, but I see merit in the life of the street person. I used the same arguments! 😄 :)

As a vagrant I used to say, "I make $2,000 a year, and $1,900 of it is profit! Is that cool or what!" Agreed. I now say, "I made $20,000, while you made $2,000 on the sidewalk in front of me. Is that cool or what!" I admit that only $5,000 is profit. I admit I work five times as long, doing things not related to making the art that is not as fun, in order to earn just twice as much money. Is it worth it? The answer is often a personal choice. So I view those like the Pilgrims, and do not complain about the street vendors. It is not hurting my business by enough to put a family out on the street. What hurts me much more is the professionals selling cheap imports and lying, calling it local handmade. Their selling in large volume hurts me. These types of sellers make huge profit to invest in connections with chain stores they have a monopoly on. But anyhow, aside from the possibility the likes of me offering boat trips to the poor...

Many read my books, hear my stories, and are happier people for it, knowing there is a frontier with trappers, homesteaders, mountain men. If a majority are not interested in being off the grid, many are or at least appreciate and understand and want to hear about it. I am there, so I take pictures for all to see, do art, write my stories, and in this way share its beauty, and even help protect wild places from bigger industry. It is the likes of me and my kind, who see what goes on and can report it.

I'm of the belief as well, there is no harm in diversity in our population, with many skills among us. We do not know what direction we may need to jump as our environment changes. I hope the price is worth the potential gain if wilderness knowledge ends up saving our species' bacon.[3] I'm fond of repeating, "People have lived subsistence for thousands of years. Modern man has been around 200 years. Most of our environmental problems have occurred in the past 200 years. So which lifestyle is more likely to be the problem?"

"Yes, Miles, but the problem is, modern subsistence people are not true subsistence when they have bulldozers, satellite phones, internet, and basically want the best of both worlds!"

I see the point. However, the very civilization those like the Pilgrims run from requires, by law, some level of advancement. Or your children will get taken by social services for example. Needing a ten dollar permit is a point I wish to make. Civilized people often say, "So you are too lazy and stubborn and wish to avoid a mere ten dollars?" Getting a simple permit requires you can read English, have transportation, have ten dollars, and be socially adept enough to conduct transactions with a government you are in terror of. (Like asking a Jew to get a permit from Hitler.)

"No thanks." Some people would rather die first.

I went through one issue I can think of. Big investigation. Karen and I had to pay for social services to fly out and investigate our living conditions in the wilderness. This cost us six months wages.

"Do the children have separate rooms?" A traditional old fashioned one room log cabin, but at least a loft, with a blanket divider to meet the legal requirements so Karen could keep her children. We had to have electricity, mail coming in, teachers showing up, permits of various kinds, papers to sign, meaning access to civilization and transportation to maintain and pay for. Fees to pay. By the time we are done, are we really subsistence? We heard a lot of, "Sign here we will bring you free food!" It is suggested we take the easy way out, put our hand out and accept welfare. It's hard to turn down. Just sign on the dotted line and goodies show up. 'No thank you,' brings a frown, not cooperating. Vague threats. Being on a list. Watched. Threat of having children taken away. An assumption of mentally unstable if you turn down free stuff. It was scary for us and had Karen in tears. Was a factor in us parting.

"Miles, my grandparents were in Germany. I heard these same stories, and I'm not going to live like that!"

Civilized laws make true subsistence illegal and those who choose this lifestyle outlaws. "Lock em up!" Becomes the public outcry. The Pilgrims feel this view is contrived, the public is manipulated, and the outcome suits the true exploiters of the land! While subsistence people in general are the opposite; the guardians of the land.

So, while I understand societies view of freeloaders who claim to support themselves living off the land as our ancestors did, there is another side to the issue. Subsistence fits in there someplace with poor. The best solution is to play your part as expected and required. What is expected and required is not subsistence as those of us who live subsistence call it. The Pilgrims believe they are not wasteful, but pictures and a description of killing a moose, fish, duck, rabbit is a graphic public image that can be focused on as wasteful and horrific. The environmental issues of civilization are not as clear with no specific person to point a finger at and charge. Yet devastating catastrophic damage is being done; but by whom?!

I recall pointing out a fact I heard thirty years ago! It took, back then, ten acres of trees a day just to print the New York times. A civilized family reads the news in this paper about people like the Pilgrims. Where does all the wildlife in that ten acres a day it takes to print that news go?

I have seen forms with questions on subsistence. Forms similar to, 'How often do you beat your wife?' Where, 'But I don't, what do you mean?' is not one of the choices. Checking a box is required; demanded. If you lie you can go to jail. Fill out all the answers; do not leave any blank. You have to check something like 'Once a month' in order to get the form approved. If you want the free windows or Social Security. "It's just a formality, no one reads this or cares. Just sign your name, it goes in a file and will never be seen again." *Will that ever get given to us in writing? Repeated in court as having been said? I think not!* Since I have had experiences similar to the Pilgrims, I can relate, and do not fault them overly much for their views. Specific questions I was asked I can think of have been, 'When will you get your subsistence moose?' and 'How much money will you make trapping this coming year?' $250,000 fine if you lie.

Getting food off the land is not like grocery shopping. We cannot usually pick and choose what we will have for dinner tonight.

"Whatever I bag today." It might be a duck, rabbit, grouse, or nothing, and it's vegetarian night. We get a moose when opportunity arises. On the one hand we know our environment and where moose should hang out. But on the other hand we often cover a lot of ground as richer sports hunters who have planes, airboats, coordinated helpers using GPS, spotters, guides, etc. scare everything in the surrounding area. The whole forest is in a panic. But mostly along the river and places easy to access that subsistence people can afford to get to.

A single hunter in a boat can cover hundreds of river miles. It becomes hard to know where anything might hide. I cannot say what day I will get a moose any more than anyone else can say what day they will see a pine grosbeak at the bird feeder. At the grocery store, you may not know a year ahead of time what day peaches will be on sale. Imagine getting asked how many pounds you will get, on what day, and how much you will pay, with a wrong answer being jail time. You ask, "Why is this your business?" Get a reply of, 'Well it helps us control supply and demand of the peach industry.' Leading to what they hope you want to hear.

"It helps keep the cost of peaches down." Who can fault that?

Trappers cannot predict the season any more than a gold prospector can tell you how much gold he will find, nor a stock speculator when and by how much his stock will increase in value. There is his dream, there is what happened last year, and there is reality. Reality finding gold can depend on depth of snow, temperatures, water depth, price of dozer fuel, etc. Subsistence living is not like a regular predictable paycheck. We do not pay predictable monthly bills. We usually have the option to hunker down and wait for better times.

As I follow the Pilgrims further, I see more of the child abuse aspect, and the subject of all that goes on, so the story gets complicated. This case becomes more an example of why the government interferes than why we all deserve freedom. The message becomes, "Watch out for those crazy bush people. They tend to be up to no good, we need to regulate them more!" I sigh. I know so many wilderness people who are simply good people. 'Good' does not make such a good read, or get in the news as much as mistakes and bad news. I smile.

"Iris, remember that news spoof about oranges?" She nods and smiles. Big headline

"*Orange Crop*" *OOOOh they must have got wiped out or something, what's it say?* The story reads, "There was no storm. The weather has been great! A cold rain came through, but the oranges just hung there." And that's the news. The roadsides are filled with crosses and wreaths honoring those who died here, hit by a car. There is no memorial for those who got rescued, those who got picked up hitchhiking in a storm, good deeds done along the road. No celebration, no 'thank you'. Doom gloom death is recorded. If we depended on knowing the reality of driving the roads by the messages along these roads, the impression would be vastly different than the views of those actually living along these stretches of highway and byways.

"Well, Iris, Papa Pilgrim ends up in jail. The kids get rescued, the land goes back to the park. It's not a best seller book. I suppose a local book about local people many of us know about. Ashes turned back to ashes and dust went back to dust."

"Like what is happening to the computer!"

I laugh. "Stupid computer!" I do not think the Feds are in my computer, for the moment. Few locals are happy with their internet connection. Stories like mine are numerous. It's not likely the Feds are in everyone's computer. Messing with us all. They could be, but why?

Or perhaps my view has to do with the fact I try to forget the bad, focus on the good. Dwell in the house of honor and light. I tend to forget the facts concerning bad news. If the government is up to what all these friends say, how can there be hope, happiness, a chance for a future?

"One answer is to give up our internet connection entirely. Go to the library with the laptop and check our mail and do business at the library." The library has about the fastest connection in town.

"Mrs. Librarian tells me it's $400 a month!"

"Miles, you'd still have the MTA bill hosting your domain name!"

"Yes, but the separate internet connection is $49 a month we'd save. I just hate paying for a service we can't use, and does us little good. I cannot even reliably get email. I wait a week and put in hours trying to upload to my web site." Some other server offers DSL connection over the phone that is supposed to be slow, but works, and is better than what I have now. My concern is I am not 100% sure all the problems are the server and would get solved by changing companies. I could drop this, pick up that, pay the various fees, and find out I went from the frying pan into the fire.

"The school has fiber optics. I think shared by the military. I am told lots of room, enough extra space for the entire community. So why can't the school share it's fast good reliable connection?" Is the military concerned if the connection is shared, it might get hacked into? It's hard to know what the thinking is. "Well, Miles, will it be ok going back to the library? It has been several years since you went there regularly after the issues with Mrs. Librarian."

"I stopped in off and on to drop off fliers that need to go up on the bulletin board. She seems to have let the past go, and I do not mention it. She's been helpful and suggested herself I come to the library to fill my internet needs. I notice there is room, it is quiet, with a good atmosphere."

I have spent a lot of my life feeling good in libraries and connected in some way. I met my first girlfriend, Maggie, at a library. I used to help people, donate time and money to libraries, so a lot of positive memories connected. Some people bring laptop computers and sit in their car outside the library and connect by Wi-Fi, check mail and do on line work. That's what it is for.

"But Miles, what if it is the Feds in the computer?"

"Well. If so, nothing I can do about it. Life goes on. What do I do? Give up? It's like in prison, you get used to people watching you shower and sit on the toilet. Including female guards." It drives some guys nuts, but is nuts going to get me anyplace? Tough it out, or perish. No one cares one way or the other. It is up to me to make a choice, am I on the train of life flying down the track, or do I get off, and get left behind.

Still a dull hum in the phone connection.

I SIGH and go check my emails

> **Hey Miles.** How have you been? It's been a long time since we connected. I have a friend that is a knife maker that's looking for a contact in Alaska that can supply animal products for knife handles. I know you stopped working with animal products, but thought maybe you could steer me in the right direction. Specifically, he's looking for walrus. Any suggestions? **Joe**

This communication seems suspicious to me. I do not recognize the guy. Most of my friends understand by now, I am out of this business and not talking about it. Understand I cannot even give advice or offer help. The web site even states so. Walrus ivory is the biggest taboo in terms of my past troubles. Offering help is the same as helping someone get heroin. I therefore suspect this is Fish and Game trying to set me up. I believe this goes on because various friends have had the same experience and remind me not to fall for it. I give a polite reply I cannot be of any help here. Not long after, a news Miner article catches my attention, having to do with wildlife rules.

> **Friday, February 26, 2016 Federal restrictions hunting**
> 'Comment period extended for rule proposal to reduce predator hunts'
> Fish and Wildlife Service argued in January that it needs to ban Alaska legal hunts...on 77 million acres of federal wildlife refuges in Alaska. Murkowski said in a written statement, "With more than 60% of our lands managed by federal government, any regulations to limit activities deserves careful scrutiny, if not outright opposition."

I see this as another example of Federal takeover of the state. The number and size of federal refuges and land holdings increases. Once obtained, Federal law rules supreme. There are no restrictions to stopping the entire state from being a refuge. I saved another article and review it.

**Federal overreach case.** John Sturgeon is a lifetime wilderness Alaskan who has been hunting most of his life by hovercraft in the Yukon flats off a remote river. He was stopped by park rangers and arrested for illegally using a hovercraft in a Federal park. He and the state argue the waterway is state regulated and not part of the park. This is a big issue setting a precedence for jurisdiction. If the Feds win, they control all waterways which effects everything happening on the water, from boat regulations, hunting subsistence etc.

The state argues ANILCA and other agreements back to the state constitution gives the state jurisdiction over its waters. This is a situation unique to Alaska. In most states, the Federal Government controls the waterways in parks. But not in Alaska. Headlines read "Alaskans deserve protection". Federal rights overreach is being applied here. Federal control could stop access to areas that amount to areas the size of the state of California.

This reminds me a lot of a previous case on the river with Jim Wilde. He was arrested, went to court and lost over Federal jurisdiction over the waterway. The argument he was given I repeat often, "If a government agent tells you to do something, you obey. It does not matter what your opinion is." In other words being right, what the law reads, is not the priority. The priority is who is talking. The presumption being, you can argue about it later in court.

**News miner opinion page**
**Disrespecting the last frontier**
"Reality TV shows set in Alaska have a checkered history. Reality TV shows found Alaska to be a gold mine. There are more than twenty. The oversimplification and distortions of reality shows have rubbed many residents the wrong way. TV shows ratchet up the drama ratings and distort the life in Alaska to snare viewers.

This article mentions the government gave tax credits to movie companies as incentive to come to Alaska to film. Bigger companies did not arrive, but the smaller teams of reality crews can make it work.

The good news right now is, I look forward to my up coming Tucson trip. I have not set up to sell in a few years and look forward to a return. I have a big mammoth tusk to bring, all my books to offer, new art, stable wood. I'm excited."

Some challenges are presented. I am not sure of the fossil market these days, since I have been gone. Likewise the country's economy. My costs have gone up. I'm going to be invested into this show to the tune of over $5,000. Can I really make enough money in two weeks to call this amount an investment I will easily recoup? At least this is an amount I can walk away from and have it not be a disaster. Last year Iris and I went to Tucson. I was a buyer, visiting regular suppliers and getting some fine cut stones, and other interesting things I can use in my art. It was good to see Dodger set up in a room. He seemed to be doing well. That is when a room opened up for this year! I was asked by the manager, who is a friend now, if I am interested in getting a selling room in a prime location. Just a few doors down from Dodger. I decided to go for it.

Mom is getting old and more often gets upset I take up too much space with my stuff. She has a small space and no room. I feel badly if my visits are more about being in the way then having a good time. I have little to say. I cannot afford to visit unless I can work. I must write it all off. To sell, I need a place to keep displays, tables, inventory. It's just how it is. I can't afford to rent a storage space. Mom has an empty shed not being used that would work fine! It used to be her husband's work space. He died, all the tools are sold. Mom says she needs the shed to store empty boxes she may need one day. All I can say is, "Ok." It's her home and shed. We could flatten the boxes, organize them for her, take up way less space. No. The boxes open, ready to use, mean more than my being able to stay. What she really wants is her freedom and space.

I feel badly. She acts like I am putting her on the spot and do not care, giving her lip service and no action. My word is no good. I had promised to find a place to store my supplies. In turn I had counted on my buddy Foil who has an empty mobile home on his lot he said is mine.

"Keep it, store your stuff, it sits empty!" Again, "You have been a big help to me in Alaska, let me be a help to you in Tucson!" Except. "Gosh golly, Miles, bummer, but something came up. Our daughter is coming back home and we told her we'd fix the mobile home up, sorry! But I'll make it up to you!"

Pass this promise on to Mom? She will roll her eyes up and not believe me. She might evict me, call the garbage man and have my things removed? She could, and might. "You were warned Miles!"

Some of the same issues I had as a child come up. Not much has changed. Mom talked about a ceramic parrot my sister broke when she was three. She has never forgiven her daughter. All those years ago. That parrot was worth more than her daughter. She tells me I was a good boy! Did as I was told! Always happy, easy going, did not want any trouble, never cried. *But while I was busy trying to keep everyone else pleased, who looked after me and what I needed? Damage was done.* I can forgive, but reality is, I'm not going to be upsetting Mom and be in her way.

"Miles, you can't keep storing your things under the trailer, someone might find out and turn me in and I could get in trouble! We have had this conversation before!" The trailer is skirted, so this is not a matter of being unsightly. Possibly related, a potential fire hazard, or rat problem. I am simply not used to such ridiculous rules. But Mom is, and it is her place.

Iris and I decided to look around for our own place to buy. Part of the land agreement on the Kantishna River with Foil was trade for help getting a good deal here in Tucson. Iris says, "Miles, I have my retirement. I want to invest instead of just keeping it in the bank!"

I have enough saved we could get a place, but this way Iris can call it hers and feel more secure. My money, her money, it's all the same if we are together. I do not mind if she buys and it is in her name. So we spent a lot of time looking around last year. It was a joy just driving around in a rented car looking at the sights that surround Tucson. We visited a place called Mammoth, then Benson, and Patagonia near the Mexican border. Four Points, and some other remote spots in the surrounding area within about an hour of Tucson. A friend is in Coolage, so we checked out this area and saw some good deals, but nothing we felt really good about.

One place we saw advertised, already sold when we got there to look. It was nice enough and we might have gone for it. A fenced in lot with small trailer on it for about twenty grand. We were bummed out it sold.

"Let's just drive around the area, and see if we can spot any 'for sale' signs!" We do like the neighborhood. We were offered another lot the first owner had with no living space for the same amount of money. We see a few 'for sale' signs up and are somewhat interested.

One place seems nice, but the owner wants more money than we want to spend. Four Points had some really low prices, but out in the hottest part of a flat desert with no view, meth labs and drugs all around. Homes that were empty had graffiti and broken windows.

San Manuel is nice, an old copper mine area where the mines gave out, but nice cement homes with good view and nice neighborhood. After talking, we feel it is too far from the airport when we'd come in. It's a $100 a person to get there from the airport. I really like the area though, and the view. It looked like game around, maybe deer. There are miles of deserted roads I could run with a dirt bike or four wheeler exploring. The drive to downtown Tucson is beautiful. But empty homes were vandalized, stripped of wire, walls busted up, maybe derelicts moving in. I liked the wilderness aspect, but Iris and I agree we can have more of the wilderness we like in Alaska. Our retirement age in the future is more about access to the doctor, an easier, more civilized life of the elderly.

Not a horrible neighborhood though, and probably a home could be protected, locked up, and have neighbors keep an eye out if we were serious and got such a place. We left Tucson without making a decision. However, the first place we visited with a realtor, Iris notices is now offered for sale on Craig's list by the owner. I know that if you hire a realtor and sell the home on your own, you still owe the realtor. So this guy might be desperate?

"Miles, maybe we can offer the guy a price we can afford, cash, and see if he goes for it!" Nothing to lose. It's not us in a bind. For us it is a buyer's market.

The place is an older small trailer that comes furnished, and has been kept in good shape. On a paved road fifteen miles from the show we do. Just outside the city limits. We saw a nice enough empty lot across the wash from the fossil show! But it is in city limits. Rules and regulations. No trailers for sure! Homes have to be designed by an engineer. Plans proposed and accepted by the city. We'd get told how many windows, how big, and looking at $50,000 minimum cost to build on this nice lot. Out of our budget. We do not want to wait to build. Patagonia was really nice! High elevation, hardly desert, twenty degrees cooler then Tucson! But it's a high end expensive tourist area with bird watching and Hoity Toity types. Well no matter where we end up, if we like any area we can simply drive to it in a hour or so to enjoy it!"

I forgot the details of the first place we saw. Two big lots, no neighbors in sight. Paved road, electric and water. No sewer, but septic tank. Low taxes in a top neighborhood. My guess is the low price is due to the huge cactus plants in the yard.

"Iris, there are laws protecting these cactus, so not easy to move them to build your dream home." Few others are interested in life in a trailer. Guessing not a lot of jobs for the poor here in this high end neighborhood. Behind this is a nice $300,000 home with swimming pool. No city rules, and no home owners association. Yet with these high end homes, probably good police protection and no meth labs. Looked like a clean neighborhood, from what I recall. We see no room for expansion. A protected desert on one side, steep mountains on the other sides. Neighbors down the way have horses, so assume someplace to go with horses, wilderness nearby. There is an art gallery not far off. The post office is in a store. It's quiet, reminds us of our own Nenana. But the economy looks better. The fire department is not so far. The airport is a $15 ride away. A hospital is within easy reach. There is a weekly swap meet nearby. Mexico is a short drive. Mom is half an hour away, and the show is about fifteen minutes away. The cactus offer shade and garden type beauty. The trailer has new air conditioning installed, and comes furnished.

The owner is a young artist from Hawaii who wants to return to go home. We think his parents sent him off to go to school and may have paid for the trailer. He is fed up! $2,000 is added to the asking price for the painting he did on the floor. A somewhat ugly painting. I could live with it, since I am an artist myself, and open minded. But in truth, you should deduct $2,000 from the value. We make our offer of thirty grand from Alaska. If the seller says no, we will come back and keep looking. He has not sold this in over a year.

"Miles, are you on schedule for leaving on our trip?" I have my passport to go to Mexico, art and everything packed. It's all exciting, and an ok retirement life. I just feel like something important is missing. Is it that I am not ready for retirement yet? I'm not sure. As if I'm looking for something. Maybe being called or pulled someplace. I cannot explain it. Sometimes how we feel does not have to do with logic and explanations. We leave in a week. Things to do before then.

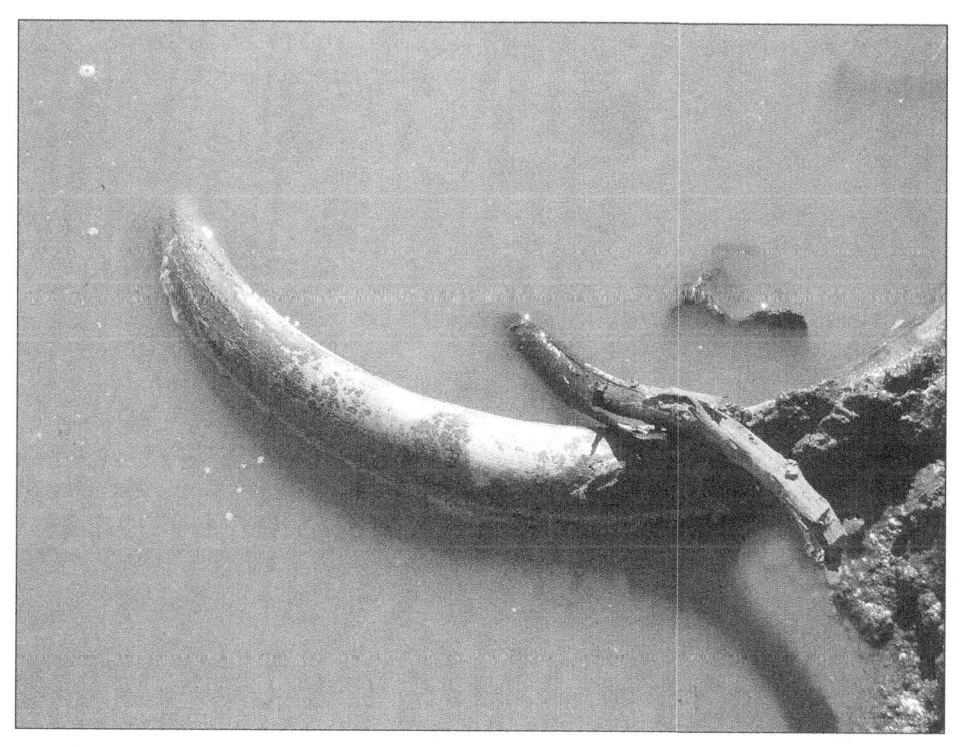

Jurisdiction over who controls the water rights affects the legality of keeping such finds as this fossil mammoth tusk.

# CHAPTER FIFTEEN

## SNOW MACHINE TRIP, GET A MOOSE

I'm gassed up and ready to go out on the snow machine. I review my list. Matches, GPS, portable Ham radio, bumper jack, rope, spare food, chain saw, snow shovel. I'm used to these trips, but as I get older I am forgetful and have to do everything slower. I love the grand touring snow machine. I have had a lot of machines in my life, and this is perfect. It's a 380, so bigger than the 200 Tundra I have, but the Tundra gets stuck a lot, especially in overflow. My 380 is much less complicated and lighter in weight than my previous Viking 550. I can at least move this if I get stuck in deep snow! I have put in several hundred thousand miles on Alaska wilderness trails. I miss the sled dogs, but in this day and age, illegal to live the lifestyle I wish with sled dogs. *Oh well! The road to happiness is not dwelling on what can't be. Need to appreciate what is.*

I'm headed for Poggy slough and the cabin I have fixed up. Kind of my get-away place. I pick up firewood on the way home so I can justify the trip out. I have always liked to do things that accomplish something. I'll spend a night, maybe two. Who knows. Iris looks up from the computer and knows by now I will be home when I get home. I step into the cabin to give her a kiss.

"Ok, Miles, have a good trip. Be safe."

I return my usual reply. "You know me! My middle name is safe! Mr. Cautious!" She rolls her eyes up. Just yesterday I accidentally lit myself on fire... again, for the dozenth time. I'm off, and this is just routine. I make these trips all the time.

The snow is deeper than usual, but not bad. The temperature is a normal ten below zero. Comfortable. The special adapted skis I put on the machine work well. Huge Skandic skis have the same bolt pattern as my much smaller grand touring!

No one else I have met has a machine set up like this. No one else I know carries a bumper jack. No one else I have met has a homemade chain saw carrier like I built, so I can grab the saw easily and do not have to balance it on the sled load of wood and risk losing it. Locals know who it is from far off when they see me on the trail. Today I see no one. Snow flies up and makes a mist in front of my face so I can't see well. I am used to this. I have a good instinct of how to get to where I want to go. All my life I called it 'a high Jesus factor' from the book, "The Right Stuff", a true story about test pilots like John Glen. I like to read. I like to daydream. I'm retired, supposedly living on just meager Social Security.

I want to find a shortcut today, so have the GPS on, watching the map and the arrow showing where I am. *The shortest distance between two points is the trail.* My unconscious reminds me of one of the rules of life we have learned. My unconscious sees I am not listening, so adds, *But of course the shortest distance is not the most fun, is it!* I remind my unconscious of a second rule that also applies. "The shortest distance is usually the curve of the universe." A shortcut through that curve can be the Twilight zone.

I discover where moose have been hanging out. No one else has been this way in a long time, maybe never. It is hard going. I had not expected tundra here. I had hoped it was more open swamps, even creeks and lakes to follow. The area is not well mapped. The maps look detailed, but were made from pictures taken from the air in 1952. This is outside the village of Nenana, Alaska, population 300.

In one place I have to cut trail through some thick woods. I suspect I am not coming back this way again! Now I know why the trail is where it is! My shortcut is of course a very long cut now. I am not sure how far I have gone… or where I am. I have a general idea of my bearing. I'm not in the least concerned, as this defines my life. Always looking for places no one else has been. The GPS says 'up ahead' someplace is Poggy. I've never been much for distance, numbers, time, dates, exactness. I'm pretty open minded about this stuff. No one I talk to knows what that means - to be open minded about time, space and matter. Nonsense. So I keep it to myself. I am simply 'out here.' Being a hunter gatherer. Exploring, as part of survival. Sometimes I day dream.

Up ahead I see an anomaly, something not natural against the backdrop. I head over that way. When I get closer, I see it is an old style sod hut that primitive people used to live in long ago. Sometimes I come across such old things, and even sometimes I meet people who prefer to live in the old way in such places! Miners, trappers, and such. I met Karen and her two children living in a sod house in the wilderness for example.[1] I assume this is such a situation. I slow down, and as I approach, there are five Natives lined up, staring at me as I shut the machine down. They look like they have never seen a snow machine before. I smile. This as well is common enough. I have in my life met homesteaders who go bushy, get lost in the past, cannot come to the present. Some have disowned time, hate 'now', refuse to acknowledge the future. Burn their money. Such people can even end up with a following in the news media. They live by choice in an eddy of time stuck in the past. This may be such a situation.

I say, "Getting any fur?" I assume this group is out trapping - the most logical reason to be here in the arctic snow and deep dark cold. I get no answer. Stoic stare, as if I am an alien. I look around. They look poor, hungry, thin. Like they have been hungry a long time, even a lifetime. This is possible. This is part of the reality of a subsistence life. I have seen this before, indeed, lived it. 'All this' that I see is familiar enough. As I get closer, the band parts and makes way for me to step into the sod hut. There is a grease lamp for light. I have built one of these lamps so know how they work. Moss in a puddle of animal grease. This one looks like an antique, related to a seal oil lamp.

I know a few reenactment people who use as many authentic things and lore as they can. Start fires with flint and steel and all that. I never saw anyone get a fire going with a bow drill before. But here is the bow, fire-stick, block with the hole in it as I have seen, and even sold before. I have friends who supply such items. The primitive Indian band is dressed in furs. I cannot make out faces or forms. I am reminded of pictures from north country, maybe Siberia, of how people lived in those old time days, that could pass for a 1,000 years ago. I ask, "Got enough supplies?" It's polite to ask. I can head to town and fetch something if there is a great need, or deal with any mail, outgoing or incoming. It is common to bring in an empty propane tank for someone to get filled, or a paperback book to read. No one speaks. As if they do not understand me.

This scene could be right out of something from the stone age. It's a good feeling. A world with no planes flying over, no fish and game telling anyone what is in season, how to hunt, what to hunt, what weapons to use, how to dress, and assigning us a number. It is nice, to stand here in a world without all that.

I see there is no food around. A stone bowl is over an open fire melting snow with no meat in it. It takes knowledge to live so primitive and survive. Just keeping an open fire in a sod hut without smoking up the place requires a lot of experience. The draft hole in the roof has to be the right size. Stones need to be set to control back draft from the breeze. What catches my attention from looking at the fire and firewood is, all the wood appears to have been cut with the stone ax setting by the entryway.

The door is smoke tanned moose hide. I mention, because such hide is worth $3,000 to Nenana beaders and skin sewers. Their clothing of furs is easily worth $10,000 in a museum. It's a little odd to have such items worth enough to buy food basics, and better, more modern supplies. They could buy a wood door like everyone else. But again. I meet people who simply prefer the old way. This is obviously not just a bunch of boy scouts having a primitive experience.

I saw moose tracks a ways back. I decide to help out these hungry people. They look at me expectantly and say something in a language I do not understand. Again, this happens. There are natives who want to keep up with tradition, speak Gwich'in. I smile, nod, and point away from here. They watch silently as I walk to my machine, climb on, and start it up. They look amazed, as the light comes on in the front. I assume they are just a bit bushy.

I make a great circle and head back the way I came, doing about fifty miles an hour. In a very short time I am further away than these trappers could walk in a week. Outside their range. All they have for transportation is homemade sinew laced snowshoes. They do not know this is where the moose are! I see sign of at least seven. I head across an un-named lake not on my GPS. In the wilds, a GPS shows where I am, but not when. Lakes have come and gone, forest fires, floods have changed the world. The map is just a basic possibility. Finding vast lakes not on the map is common. I cross such a lake to a willow patch across the other side.

I know this is where moose will find feed, and should be hanging out on this side away from the wind, facing what little light there is. Sure enough, a bull is feeding along the edge of the willows. A rifle is part of my normal gear, but I carry only black powder now as I'm a felon. The good news is, this is my modern in line - more accurate than my old Hawkins only because I choose to spend four dollars a shot using modern solid copper, Teflon coated projectiles and modern pyrodex powder, instead of black powder. I have made black powder, made my own caps, and cast my own lead balls before, so I have the knowledge. I had my fun in my youth, but now, why bother? It's probably a 911 Homeland Security issue to make your own gunpowder without a permit anyhow. The assumption would be I'm making bombs to start a revolution.

It is a wonderful day. Off in the far distance is Denali. Between here and there is just one road in a hundred miles. Nothing but tundra, un-named lakes; a vast expanse of heaven. Flying snow sparkles in the light of the pink sunrise. The moose does not look up until I am within shooting range. I come in even closer, since the moose is not spooked yet. As if he has never heard a snow machine or seen a human. This as well is possible. It is a fact, there is a lot of wilderness here. I have only one shot, but am not concerned. It has been many years since I needed more than one shot to bring down any large game. I'm a good shot. I estimated I have killed over fifty moose in my lifetime. This is very much routine. As routine as the civilized person going to the grocery store. I know exactly where to aim from the position the moose presents me with. It's not a perfect profile. I wait. The bull is facing me. A heart shot is made best with a more broadside profile. I prefer heart shots. I know the moose will decide to leave, and will swing his head before he turns. There will be a moment when there will be a two second pause, and a good clean perfect broadside shot. I wait for this moment. The black powder rifle is rested on the handle bar of the snow machine so I have a steady aim.

The perfect moment arrives and I take the shot. "Whump!" Like dynamite going off. Five feet of flame, a plume of smoke that would gag a maggot. All quite familiar. I see the moose sag in confusion. No pain, no movement. This as well is expected. I know the symptoms of a heart shot. He never heard the gun go off. I'm not excited. It's just making meat.

I pause, as is often a good idea. Give such a large animal time to know it's dead. It is possible for the heart to be stopped and the moose, bear, or other large animal to not know that, and still have the ability to lash a hoof out, twist a head and snag the small human body. I smile. When I was young, I'd dash right in! Feel a need to slit its throat, or simply go one on one because it was exciting. I'd get clipped by a lashing hoof, tossed off balance and laugh. But that was forty years ago. I'm a senior now. Waiting a minute or two is safer.

*May as well re-load this smoke pole while I wait.* This will take a couple of minutes. I keep all the fixings in the hollow rifle stock. I made the butt plate a swivel cover. As I reload, I daydream.

I remember when getting a moose was exciting. Worth writing down on the calendar with an explanation mark. Telling of it was big news! I'd spend pages describing in my writing how it was, step by step and what it felt like. The thrill is gone. It's like the civilized person grocery shopping. How thrilling is it?

"And then I merged to the left turning lane as I got close!" Will there be a holding of breath, Oooh and Awe of a captive audience? "Then guess what! A red light! I had to wait. When the light was green I turned slowly and snuck to the back parking lot because I knew there would be good parking!" Will an audience really whisper how intelligent that is, how wise you are, and how you need to write that in your next book? Will you hear, 'That was so brave, how do you do it? You could get hurt!' Will you go on describing the vegetable isle and then the canned goods? Is it worth the telling? I think not.

So this is how I feel. If anyone finds it exciting, well, I'm not impressed. I'm even less impressed when people are disgusted. How would the civilized person feel if I got angry and reprimanded them for going shopping. "You polluted the air with your filthy car exhaust! Look at this chemical injected crap you got that you call food for the body! It's disgusting!"

I run a lightly oiled rag down the barrel with a ramrod to clean out the salt based corrosive explosive residue, then drop three large powder pellets that have already been pre-measured and shaped to the exact diameter of the barrel. The 350 grain copper projectile already has a plastic wad on the back that seals the gasses. I need the ramrod with a different tip to force the tight copper down on top of the powder.

A mark on the ramrod tells me I have everything seated when it is at the lip of the barrel. A 202, twelve gauge primer is set on top of the nipple. I close the breach and check to see the rifle is on safety. This process can be accomplished in ten seconds by an expert in a hurry. It can be a special challenge when the first shot is a miss or an exciting situation is created with a wounded large animal. I do not believe it is the excitement I miss, to where I'd deliberately create challenging situations. I carefully put the rifle back into its leather sheath. This is an antique smoke tanned soft leather scabbard I acquired at a pawn shop. Someone had used it for a long time. It does a good job protecting the rifle from snow, getting dirty, or banged around in my lifestyle. It's as old as 100 years.

I start the machine and skip across the deep snow as lightly as a snowshoe hair. The moose is dead, and I gut it, saving the heart and liver. This is an hour of work. Not even work. Just something you do when you get a moose. In civilization there would be questions of the legality of this! The legal question does not even cross my mind. I am after all, on my own planet. I have never met Fish and Game in the field. For all I know they do not exist here. *Because I don't want them to.* My logical mind tells me of course, they must exist, even here! My unconscious tells me we are in another world now. I have been saying this for years. I feel just the same as I have for forty years.

I get some rope out of my survival gear and unhook my sled, which I will come back for. I lift the head of the bull and set it on the back of my machine. I have done this before. I know it works well. I know I cannot lift, or even roll this heavy moose onto my sled. I lash the head to the backrest bar, and the 1,000 pound bull pulls like a sled. The feet slide behind and the fur is slippery. There are no trails out here, no planes overhead, no human sign. This is a scene that has not changed in 10,000 years. The only difference is, I have inserted a snow machine into this movie of life.

I head back to the native camp. They are all lined up again expectantly. I pull up closer this time and stop the machine right in front of the hut. I get off, untie the moose, and bow, point, indicating this is for them. There is a lot of excited babble, crying, bowing down, followed by diving into the meat like animals. This is the first time I wonder who they are, what they are doing here, as this is beyond what I have seen before. They act like they are starving, ripping off hunks of meat and hide with their teeth, and eating it raw. *Who among the civilized have jaws strong enough to rip a moose hide apart?* I feel a little embarrassed for them. None of the faces look familiar as anyone I see around Nenana getting mail. I know most of the homesteaders and remote people. I have met families before who have never ever come into civilization. I know enough to comprehend this is a temporary seasonal nomadic camp. I understand in the lifestyle they have chosen, it is not possible to stay put here because they would run out of dead trees to drag to the hut to stay warm. They have already run out of game they can easily reach within snowshoe distance.

These look related to the distant Minto tribe of Athabascan Indian. Somewhat related to the Nenana tribe, who I am more familiar with. Maybe shorter. Elders tell me they used to be called 'Pike Eaters' when other tribes ate salmon. A small band was run out of the Arctic country on the other side of the Brookes range. Maybe 500 miles from here. I'm guessing a thousand years ago. Chased by the feared Eskimo. The Eskimo believed the Indians were deliberately stopping the caribou from crossing the mountain range. Possibly, but inadvertently, true. The Indian had discovered making stone figures in a line to steer the caribou into good killing zones. The Eskimo saw it as some kind of voodoo hex stuff, Uclainy as they call it.

So the Indian band on one side of the mountains was run off by the Eskimo on the other side of the mountains. The savages settled in the interior away from the major rivers where the salmon are. They were shunned by other tribes, yet left alone to live their way. I know a few of them. One of my best friends, Josh, is from Minto. The winter trail is only fifty miles from Nenana. Josh still goes there with sled dogs. I would not think any of the Minto Indians still live in the old way, I mean really old way of 200 years ago. It would be illegal. Children, babies, would be taken away as living an inhumane unfit cruel life.

In the news was a family who had their children taken from them because the family was living in a tent in the winter outside a wilderness village. The children were happy, well fed, warm, not abused, and going to the village school. There appeared to be no other circumstances beyond, "Those poor dears having to endure tent life!"

In this day and age you need an ID. Report for jury duty, and all that. Been there, done that, been arrested, know all about it. This tribe has my respect. I may even envy them.

Josh has told me what life was like before White man.

"Few who remember wish to go back!" I understand! They did not even name their babies till they were three years old because most died! A band of twenty was a large group! The land would not support a bigger group. Here is an example of trying to reenact this life. "Welcome to the wonderful world of starvation!" These Natives do not understand me, and only look up and nod.

One seems to be older and in charge. I notice a grizzly claw around his neck. Tattoos surround scars on his face and neck where it appears a bear clawed him. Old style tattoos done with fire soot. I see what would pass for a medicine bag. I studied a little on such old ways and recognize things. It looks to me like the band had been in prayer. They are acting like their prayer has been answered, as they show reverence to the head Grizzly man, giving him thanks, then smile at me as if I am something conjured up, but not real. The mood appears to be, previous doubters, ready to make amends to a powerful Shaman.

I see a bic lighter on leather around his neck! *Is this the shaman from Flowers's tribe!*

I am not sure if he or anyone else recognizes me. How could they not? Or recognize me, but are not pleased? I had warned Flower about the Shaman and the Bic lighter. I suspect the Bic lighter stopped working. The Shaman controls his band with fear. Maybe told them not to treat me as a friend? Is one of these fur covered people, Flower or her father? If the same tribe, maybe a different time, a decade or more different, a whole generation later?

It is common for me to be in situations and use my imagination! So common, not much surprises me. *Nothing can be reality if you cannot even imagine it or accept it.* My unconscious reminds me what we preach to others. *We spell bind people at shows selling our art, books, stories, fossils and artifacts.* I accept what I am seeing. I am not afraid, or concerned. I smile and nod. I blend in, play my part. It was easy for me to get a moose, and it meant a lot to these poor people. I feel good. They care. They are grateful. 'Pay it forward' is something I believe in. One day they can return the favor, maybe not to me, but they may meet another White human someday and have compassion. Remembering the day Wild Miles saved them! Described in the spoken history as, "The day we were saved by a wild beast with one eye that a light came out of, ridden by a man who was all White, dressed..." And so on and so forth. Bible type stories. *What an imagination!* 😄 ·)

Of course living in the old days with the conveniences of modern day would be wonderful! No starving. No snowshoeing for days and weeks. All the advantages of civilization, without the price to pay! I turn to go, and the Shaman hands me a leather pouch. I understand this is a thank you, or payment of some kind. I do not know what is in it. I have been here and done this before. Well not exactly like this.

There had been that guide, 'what's his name' Lynus, that ran my trapline to get to his, and liked to use a hide toboggan sled as done 200 years ago. His dogs would be half dead, and everywhere he went it was an emergency. I had to help him, offer my own meager supplies, share what I had, then do without later on. He would believe he makes it all right by offering trade goods. The old time antique lantern for example. I'd smile politely and accept his gift. Things I really had no use for. What I really wanted was supplies he needed and took if I was not home. He helped himself to my expensive hard to come by dry fruit, things he himself could not afford to buy. He did not take the survival 'all you want cheap' rice I have 300 pounds of. I do not like being between a rock and a hard place, obligated to help people, trapped. If I help, it is on my terms, a favor, not an obligation. A clan like this could suck me dry if I share all I have openly. It is common to find that people create the very mess they find themselves in, and helping becomes enabling.

People who choose to burn money do not deserve anyone else's hard earned support, beyond a onetime emergency hand out given voluntarily. *I have received gifts, hard to explain in the past.* Where did you get that? Where is the receipt, who did you get it from?" If I cannot produce what is required by the authorities—basically

'Big Brother'— the assumption is, it is stolen or illegal. Legal is a paper trail! Someone with an address, a business name. Not going to happen! I smile. *Could you imagine asking this tribe for a receipt?* Ha! What a joke! They do not know what paper or pencil is. It has not been invented yet. I know that! But I tried to explain to authorities in the past, and I did not get very far. Aliens from Mars gave it to me!" "People who were out hunting mammoths gave it to me, in a trade deal we worked out!" I have no answer. I do not need an answer. It doesn't matter when, who, where or when. It simply 'is'. But then, this that I am doing, needs to be about keeping secrets. Making sure I have no proof, and cannot be believed. I'd be heartbroken to do or say something that gets these people arrested. *Go away! Leave me alone! I know nothing.*

I've described this before, as living in the days of the sheriff of Nottingham and meeting Robin Hood and his merry men in the Kings woods. Getting involved in poaching the Kings deer. Or doing a trade deal with the Hood. The last thing I want is to see the Sheriff escort Robin out of the forest to be hung for trying to eat. Me along with him. If choices have to be made, I side with the savage. I also talk about Tarzan. Imagine Tarzan in civilization. No permit for the monkey, wearing a loin cloth in public, barefoot, illegal campfire, etc., etc. Crocodile Dundee meets some civilized lady who takes care of him, but imagine walking the streets of New York with an open blade eight inches long and see how far you get.

I stop at Poggy and spend a night reading by kerosene light by the wood stove. My blood pressure comes down. I have visited my man cave as it is explained and understood to civilization. When I return to Nenana and my home, Iris greets me, "Did you have a good trip?" As usual I answer politely, short answers to standard boring questions. Iris is not an outdoor person. Nor interested in imagination. This is fine. *Possibly no one could come with me where I have been.*

"Yeah, ok. Ran into some trappers I did not know." And that is that. I am not sure I refer to it as 'time travel.' I do not call it anything but 'life.' It's not a secret when no one is interested.

No one knows where I get all my cool artifacts. Rooms full of the odd, the rare, the collectable. No receipts. No names. I do not recall." What did I pay, how do I do my taxes? I blend it in with what is acceptable and deemed as normal, hopefully not raising any red flags. I nod my head when everyone else does, smile when everyone else does. Laugh when it is expected. I do not like visitors and hardly get any. Hardly anyone comes in my work area and shop. Not even Iris, as it doesn't interest her and it's far too dirty. Old useless crap. I smile. I nod. All is well and as it should be.

Now and then a puzzled, "But? Where could you possibly come across this? It's 10,000 years old!" I honestly do not recall. I am not in a position of having to lie or cover anything. My unconscious takes care of certain details and protects me. I trust my unconscious. Those who feel I am a total fraud, make it all up, Looney tunes, are sometimes puzzled that I can come up with so much 'stuff!' Where do you get it, I mean really?" I try to honestly explain, but get a distant look, and I know the person is bored. So I stop explaining.

I have pat short answers like, "Magic! Talent! I know people!" *If I am so nuts, would it make sense I am this functional and independent? Isn't one of the definitions of being nuts the inability to take care of yourself? Maybe not 'nutty', as in 'fruitcake', but 'nutty' as in John Lennon, Edison, Einstein, Picasso, Hemingway, Poe, Mark Twin, and so on and so forth.*

I absolutely do not want anyone to believe me! Goodness! I know by instinct this would not do! Maybe I do not need the competition? Or, where I go is paradise. Why would I want to invite civilization who would regulate and ruin it? The people I visit, the things I see and am able to do, feel, be part of, is sacred. Offered to me in trust. Taken away if betrayed. It is not up to me to invite you to heaven. You have to earn it. *First you must die.* Would these primitive people show themselves if I had company? Do we want university people in helicopters flying grid patterns over the area? An ending like, 'Brave New World?'

Are there others who do the same as me, know how to meet primitive people? I am not alone? Could be, possibly, even most likely! Still, people say all kinds of things about 'understanding' and 'being there!' People who claim to talk to the dead, meet Sasquatch, read sign, speak to animals, the spirit world and what not. Being Shaman, because that is cool. How would I separate wheat from chaff? I do not want that responsibility.

I have already made serious mistakes trying to introduce people I trusted to sacred places. People I assumed would appreciate it. People who became a serious, very serious problem! Like Foil who I am still dealing with. It's like fisherman not revealing their favorite stream. It's not paradise anymore when friends tell best friend, who tell, their best friend. You show up one day and there is a party, the trees have been cut, garbage in the water, no fish, a homestead nearby. The footpath now cut open by four wheel drive ruts. All because I told my best friend. Been there and done that. Seen the results.

I play my part. I do not want any problems. The prize I am handed is added to hundreds of others in a shop no one who knows what they are looking at has seen. Does this have anything to do with why the Feds spent over a million dollars trying to find out what I am up to, and locked me up? My phone is tapped, my computer monitored. It's enough to drive a sane person bonkers.

It all becomes part of my story of magic, The life of Indiana Jones, the man of mystery.

"The government is after me!" Clark Kent by day, Superman when required. "Now you see me, now you do not! Step closer, and hear the story! Step right this way, for $10 this piece of art can be your memory of meeting a savage! It's a secret between us, do not reveal your source!" A sly smile crosses my face when everyone leaves and I close the full cashbox.

It's been a delicate balance act. On the one hand I can't be believed enough to be checked out. But on the other hand, how can I make a living, receive respect, help others, if I'm not believed at all, by anyone? So I gather pictures, 'things' that can pass as evidence, but nothing absolutely foolproof. While no one comes in my shop, it is a lonely life.

People like Foil call me nothing but a con artist! Tricking people to suit my own selfish needs! Just like him, no different! I see a difference. I have the ability and could potentially talk people into and out of things that suit my needs. Sometimes I fool with that ability! However, I never talk people into things they can't afford, or do not want. Some of my deals are border line shady, but as the judge said, "I see no victim here." I do not take people to the cleaners and leave them with their life savings gone and then feel happy about it.

I may blend truth and reality into the shell game with fantasy, except for exclusive in the know customers and friends. But even then! Like the life of a gypsy, all dust in the wind when examined under a microscope. A necessary precaution of survival. Those who believe in out of body travel, or ghosts, do not try to seriously prove it. Few devout Christians spend their life trying to prove to doubters there is a God. They attend church, bow, pray, with other believers.

There is a long trail of those who tell me how much I helped and touched other lives as a Medicine Man. Or simply as someone who offered something they had no other way to receive.

---

I'VE REACHED a point in life of being unhappy, disillusioned, trapped, with nowhere to turn.

It is simply harder as I get older to keep up the 'Ra ra life is so good!' mentality and trying to pass that on to others as an optimist. I can face only so many rolled up eyes and snickers! I keep up the faith, for days, weeks, months, and years. Believing there is a reason. Believing I have a purpose.

# CHAPTER SIXTEEN

## ANCIENT LIFE, HUNTING MAMMOTHS

Shaman had a lot of thinking to do. This is always done alone. *I'm a Shaman, one who has a foot in two worlds, but belongs in neither.* Much of his work is a delicate shell game, mixing the real with imagination and tricks. Much that has to do with healing the sick, time travel, solving problems of his tribe. Much has to do with the mind and believing. The mind must be prepared, ready to believe and accept. Sometimes it requires sleight of hand. Magic lights, glitter dust, awful tasting potions made of inert materials that all come to life with his words and chants. There is a lot of responsibility involved when dealing in hypnosis, mass hysteria, brainwashing, and expecting positive results. His people trust and depend on him to cure them, keep them from starvation. His role is as important as the chief. Failure is tough. People expect miracles.

Sitting on an uprooted tree stump he examines the tiny piece of copper Flower's father dug out of the beast the God brought, in answer to Shaman's prayer to save his people. *I know this copper material. We do not have any, but a rich man from Siberia brought an arrowhead made of this. He said it came from the fire underground. Hot molten rock. I could not afford to trade for any. This looks more pure, and is not very big. How did it kill the beast?* Shaman wonders about the beast. It is already gone—eaten—but it looked like an elk with big flat spread antlers which had more points and the fur is shorter, different. The nose is different. If he had to give an answer, he would say it came from the future. *My magic must be getting stronger to be rewarded in this way!* Surely Shaman will be respected in his tribe more now! A little late in life. Often there is a heavy price to pay from the Gods when magic is performed! It is easy to make a deal with the Devil, harder to deal with God.

*Did God or Devil bring the magic?*

In spoken history another deal was made with the Gods. After the food, came the sickness! Half the people died, not of the starvation! Dead by the dust of the God who brought the food! Shaman is wise, and had everyone stay upwind of the God. After the God Soonshine left, they all washed, to make sure no dust of the God stayed with them. Everything got blessed in the holy yellow dust of the special flower that kills small creatures living on the body. *I looked into the eyes of this God and he looked friendly. He brought food and did not demand anything in return. Not like I expected.* It is hard to believe such a God would bring sickness or death.

His eyes go once again to this small piece of copper. Where did the God get it? How could such a small thing kill a big beast with only one stab? Where is the rest of the spear? This small object raised havoc in the beast! Broke bones, penetrated deep! The copper was pressed so hard it deformed and bent! Such powerful arms to stab this hard! Or did he throw it? Even more impossible! Such powerful magic! Way beyond Shaman's understanding. Now, more decisions to make.

Weather has cleared. Mammoth have come out from protected glacier crevices. Moving North. His tribe must follow them. These beasts are feeding at the base of the vast glacier that covers most of the world. Shaman leads his people through a pass and to the top of the glacier 200 feet above the feeding mammoths. For some reason Shaman had been compelled to keep his tribe moving to get here. Was it for a meeting with destiny, and this God who calls himself Soonshine? The chief has lead his people before, but rarely the Shaman.

The Chief has his people dig a hut size chunk of ice loose, propped up with a stick. A braided leather rope is fixed to the stick holding the ice. This is a sort of deadfall. When a mammoth is directly under this heavy weight, 200 feet on top of an ice cliff, the rope is pulled and the ice falls. The rope is wrapped around the stick, so when pulled, the stick turns. The top of the bent stick in uneven so in turning, can no longer balance the heavy weight.

A great scream of joy goes out as the ice hits the mammoth and kills it! The mammoth utters a moan and the job is done. The people of the mammoth can now eat for over a month. Camp will be made here near the mammoth. Blood runs across tan sedge grass and flows into snow across the trail.

The chief, Grizzly-scars, points to Firemaker, and tells him where the fire pit will be. There is a ritual that involves the fire pit as the basis of orienting everything else in the physical world. A Quonset type hut will be built around the pit. The women will be to the east, the rising sun. Men will go to the west, the setting sun, the direction they come home from at the end of the day because the last light is here.

They will live here a month till the meat is gone. This works out well because in a month there will be no more wood easy to reach to keep warm, and the waste put out by the group will be unsanitary. Half the battle of avoiding disease is to be mobile, with a small group in a clean environment. There are, in fact, few sicknesses that pass between species in this environment.

Weapons are organized to the south, the direction the hut entrance faces, away from the north wind. There is a place for food items, clothing, everything made of wood. In this way, everyone knows how to find the necessities when needed, and where their spot is. There is no rushing around in a crowded dark place, less gets lost, stepped on, fewer arguments. The ability to have designated personal space, mixed with being able to find anything in an emergency.

Firemaker gets out his birch box filled with ashes from the last fire. Digging down into the center of hot ashes, he finds the fist size hot coal for this morning's fire. It can keep in the ashes under low oxygen for three days of travel. The hot charcoal is laid on top of shredded bark and twigs. Firemaker blows, and the red orb gets hotter, till suddenly flames shoot up. "Bless you for giving us heat and light." Firemaker smiles to the east, the direction the fire in the sky rises. *It is a small piece of the sun I carry in my box. Once there is fire, the people can relax.* All else will be well, as long as they are warm.

Knapper hands out flint tools to the women who begin the job of getting a piece of hide off the mammoth and at the meat. Knapper, like Firemaker, sees himself as a very valuable member of the tribe, maybe the most important! He is the expert tool sharpener in charge of weapons! Much has to be done before the mammoth freezes. The wind along the base of the glacier blows the snow away, and exposes the dry grasses the mammoth eat in winter.

The snow free space is also good for the people. Sticks are heat formed over the burning fire - long 20 foot poles of the tamarack tree. As the wood warms, steam is made from the moisture inside, making it easy to bend the wood. The pliable steamed wood is bent in a pattern to form a dome of hoops in a Quonset type hut pattern the people traditionally use. Before nightfall there is enough mammoth hide to cover the poles and make a cozy warm shelter for the family size tribe of five. The base is banked up with loose snow to keep out drafts. On longer stays, especially in summer, there can be three to four huts.

A smoke hole in the roof can be made larger or smaller with skin flaps that also can be moved with poles to control the draft as the wind shifts or increases in strength. Firemaker tends to the smoke hole. The fire pit offers warmth as the smoke rises and leaves out the hole above. The air is more pure down low, so the people tend to walk stooped over when in the shelter. This is fine. They are in the hut to rest, get warm, sleep, or eat. The rest of the time is spent outside the shelter. Firemaker spends the most time in the hut tending the sacred fire. The fire is always in use for something, not just warmth! There is snow to melt for water, mammoth fat to render into lard that will not spoil for lamp light. Lard is used in medicines and pemmican travel foods, as well as waterproofing leather.

Firemaker is joined by one of the two women who spends a lot of time doing clothing repairs for everyone in the clan. The fire is used for light, so projects are done around the fire. In this far latitude there is only two hours of light a day right now. Outside, the dogs are sleeping after gorging themselves on the gut pile of this mammoth. The clan knows if an enemy comes, or predator, these dogs will alert the camp and begin a defense, till tribe members arrive with spears.

The tribe can afford to keep a few camp dogs. Mostly wild they can be put to use as pack animals, guards, and sometimes a smart one is adopted into the tribe and treated special, one who understands commands. They follow the tribe as it suits them. Sometimes they get fed, sometimes they can steal. Other times they wander off on their own for weeks at a time. The wild dogs know not to enter the huts. Sometimes a starving tribe eats them. The dogs left but are now back, word having traveled a mammoth is down. Perhaps the whoops of the tribe alerts the dog sentries on the hill. Both the tribe and the dog benefit from this loose arrangement. Symbiotic without total dependence.

"A woman's work is never done!" Flower giggles as she positions herself on a rock where she can sit and lean back against one of the hut poles. The shelter is built the same way each time, so she knows her place and this is her pole and spot to work. All her repair tools are in a leather bag tied with a draw string hanging from a cut off branch on this pole above her head. Every camp is set up so. She gets out a bone needle she uses as an awl and her bundle of dried sinew for thread. "The mooz had good sinew, better than elk!" The beast the God brought is being called a mooz, because this is what the God called it. Today there are skin pants to sew for Shaman. Maybe he will do her a favor if she does a good job; give her some medicine made from the arctic poppy. However, she knows it is simply her role in the tribe to do clothing repairs. It's better than being out in the dark wind and cold being the one who gathers wood! The tribe operates on exchanges of favors, sharing of roles.

Flower had been ordered by Shaman not to speak to Soonshine. In her culture woman are owned by the men. Not quite, but women let them think so. It was no big deal not to talk to Soonshine. She has other instructions from Chief Grizzly that involve a distant future, and she has had her own dreams since she was a child concerning the arrival of Soonshine and his fate. All is happening as it should. Soonshine had told Flower what would happen with Shaman and his Bic power. She watched the prediction unfold in silence.

Shiny bits of shell, pretty pebbles, herbs, dried berries, all varieties of promises, favors, positions to sit, amount of food, are all in flux, exchanged, shared, denied, in a world without money. There is an understanding of credit. The, "I will gladly pay you Tuesday for a hamburger today," concept. *Says Wimpy to Popeye.*

"Where do you think the Mooz God came from, and do you think we will see him again?" Firemaker can only speculate. "I doubt it. It took a lot of magic to bring him here once. Would you want to see him again, Flower?" Sometimes women who are part of such a small tribe look to men outside the tribe as a way to get new bloodline into the tribe, to make the tribe stronger and more diverse. Flower would not say this out loud, as other men may not like this thought. She gives another reason, "I am curious about the kind of clothes he wore. Not skin like ours, not as heavy. It was ugly, but maybe very practical! I'd like to look closer at those fine threads!" She also decided this God might be a person, just from another time. *He seemed to have understanding and kindness in his eyes.*

It is Flower's job to cut and take care of the sinew from the mooz for her sewing work. She had been taught by her mother how to dry, then separate the fibers with her teeth. Thread, anything to do with clothing, is of interest to her.

Firemaker frowns, "What else would you wish to look close at!"

"It does not matter. We will never see this God again, as you say. Even Shaman says no, it was a onetime visit to deliver us from starvation." Long ago there was a visit, I hardly recall, like I forget for a reason, not supposed to know. *Shaman made us all forget.* "It took all Shaman's special magic powder. Now he is out of the right herbs." "Do you think Shaman can really leave his body and time travel?"

Firemaker suspects Shaman feeds them all a mind altering drug. However, if the visitation was an hallucination, where did the meat come from? There is, of course, only one right answer. Even though she shivers before replying, she answers as she should. "Yes! Of course! Aren't you a believer?"

Firemaker quickly answer, "Yes of course!" They are both a little afraid of such power and strange events that surround him. Shaman is one to be pleased, scary to contemplate displeasing him. *I mean who wants to be turned into a frog!* Can Shaman ever be fully trusted, ever be one of us? He is necessary as part of the tribe, but more respected out of fear, than love.

Firemaker still wonders how Flower really feels, and if she'd trust him to speak about such matters truthfully. He has been interested in her for some time now. He watches her by the firelight. She is naked from the waist up, as is the custom in a warm shelter with their people. Flower is in her prime at 16. Hair down to her waist, very smooth unscarred skin. She has a happy disposition, always laughing, seeing the good side of everything. She would make a good mother, the way she is thoughtful of others, and carries the doll around that her mother gave her. The mother is now dead, and Flower grieves for her, clinging to the doll. She doesn't speak about it much. Flower also thinks this visitation from God was the same God who visited in the past, years ago when she was 12. So strange looking compared to her people, it is hard to tell! *All White Gods must look alike!* No one else made the connection, or will speak of it out loud. Now that the group has a mammoth and will have food for a month, the tribe can relax, hang around, tell stories and discuss possible futures.

Firemaker has to go out of the hut to relieve himself. There is a designated area at every camp. Always downwind. Men go the furthest. Anyplace over 100 feet. Winter is easiest when everything freezes. Firemaker uses his foot to push snow over his leavings. This will be his designated spot others will stay away from. This makes for a generally sanitary condition at camp. A piece of rabbit skin is used for wiping. Firemaker always looks around to make sure there are no predators lurking! There is the cave cat, the Smilodon to watch for. Wolves are more shy, usually not a problem due to semi-domestication of camp wolves. The short face bear can be 14 ft. tall, but usually feeds on mammoth, not people! These bears are smart, and usually leave people with the pointy sticks alone. But even so, all animals are opportunists and cannot be totally trusted. Any starving or wounded wild thing will do what it can to eat! Or in spring, protect the young or a food cache.

Still, the situation is not so scary and horrible that life is miserable. It's a good life overall. The tribe is mostly happy. There is a lot of laughter and storytelling around the fire. There are extremely few fights, anger, hurtful things happening. *When people are busy, working, physically healthy, not bored, lots of room, there is little strife.* So says the Shaman, who thinks upon such spiritual matters, and the welfare of all. Shaman is the one who gives advice to the chief. So far, good decisions have been made. *Maybe the tribe can build back up again to the 20 it once had.*

Flower is not interested in getting pregnant yet. Firemaker is not really her mate by choice. The chief and Shaman consulted and decided for the better of the tribe, it would be good now for Flower to help build the tribe back up. Usually an alpha male breeds. In this case, either the chief, or the Shaman. Since the band is so small the two get along well, work together! It was decided! The Shaman made a prediction! Not all predictions become reality. Just as all hunts do not end up in meat at the end of the trail.

Shaman knows, if Flower believes strongly enough, she can more easily become pregnant! If she believes even stronger, her energy will go in the right direction to create a strong son. If the tribe is already big, the women do not get pregnant and never have twins! When the tribe needs babies, behold! There are twins! And behold again! Most are males who can take care of the tribe! Even animals know this! The female must feel safe and optimistic. She controls the chemicals in her body. With a little help from Shaman. This is part of 'Shaman knowledge' that makes him look more like the man of magic who can predict the future. All good Shaman, witches, people of magic, and politicians, good leaders, know this.

Likewise, Shaman continues the tradition of taboo. Many and complicated. In this way, if a prediction fails, it is not his fault. If Flower does not have a son in spring, it must be because she ate out of a wooden bowl on a Thursday! The focus of attention is now on someone else's indiscretion! But odds are in Shaman's favor his predictions will be correct.

Flower tells Shaman, "Your powers get stronger! You leave your body and learned to go further. Perhaps to another time. Our ancestors spoke of this, but who of us has ever been a witness? You saved our lives!" But the real unasked question Flower is leading to is, *"Can you bring this man back again? He seems interesting, and might be a big help to the tribe."* It might be good to know such a person better. The Shaman is not sure, but wonders more, if this God would want to come back, for why would he, if he is so much more advanced?! What would Shaman or his group have to offer such a God? Shaman already knows this, 'Man of the Future,' is called Wild Miles by some, and Sunshine by others. He keeps this knowledge to himself, so the focus can be on him.

Shaman sees this in dreams, as if there is a connection between the two of them. Flower feels a similar connection, maybe stronger. It is Flower's influence on Shaman that has more to do with crossing the time line. Shaman would never admit this. Nor does Shaman know Soonshine trusts Flower, her father, and the chief more than Shaman. But Flower knows, and simply flatters Shaman. Even in a group of five savages, there is politics.

It is hard to know if Sunshine will ever be back, and what the relationship is, if there is some reason for this connection.

## THE END OF BOOK 8

**A personal note—**

Reviews help! If you enjoyed this book, please leave a review where you purchased it—it would be greatly appreciated!

Sign up for my newsletter, "Keeping Up With Miles," @ www.milesofalaska.com Deals, new books, comments, links to YouTube. Stay updated!

## The Alaska Off Grid Survival Series Summary

**Book 1 - Going Wild**

In 1973, I am 22 years old, and a city kid. I enlisted in the Navy and got out after the Vietnam War.

I travel to interior Alaska, a 'Cheechako' (Greenhorn) by Alaskan standards. But I have been raised on Walt Disney and feel qualified to be a mountain man!

I arranged with a pilot to drop me off in the wilds of Alaska. I do not have everything I need and have things I do not need. I learn about guns, trapping, and the loneliness of living in the vast wilderness with no other humans around.

I do not see anyone for many months, then walk out of the wilds to civilization in the spring. After working odd jobs to make supply money, I return to the wilds in the fall and have a hard time my second winter. I almost die, and need to be rescued.

I decide to build a houseboat so I can travel around without having to build another cabin. I have to accept summer work in Fairbanks to pay for the boat materials and work under a builder. The boat takes much longer to build than expected.

I live as a street person much of the time to keep expenses down.

**Book 2 - Gone Wild**

I have many adventures on the houseboat and acquire a dog team. There are issues with the police, a bear on my boat, and a trip to see my family who live a civilized life.

My houseboat sinks. I get lost and learn other hard lessons. I start doing artwork and end up on TV. I win a land lottery and start my first homestead.

There are mail order women, and I live with a woman and her kids. Ten people are murdered in a village we visit, and myself and the family are almost among them. Family life is more difficult than I imagined.

Fish and Game becomes a concern.

I head back into the wilderness, which leads into book 3.

**Book 3 - Still Wild**

I acquire a couple more homesteads and cut more trapline.

I give up sled dogs and enter the world of snow machine adventures.

I winter in Galena and visit many native villages. There are bear encounters, and many survival situations to learn about.

I become a serious mammoth hunter and find fossils as part of my living. I work with a land surveyor specializing in homesteads and wilderness surveys, getting paid to use my boat.

My art sells well, so I do some big shows. I become more social and understand civilization better. I see the wisdom of being accepted by others. I learn. I grow. I try to change, as the world does.

The economy changes. It is less acceptable to be a trapper. I never become totally civilized as a city person defines it, but maybe I do, relative to the life I had in book one.

**Book 4 - Beyond Wild**

I am getting past just survival and doing well, even prospering. I own more than the houseboat can easily haul. Gas gets expensive. I need a new houseboat engine.

There is a homestead and trapline that keeps me in one place now. There are more bear stories and adventures into the wilds, including a 300-mile boat trip looking for mammoth tusks, which has disastrous consequences.

I find where I want to live on the Kantishna River. A river 300 miles long with about five people on it. I hang out in the native village of Nenana, spending a lot of time here.

I get my first computer and learn to build a website. People are looking at the pictures and buying my raw materials and art. This is a chance to make a difference.

Life is beautiful. Life is precious. I Dare to live it.

**Book 5 - Back To Wild**

I acquire a home in Nenana and start a web store. I am forced out of my subsistence lifestyle, partly because of changes in the laws. I do some serious mammoth hunting.

Unstable power causes a lot of computer data loss. I learn by punching keys to see what happens. It takes a long time to get good enough to create a book.

I continue the Mammoth hunts. The Tucson fossil gem show and State fair do well for me.

This period of 'being civilized' that I am trying out, has advantages, but also a price to pay—a big change from the wilderness life and being alone!

I am a suspect in a murder investigation. Another trapper tries to move in on my territory. There are neighbors and infringements on my property.

I fear I cannot change who I am. There is difficulty blending the two lives and ways of thinking. There are mail-order women coming and going, as well as the usual adventures and situations I manage to get myself into.

**Book 6 - Surviving Wild**

Iris is my partner. Business grows, with money coming in, but causes 'complications.' I understand why I left for the wilds in the first place.

I get better at fossil hunting and have some exciting trips getting mammoth tusks and other ancient treasures. I am viewed as an expert on a few subjects and Discovery TV and reality shows contact me several times.

The new life in town causes legal issues that have been nipping at my heels off

# ABOUT THE ALASKA OFF GRID SURVIVAL SERIES

and on throughout my time in Alaska. Fish and Wildlife ask, "Why are you alone out here where we cannot keep an eye on you? We know you are up to something. What is it you have to hide? We will find out!" This mentality is that different is bad and of concern. I end up being investigated. A SWAT team shows up at my property with a dozen cars and 20 cops.

My arrest makes headlines. I'm sentenced to Federal Prison for six months as a felon. This is a stark contrast to 'Book 1-Going Wild,' where I have as much freedom as it is possible to have.

*How did I get from there to here?*

### Book 7 - Secretly Wild

I am a convicted felon, describing life in prison from the viewpoint of someone used to freedom and the wilderness life. The same feather in the hat I wore on the cover of Ruralite magazine in 1979, is now worth five years in prison.

What do I need to do to survive here? There are classes to take, books to read, farm work to do, and people to help. There are interesting felon stories.

I observe more crime within the prison system by the system than I am accused of committing. "The prison could not survive if we operated legally," I am told by officials. I do my time. Now what? Am I a better person? I see the error of my ways. I am saved. Society is safer now.

### Book 8 - Retiring Wild (This Book)

I talk about news relevant to living off the grid as an individual in the wilderness that few citizens are aware of. I adapt my business, and still have adventures, depending as much as I can on the subsistence life I love and understand that is now becoming illegal as a white man.

I ponder whether the end of my life is in agreement with the views I held dear from the beginning. I have hope that even in times of control and suppression, I can still focus on the plus side, and continue to find ways to enjoy personal freedoms and individuality.

I continue to explore choices, how to have better control of my destiny, happiness, and success. I refer to this as 'Survival.' I have few regrets, and hope my life's path as written can provide entertainment and insight.

As someone who is interested in being different, not one of the sheep, I look realistically at the rewards that choice offers, but also the price that has to be paid.

---

Please visit www.alaskadp.com for links to the books.

Visit www.milesofalaska.com to find a bio of Miles, additional photos, stories, how-to videos, handmade artwork, and raw materials for sale.

### Magazine and News Stories

*Alaska Magazine*

Alaska Magazine July 77—Survive by Miles Martin two pages, Photos. By Miles about my rescue, walk out on the Yukon River, five days at 50 below zero.

**Nomadic House Boater Have Cabin Will Travel** January 81—by Miles. Three pages, four color photos, a map. About life living on a houseboat, trapping and selling art (photo of my art), and all the adventures I have had on the river.

**Would You Make A Good Bush Homesteader?** June 86—by Miles four pages, six color pictures (One shows my custom knives.) A story I wrote about what it takes to be a homesteader.

**Surviving The Big Lonesome**— March 98—by Jim Rearden five pages, two color photos, one double page photo of Miles. Photos by world-famous photographer Jean Erick Pasquier. Describes life in the wilderness.

*GEO Magazine*

GEO in Germany is like "National Geographic" in the US.

**Life in The Wilderness Alaska Special—87** by Miles Martin ten pages, sixteen color photos, a map

Photos by Jean Erick, one of the best photographers in the world, I Wrote it myself, winter life in the wilderness.

**Alaska Special - 95** Einer gegen den Rest der Welt

Eight pages, seven color photos, three are double page. A follow up story to the first, written by New York Times reporter Ted Morgan, with Brigitte Helbing, photos by New York Times photographer Rex Rystedt. My fight for a lifestyle.

*The New York Times*

**New York Times Magazine** an insert to the paper, April 17, 1994, section six, The Vexing Adventures of the Last Alaskan Bushrat.

Six pages, four color photos, one is a double page Written by New York Times writer and bestselling author Ted Morgan. Photos by Rex Rystedt (World-renowned photographer). Facing twenty years in jail and a $10,000 fine for putting artwork on a bear claw and selling it.

**Book-- A Shovel Full of Stars** 95—Published by Simon and Schuster — New York

By Ted Morgan about ten pages with Miles. About one of the last homesteaders,

# MAGAZINE AND NEWS STORIES

and the lifestyle I live, of a Subsistence person.

*Ruralite Magazine*

Put out by Golden Valley 180,000 circulation
   **Wild Miles August 7**9, two pages, four black and white photos, Full cover page photo of Miles doing artwork. Story and photos by Margaret Van Cleve — Mostly about my artwork, some about my lifestyle on a houseboat

*Newspaper, Daily Newsminer, Fairbanks Alaska*

Associated Press, date unreadable, think a Thursday, and think spring of circa 74 **'Trapper rescued by Chopper**; Vows to Return to the Bush' headline, one column, National news, about my rescue after five days walking at 50 below.

*Alaska Trapper Magazine*

Put out by Alaska Trappers Association, a cover photo of me with Wolf Five-page story by Miles comparing snowmachine and snowshoe trapping Nov. 99—four pages. Over the years, another six-seven articles on various trapping and related issues. Contact organization for exact issues.

Me in 1975.

# OTHER TITLES AVAILABLE FROM ALASKA DREAMS PUBLISHING

Visit www.alaskadp.com to see these titles.

**Books by Miles Martin:**

- Going Wild
- Gone Wild
- Still Wild
- Beyond Wild
- Back To Wild
- Surviving Wild
- Secretly Wild
- Retiring Wild

**Titles by other ADP authors:**

- Rookie
- Alaska Freedom Brigade
- Apache Snow
- In Search of Honor
- A Coming Storm
- Arizona Rangers Series – Blake's War
- Legend of Silene
- Inspiring Special Needs Stories
- My Life In The Wilderness
- All Over The Road
- Ghost Cave Mountain
- Inside the Circle
- The Silver Horn of Robin Hood
- Alaskan Troll Eggs
- Through My Eyes
- The Professional Ghost Investigator
- The Adventures of Jason and Bo
- Seeds Of The Pirate Rebels

# NOTES

### CHAPTER 1

1. 'Past Flash' = A past moment frozen in time we remember as if it were yesterday. Taken from the music industry, "Knock out nifty of the past." "Golden oldie."
2. More than one of these 'customer friends' tells me the Feds bothered them for a long time, trying to bait them into illegal deeds. The problem originated with having done business with me. Customers who do not wish to be harassed reply, "I don't know Miles!" And cut off all ties. Few want to be profiled, on the 'person if interest' list.

### CHAPTER 2

1. For decades I collected news articles, saving times, dates, verification of what I speak of. I 'm not sure why I bother. Who is going to look it up? Who is looking for verification and proof? We believe what we want. Facts? Is the newspaper even factual?
2. In book one I entertain tourists who treat me to dinner to hear 'stories about the majestic wolf.' I had to tread lightly knowing they do not want to know the truth of what I saw.
3. Josh has been part of my story since the beginning and I have a lot of Josh events, some of which are 'different.'

### CHAPTER 4

1. Just one example, four out of five days have front page headlines such as, "Police Chief to be investigated," Daily News Miner Sept. 24th. Experience has shown me, if there are issues at the top, there will be other issues on down the chain of command. Other headlines reflect this. Even Trump, running for presidency, is saying the entire system is corrupt.

### CHAPTER 5

1. In book one, my rescue
2. Well described in book one.

### CHAPTER 6

1. I have been stopping here for forty years, and mention Tolovanna in most of my books.

### CHAPTER 7

1. Written about in an earlier book concerning the boss stealing government furniture and selling it etc., and 'everyone' was involved, if you wanted to keep your job. No thanks. But to make one person out of the bunch, a felon is not understanding the situation.

# NOTES

## CHAPTER 8

1. Covered well in previous book in the series

## CHAPTER 9

1. Later related scandals involve, as I guessed, Facebook owner, large companies, and President Trump.
2. Two years from now no one at Denali ever even heard of Glitter, out of business. Last I heard crying the blues, wife got her citizenship and left him, in and out of court over child custody. Broke. I Would not wish to be him.

## CHAPTER 12

1. A time already arrived when on several occasions I have been asked how many guns I own, that they might be confiscated. Black powder has no records kept. Bullets have been hard to acquire for months at a time.
2. The main instigator died in a motorcycle crash at 19. All of the other five died before they were forty. I suspect as a result of a lifestyle they chose. Thus, the pain is in knowing this, what a waste of a life.

## CHAPTER 13

1. A Geek friend investigated my code in Web Studio, my web building program, and told me this low budget outdated program does not allow for resizing a page I create, and my pages are unusually wide. This creates issues with some viewers. Better programs allow for varieties in monitor settings. My program came out before cell phone viewing for example. My old program is not something to be proud of, like my 86 Ford with real crank up windows! Outdated could mean, "No can do business."
2. As I edit, in the news, potential presidential vote manipulation, maybe Russians, maybe Trump. Rumors are not getting explained, or going away, but becoming more solidified into facts. Other hacking and internet privacy issues.
3. A year from now I disconnect this computer from the internet and plan to never have it on line again, nor share disks or drives that do. It runs great!

## CHAPTER 14

1. I keep bringing up this ANILCA agreement because it is complicated. I grasp only part of it and its implications one at a time. Were the Natives involved in its design? It seems like it was rushed through hush hush, is that so? How does it affect me? It is taking forever to deal with just the tip of this iceberg.
2. In fact, after the permit is issued, it is later decided legal access should only be allowed in winter, when the land is the least impacted. Once a permit is acknowledged, restrictions can, and usually are, put on it. The purpose of a permit is control.
3. In 2017 there are a series of hurricanes typhoons, earthquakes that, for example, cut off power to all of Puerto Rico. 2020 has a COVID virus event.

## CHAPTER 15

1. Written about in, I think book two or three. You know me and numbers and names! Are you following the series?

Made in the USA
Monee, IL
05 August 2022